# INTRODUCTION TO MODERN MODELLING METHODS

# THE SAGE QUANTITATIVE RESEARCH KIT

*Introduction to Modern Modelling Methods* by *D. Betsy McCoach* and Dakota W. Cintron is the 9th volume in *The SAGE Quantitative Research Kit*. This book can be used together with the other titles in the Kit as a comprehensive guide to the process of doing quantitative research, but is equally valuable on its own as a practical introduction to Conceptual Grounding and Concrete Practical Guidance on using Modelling Methods.

**Editors of The SAGE Quantitative Research Kit:**

Malcolm Williams – *Cardiff University, UK*

Richard D. Wiggins – *UCL Social Research Institute, UK*

D. Betsy McCoach – *University of Connecticut, USA*

**Founding editor:**

The late W. Paul Vogt – *Illinois State University, USA*

# INTRODUCTION TO MODERN MODELLING METHODS

## D. BETSY MCCOACH
## DAKOTA W. CINTRON

Los Angeles | London | New Delhi
Singapore | Washington DC | Melbourne

THE SAGE QUANTITATIVE RESEARCH KIT

Los Angeles | London | New Delhi
Singapore | Washington DC | Melbourne

SAGE Publications Ltd
1 Oliver's Yard
55 City Road
London EC1Y 1SP

SAGE Publications Inc.
2455 Teller Road
Thousand Oaks, California 91320

SAGE Publications India Pvt Ltd
B 1/I 1 Mohan Cooperative Industrial Area
Mathura Road
New Delhi 110 044

SAGE Publications Asia-Pacific Pte Ltd
3 Church Street
#10-04 Samsung Hub
Singapore 049483

Editor: Jai Seaman
Assistant editor: Charlotte Bush
Production editor: Manmeet Kaur Tura
Copyeditor: QuADS Prepress Pvt Ltd
Proofreader: Derek Markham
Indexer: Cathryn Prithcard
Marketing manager: Susheel Gokarakonda
Cover design: Shaun Mercier
Typeset by: C&M Digitals (P) Ltd, Chennai, India

**Library of Congress Control Number: 2020945954**

**British Library Cataloguing in Publication data**

A catalogue record for this book is available from the British Library

ISBN 978-1-5264-2403-7

# CONTENTS

| | |
|---|---|
| *List of Figures, Tables and Boxes* | xi |
| *About the Authors* | xvii |
| *Acknowledgements* | xix |
| *Preface* | xxi |

**1 Clustering and Dependence: Our Entry into Multilevel Modelling** — **1**

| | |
|---|---|
| Nested Data and Non-Independence | 3 |
| Intra-Class Correlation Coefficient | 6 |
| Effective Sample Size | 8 |
| Computing the Effective Sample Size n^eff | 9 |

**2 Multilevel Modelling: A Conceptual Introduction** — **15**

| | |
|---|---|
| Review of Single-Level Regression | 16 |
| *Regression Model With No Predictors* | 16 |
| *Multilevel Model With No Predictors* | 16 |
| Random Effects and Variance Components | 19 |
| *Residual Variances in MLM* | 19 |
| Intercepts as Outcomes Models | 20 |
| *Adding Level-1 Predictors* | 20 |
| Full Contextual (Slopes-as-Outcomes) Model | 23 |
| Variance–Covariance Components | 23 |
| Advice on Modelling Randomly Varying Slopes | 24 |
| Centring Level-1 Predictors | 26 |
| Important Guidance on Group Mean Centring | 27 |
| Estimation | 29 |
| *Conceptual Introduction to Maximum Likelihood* | 30 |
| *Maximum Likelihood Estimation in MLM* | 31 |
| *Reliability and Estimation of Randomly Varying Level-1 Coefficients* | 32 |
| *Empirical Bayes Estimates of Randomly Varying Parameters* | 33 |
| *Deviance and Model Fit* | 35 |
| *Akaike Information Criterion and Bayesian Information Criterion* | 37 |

*Using Model Fit Criteria for Model Selection* 39
*Proportion of Variance Explained* 41
*Effect Size* 43

**3 Multilevel Model Building Steps and Example** **45**

A Review of Traditional Model Building Steps in MLM 47
*Raudenbush and Bryk's Proportional Reduction in Variance* 48
Model Building Steps for Confirmatory/Predictive Models 50
An Example of a Two-Level Multilevel Model 54
*The Random Intercept-Only, One-Way ANOVA or*
*Unconditional Model (Model 1)* 56
*Random Coefficients Models (Random Intercept and Slope Models)* 58
*Evaluating Variance Components/Random Effects* 61

**4 Introduction to Structural Equation Modelling** **67**

Advantages of SEM 68
Assumptions and Requirements of SEM 70
Understanding SEM 71
What Is Path Analysis? 72
*Path Diagrams* 72
*Exogenous and Endogenous Variables* 74
Estimating Direct, Indirect and Total Effects in Path Models
With Mediation 75
*Direct Effects* 75
*Indirect Effects* 76
*Total Effect* 77
*Mediation and Bootstrapping* 77
*Mediation and Causality* 78
What Are Latent Variables? 78
*Measuring/Modelling Latent Constructs* 79
Factor Analysis 80
*Types of Factor Analysis: Exploratory and*
*Confirmatory Factor Analyses* 81
Measurement Models Versus Structural Models 81
*Disturbances and Measurement Errors* 83
Estimation 84
Model Fit and Hypothesis Testing in SEM 84
*Alternative Measures of Model Fit* 86
Model Comparison 87

*Chi-Square Difference Test* 87
*Fitting Multiple Models* 87
Model Modification and Exploratory Analyses 88
*Modification Indices* 88
Model Fit Versus Model Prediction 89
When SEM Gives You Inadmissible Results 90
Alternative Models and Equivalent Models 90
A Word of Caution 91

**5  Specification and Identification of Structural Equation Models    95**

Computing Degrees of Freedom in SEM 96
*Model Specification of Measurement (CFA) Models* 100
Model Identification: Measurement Models/CFA 102
Identification in CFA Models 103
*Freely Estimated Parameters in CFA* 103
*Degrees of Freedom for Hybrid SEM* 105
*Equations for a Measurement (CFA) Model* 106
Introduction to Systems of Equations Using Path Diagrams 107
*Getting Started With Wright's Rules* 107
*Standardised Tracing Rules* 107
*Example of the Standardised Tracing Rule* 109
*Standardised Tracing Rules for Measurement Models* 111

**6  Building Structural Equation Models    117**

The Two-Step Approach to Fitting Latent Variable Structural Models 118
*Why Fit the Measurement Model First?* 118
*Structural Model* 119
Suggested Model Fitting Sequence for Hybrid Structural Equation Model 120
*Phase 1: Fit the Measurement Model* 120
*Phase 2: Specify and Fit the Structural Model* 121
Example: Mediation Example Using a Structural Equation Model
With Latent Variables 126
*Data* 128
*Measurement Model* 128
*Conceptual Model* 130
*Model Building Steps* 130
*Measurement Model* 130
*Just-Identified Structural Model* 135

*Conceptual + Necessary Conceptually Omitted Paths Model*                138
*Direct, Indirect and Total Effects*                                      140

**7  Longitudinal Growth Curve Models in MLM and SEM**                    **145**

Conceptual Introduction to Growth Modelling                              148
*The Importance of Theory and Data*                                      148
*Requirements for Growth Models*                                         150
*Time-Structured Versus Time-Unstructured Data*                          152
*Growth Modelling Using Multilevel Versus SEM Frameworks*                153
Multilevel Model Specification for Individual Growth Models              154
*The Two-Level Multilevel Model for Linear Growth*                       154
*Variance Components*                                                    157
Estimating Linear Growth From an SEM Perspective                         159
*Means and Intercepts*                                                   159
*SEM Specification of Growth Models*                                     161
*Degrees of Freedom for SEM Growth Models*                               161
*Interpreting the Parameters in SEM Growth Models*                       163
*Model-Predicted Scores*                                                 163
Incorporating Time-Varying Covariates                                    164
*Incorporating Time-Varying Covariates in MLM*                           164
*Variance Components*                                                    166
*Incorporating Time-Varying Covariates Into SEM Latent Growth Models*    168
Piecewise Linear Growth Models                                           169
*How Many Individually Varying Slopes Can My Data Support?*              169
*The Mechanics of Coding Time for the Piecewise Model*                   171
Polynomial Growth Models                                                 174
*Intercept-Only Model*                                                   175
*Linear Model*                                                           177
*Quadratic Model*                                                        177
Fitting Higher Order Polynomial Models                                   180
*Cubic Model*                                                            180
Building Growth Models                                                    183

**8  An Applied Example of Growth Curve Modelling in
MLM and SEM**                                                            **187**

Model Building Steps for Growth Models                                   191
Model 1: The Unconditional Linear Growth Model
(Age 11 – No Covariates)                                                 192
Model 2: Piecewise Unconditional Linear Growth Model
(Three-Piece Model)                                                      197

Model 3: Piecewise Unconditional Linear Growth Model
(Two-Piece Model)                                                    202
Model 4: Unconditional Polynomial Growth Model                       206
   *Model Comparisons and Selection*                   212
Model 5: Piecewise Conditional Growth Model
(Two-Piece With Gender Covariate)                                    214
Conclusion                                                           217

*Appendix 1: Brief Introduction to Matrices and Matrix Algebra*      221
*Appendix 2: Linking Path Diagrams to Structural Equations*          227
*Appendix 3: Wright's Standardised Tracing Rules*                    231
*Appendix 4: Wright's Unstandardised Tracing Rules and Covariance Algebra*   241
*Glossary*                                                           249
*References*                                                         259
*Index*                                                              273

# LIST OF FIGURES, TABLES AND BOXES

## List of figures

1.1 Effective sample size for a sample of 1000 as a function of
intra-class correlation coefficient and average cluster size     11

1.2 Effective sample size for an original sample of 1000 as a
function of intra-class correlation coefficient and average cluster size.
This figure demonstrates that even a seemingly small intra-class correlation
coefficient can have a large effect on standard errors and tests of statistical
significance if the average cluster size is very large     12

2.1 (1) Randomly varying intercepts, (2) randomly varying slopes and
(3) randomly varying intercepts and slopes     22

4.1 A path model in which growth mindset predicts academic persistence,
which in turn predicts academic achievement     73

4.2 A simple mediational model     75

4.3 A simple unidimensional factor model     80

4.4 A structural model in which growth mindset (measured with a latent
variable) predicts academic persistence, which in turn predicts academic
achievement     82

5.1 A simple path model with four observed variables     98

5.2 A standard confirmatory factor analysis (measurement) model for
math and reading ability     101

5.3 A hybrid structural equation model: A path model that includes
latent variables     106

5.4 Tracing rule: You can go backward and then forward. This figure
illustrates a linkage due to common causes (upstream variables)     108

5.5 Tracing rule: No going forward and then backward. There is NO linkage
due to common effects (downstream variables)     108

5.6 Using the tracing rules to compute model-implied correlations     110

5.7 A two-factor CFA model with standardised path
coefficients and correlations     112

5.8 A two-factor CFA model with standardised parameters
and a correlated error     114

6.1 A conceptual model in which motivation completely mediates the pathways from cognitive ability and parental expectations to achievement. This model does not allow for direct effects from cognitive ability and parental expectations to achievement     122

6.2 The conceptual model in which motivation completely mediates the pathways from cognitive ability and parental expectations to achievement. The dashed lines represent conceptually omitted paths. The just-identified structural model includes all paths in Figure 6.2     122

6.3 Flowchart for model building steps with hybrid structural equation modelling     125

6.4 The conceptual hybrid structural equation model for our example. Note that there are three conceptually omitted paths in the structural portion of the model: (1) the path from ITBS-PR to classroom post-test mathematics achievement, (2) the path from classroom pretest mathematics achievement to ITBS-PT and (3) the path from ability to classroom post-test mathematics achievement     127

6.5 The measurement model. The five structural (conceptual) variables, Ability, ITBS-PR, Classroom Math Achievement (pretest), Classroom Math Achievement (post-test) and ITBS-PT are exogenous variables in the measurement model, and the measurement model freely estimates covariances among all five of these structural (conceptual) variables     131

6.6 Just-identified structural model     136

6.7 Conceptual + necessary conceptually omitted paths model     139

7.1 Growth in the number of articles in PsycINFO that mention growth modelling (1980–2019)     147

7.2 Individual growth plots for a sample of 20 students' scores on a reading test administered at ages 6, 8 and 10     150

7.3 Four different quadratic growth trajectories     151

7.4 A linear growth model in the structural equation modelling framework     159

7.5 A piecewise growth model with two slopes. The first slope captures growth from waves 1 to 3 and the second slope captures growth from waves 3 to 6     172

7.6 Polynomial order for polynomial growth models     175

7.7 Two cubic functions where the $a$, $b$ or $c$ in the function $ax + bx^2 + cx^3$ are either positive or negative     181

7.8 Latent growth model with a cubic growth trajectory in structural equation modelling     182

8.1   Mean *ATTIT* by age                                                                                      189
8.2   Mean *ATTIT* by age and gender                                                                           190
8.3   Individual trajectories of *ATTIT* change for nine individuals in the data.
      For the purpose of clarity in presentation, not all individuals are shown    190
8.4   Unconditional linear growth model                                                                        193
8.5   Three-piece unconditional growth model                                                                   198
8.6   Two-piece unconditional growth model                                                                     205
8.7   Predicted values by model type. The plot includes the sample means
      along with the predicted values. Note the sample means are equivalent to
      the predicted values for the three-piece and two-piece models                206
8.8   One unconditional cubic growth model                                                                     207
8.9   Two-piece conditional linear growth model                                                                215

## List of exhibits

4.1   Symbols in a path diagram                                                                                72

5.1   Systems of equations corresponding to Figure 5.2                                                         107

7.1   Variance-covariance matrix that includes four variances and
      six covariances                                                                                          171

8.1   Comparing the MLM and SEM approaches for the linear growth model    195
8.2   Comparing the MLM and SEM approaches for two-piece piecewise
      growth models                                                                                            203
8.3   Comparing the MLM and SEM approaches for the cubic growth model     208
8.4   Calculating degrees of freedom (*df*) for the unconditional linear,
      piecewise and polynomial growth models for growth in *ATTIT*,
      measured across five time points                                                                         209

## List of tables

2.1   The total penalty imposed by each of the model fit criteria for
      models that differ by 1, 2, 3 and 4 parameters                                                           39
2.2   The per-parameter penalty imposed by each of the model fit criteria
      for models that differ by 1, 2, 3 and 4 parameters                                                       40

3.1   Results of two-level analysis evaluating the effectiveness of
      a supplemental vocabulary instruction intervention with
      kindergarten students                                                                                    57

5.1    Model-implied correlations among the four observed
variables in Figure 5.6    110

5.2    Model-implied correlations among the six observed variables
in Figure 5.7    113

6.1    Descriptive statistics for the observed variables in Figure 6.4    128

6.2    Correlation matrix for the observed variables in Figure 6.4    129

6.3    Model fit information for the measurement and structural
hybrid models    132

6.4    Measurement model results (step M1)    132

6.5    Standardised results for the just-identified structural model
and conceptual + necessary conceptually omitted paths model    137

6.6    Table of conceptually omitted paths    137

6.7    Standardised total, total indirect, specific indirect and direct effects
for Model S2 (the conceptual + significant deleted paths model)
with 95% bootstrap CIs    140

7.1    Demonstration of how centring using age controls for age    157

7.2    Demonstrates how the means and loadings for the intercept
and slope are used to compute the expected mean at each time point    164

7.3    Coding scheme for two distinct growth rates    173

7.4    Coding scheme for two separate growth rates with baseline
growth slope    174

7.5    Computation of first-order, quadratic and composite
growth parameters    179

8.1    Descriptive statistics for tolerant attitudes towards deviant
behaviour (ATTIT)    189

8.2    Model 1 results using the linear growth model in MLM and SEM    194

8.3    Predicted values for the linear growth model    197

8.4    Coding for the three-piece linear growth model    198

8.5    Model 2 results using the three-piece growth model in MLM and SEM    199

8.6    Predicted values for the three-piece piecewise growth model    201

8.7    Coding time for the two-piece piecewise model    202

8.8    Model 3 results using the two-piece growth model in MLM and SEM    204

8.9    Predicted values for the two-piece piecewise growth model    206

8.10    Coding time for the cubic polynomial growth model    210

8.11    Model 4 results using the cubic polynomial growth model
in MLM and SEM    210

8.12   Predicted values for the cubic polynomial growth model            211
8.13   *ATTIT* sample means compared to predicted *ATTIT* values for
       each longitudinal model                                          212
8.14   MLM model fit indices                                            213
8.15   SEM model fit indices                                            214
8.16   Model 5 results using the two-piece piecewise conditional
       growth model in MLM & SEM                                        216

## List of boxes

2.1   A Note on Multilevel Versus Combined Equations                     18

7.1   How to Specify Means Within a Standard Latent Variable Model      161
7.2   Model Building Steps: Growth Curve Models                         185

# ABOUT THE AUTHORS

**D. Betsy McCoach** is Professor of research methods, measurement and evaluation in the Educational Psychology Department at the University of Connecticut, where she teaches graduate courses in structural equation modelling, multilevel modelling, advances in latent variable modelling and instrument design. She has co-authored more than 100 peer-reviewed journal articles, book chapters and books, including *Instrument Design in the Affective Domain* and *Multilevel Modeling of Educational Data*. She founded the *Modern Modeling Methods* conference, held annually at the University of Connecticut. She is co-principal investigator for the National Center for Research on Gifted Education and has served as principal investigator, co-principal investigator and/or research methodologist for several other federally funded research projects/grants. Her research interests include latent variable modelling, multilevel modelling, longitudinal modelling, instrument design and gifted education.

**Dakota W. Cintron** is a postdoctoral scholar in the Evidence for Action (E4A) Methods Laboratory. His research focuses on the application, development and assessment of quantitative methods in the social and behavioural sciences. His areas of research interest include topics such as item response theory, latent variable and structural equation modelling, longitudinal data analysis, hierarchical linear modelling and causal inference. He earned his Ph.D. in educational psychology from the University of Connecticut. He has previously held research positions at the Institute for Health, Health Care Policy and Aging Research, the National Institute for Early Education Research, and New Visions for Public Schools.

# ACKNOWLEDGEMENTS

First, thanks are due to Sarah Newton and Joselyn Perez, who read and provided feedback and edits on several of the chapters in this book and Faeze Safari, who helped check and format the references. Thanks also to the two co-editors of this series, Dick Wiggins and Malcolm Williams, for inviting us to write this volume, for providing helpful editorial feedback, and for being patient with and supportive of us throughout this process. Thanks to the editors and publications staff at Sage, especially Jai Seaman and Lauren Jacobs.

## From the first author

I would also like to thank David A. Kenny, who has provided exceptional mentorship to me as I have learned, used, and eventually mastered these techniques, and who provided me with sage advice about the book. (However, please do not blame Dave for anything that I have not yet mastered.) I would also like to thank Ann O'Connell, who was my partner in teaching a week-long multilevel modeling course for over 10 years. And although it may seem unusual, I would also like to thank my co-author, Dakota Cintron for his support and patience throughout this process. We started this project as he began his Ph.D. with me, and we completed it as he finished his Ph.D.

Thanks to all my current and former students in my SEM and HLM classes, especially those who asked difficult questions. Thanks to my exceptional faculty colleagues in the RMME program, the many *Modern Modeling Methods* (*MMM*) conference keynoters and speakers, and my conference friends from MMM, AERA, and SREE, all of whom have taught me so much about modeling.

Finally, I would like to thank my family and friends for their support and encouragement throughout this process. They have been very patient, waiting for the book to be finished. My husband, Del Siegle, babysat our children, Jessica and Del, on seemingly countless weekends so that I could write and edit this volume. The book took so long to write that by the time I finished, our children no longer needed a babysitter!

# PREFACE

Welcome to the wonderful world of modelling. One might quibble with our use of the word 'modern' to describe the modelling methods that we introduce in this volume: multilevel modelling, structural equation modelling, and growth curve modelling techniques have existed since the last century. In fact, Willett (1997) referred to growth curve modelling as a modern method for the measurement and analysis of change over 20 years ago! However, advances in computing capability and software accessibility have led to their meteoric rise in popularity in the 21st century.

The title of the book is inspired by an annual conference, held every year at the University of Connecticut. The *Modern Modeling Methods* ($M^3$ or MMM) conference is an interdisciplinary conference designed to showcase the latest statistical modelling methods and to present research related to development and implementation of these methods. The first author founded the MMM conference in 2011 to create a home for methodologists from different disciplines in the social sciences and statistics to share methodological developments and methodological research related to latent variable and mixed effects models. These models are used in a wide variety of disciplines across the social and behavioral sciences and education. The substantive areas in which these models are applied vary a great deal across disciplines, and the terms and notation may differ across disciplines. However, the fundamental modelling techniques employed are the same (or quite similar). Therefore, this book provides a foundation for two basic modelling methods: structural equation modelling and multilevel (mixed) modelling. Afterwards, we demonstrate how each can be used to fit growth curve models. Because we are educational researchers, our examples are educational in nature. However, the modelling methods that we present are applicable across the social, behavioral and natural sciences.

This book contains three sections. Section 1 (Chapters 1–3) focuses on multilevel modelling. Chapter 1 introduces the notion of dependence. Chapter 2 provides an overview of multilevel modelling (MLM). Chapter 3 outlines a sequence of recommended model building steps for MLM and provides a data analytic example of MLM. Section 2 (Chapters 4–6) introduces structural equation modelling (SEM), including path modelling, measurement models and latent variable structural equation models. Chapter 4 provides an overview of many of the core concepts in SEM. Chapter 5 covers several foundational aspects of SEM specification, identification and generation of model-implied variances and covariances. Chapter 6 outlines a sequence of recommended model building steps for SEM and provides an example of SEM. Section 3 (Chapters 7 and 8) introduces growth curve modelling and uses growth curve models to integrate the fundamental techniques covered in the first two sections. As such, the third section helps explicate the correspondence between MLM and SEM. Chapter 7 discusses how to fit and interpret growth models in both a multilevel framework and an SEM framework and provides a recommended model building sequence for fitting growth curve models (in either SEM or MLM). Chapter 8 provides an applied example of a non-linear growth curve model. We present and compare the results of the structural equation and multilevel models. Thus, the growth modelling chapters serve as a finale of sorts, pulling together many of the topics and concepts from across the entire book.

Given the length of the book and the scope of the content that we cover, we have had to curtail potentially interesting conversations about the methods that we introduce. We also had to eliminate many of the technical details about fitting the models, and we are virtually silent on the actual computation of the models. We also neglect to present most of the mathematical foundations of the models. For those who are interested in a more theoretical and mathematical introduction to structural equation models, we have included four appendices that introduce matrix algebra and covariance algebra and present the standardised and unstand-ardised tracing rules in a more formal fashion. Students who would like to pursue these methods in more depth should familiarise themselves with the content of the appendices. For example, the appendices represent content that our Ph.D. students in research methods, measurement and evaluation are expected to master. However, for students who want a more conceptual introduction to SEM, the appendices are optional: it is not necessary to read the appendices to understand the material presented in the book.

This has been a challenging book to write because we try to cover so much ground in such a short book. The topics for each of these three sections (SEM, MLM and growth modelling) can easily fill an entire 3-credit, one-semester graduate course in quantitative methodology, and we could have easily devoted an entire volume to each of these three topics (MLM, SEM and growth curve modelling). Indeed, there are entire full-length books on each of these topics. Our goal for this book is to provide a solid foundational understanding of these modelling techniques and to provide the necessary preparation for more advanced treatments of these topics. However, this book is not meant to be the last word or the most comprehensive treatment for any of these three modelling techniques. We hope that this introduction to modelling excites and stimulates your curiosity enough that you want to tackle whole books on these topics after reading this book, and we provide suggestions for supplemental texts throughout this volume.

This book is designed for applied researchers, and we presume no prior knowledge of SEM, MLM or growth modelling. We have tried to minimise equations and technical details, and the book has no prerequisite mathematics requirements (i.e. this volume is accessible to people without any college-level mathematics knowledge, such as calculus). However, we do assume that the reader has a solid foundation in and strong understanding of multiple regression. Therefore, if you are not completely comfortable with multiple regression, we suggest that you read Book 7 (*Linear Regression: An Introduction to Statistical Models* by Peter Martin) before embarking on this journey.

Although we do present tests of statistical significance and hypothesis tests throughout the current volume, we rely on logical, systematic and reproducible decision criteria when building, choosing and interpreting models. Further, we emphasise the importance of effect sizes, practical significance and interpretability over blind adherence to an arbitrary dichotomisation of results into two buckets: $p < .05$ and everything else. We agree wholeheartedly with Goldstein (2011), who said 'the usual form of a null hypothesis, that a parameter value or a function of parameter values is zero is usually intrinsically implausible and also relatively uninteresting. Moreover, with large enough samples, a null hypothesis will almost certainly be rejected' (p. 39).

When using any modelling techniques, a coherent conceptual base should inform and guide the statistical analyses. The model theory and included variables need to be consistent with the purposes of the study and the research questions or study hypotheses. Modelling techniques provide research tools to develop and test substantive theories. The models that we present in the pages that follow allow us to ask and answer interesting and nuanced research questions. However, the role of theory becomes even more important when using modelling techniques such as SEM and MLM. No statistical model, no matter how 'modern' or fancy, can replace the need for strong substantive theory. Instead, when carefully and thoughtfully developed and applied, modelling techniques provide a complement to strong substantive theory. One of the challenges of writing about methods is that the focus is on the methods, rather than on the substance. Therefore, descriptions of methodological decisions often seem canned or trite, and interpretations feel shallow. We have tried to present thoughtful guidelines for how to build models. However, we fully recognise that in trying to make our examples accessible and digestible to a wide audience, they necessarily lack the depth, complexity and nuance that you are likely to encounter in your own research. Thus, understand that our guidelines are **suggestions**, not commandments. Likewise, our examples represent merely one simple, constrained set of exemplars rather than an exhaustive set of potential analyses. We walk through a small set of simple models to provide a sense of the process of building, evaluating and interpreting data using the models that we highlight.

We have deliberately chosen not to present syntax or code or gear the content of our chapters towards particular software packages. The first author has spent more than 20 years using SEM and MLM and more than 15 years teaching others how to use these two techniques. Over the course of that time, she has used AMOS, LISREL, EQS, Mplus and lavaan to fit structural equation models, and she has used HLM, MLWin, Mplus, several packages in R, SAS, Stata and SPSS to fit multilevel models. The popularity, availability and functionality of statistical software packages for MLM and SEM will certainly change across time, but the fundamental concepts that we present in this book are timeless. However, we do recognise that learning new statistical packages can be challenging and frustrating. One of the obstacles preventing some researchers from using these techniques is the fear of needing to learn a different software to set up and run multilevel and/or structural equation models. Therefore, on the companion website for this book, we have provided the data for the models that we present in this book, as well as the syntax/code to run the models in Chapters 3, 6 and 8 in one or more software packages. These days, we prefer to conduct our latent variable models in Mplus or lavaan (an R package) and our multilevel models in HLM8, lme4 (an R package), Stata or Mplus. In Europe, MLWin is the most historically important MLM program. In the USA, it is HLM. In fact, HLM was such a popular package that in the 1990s and early 2000s, it was not uncommon to hear Americans refer to multilevel models as HLMs. There was a joke that you could tell where someone was from based on the way that they referred to these models. Those who called them HLMs were almost surely Americans (and probably used the HLM software); Europeans conducted MLMs, most likely using MLWin. These days, it seems as if R has captured a lion's share of the statistical software market, and that is true in modelling as well. The great advantage of R is that it is free and has open access. (The first author often jokes that R is the Wikipedia of statistical software packages, and frankly, the analogy is quite apt.) If you visit our companion website, you will find R code for all of the examples and Mplus syntax for the SEM and longitudinal models. Without further ado, let's dive into the wonderful world of modelling!

# 1

# CLUSTERING AND DEPENDENCE: OUR ENTRY INTO MULTILEVEL MODELLING

## Chapter Overview

Nested data and non-independence ........................................................... 3

Intra-class correlation coefficient ................................................... 6

Effective sample size ................................................... 8

Computing the effective sample size n^eff ................................................. 9

Further Reading ................................................... 13

Frequently in the social sciences, our data are nested, clustered or hierarchical in nature: individual observations are nested within a hierarchical structure. 'The existence of such data hierarchies is neither accidental nor ignorable' (Goldstein, 2011, p. 1). Examples of naturally occurring hierarchies include students nested within classrooms, teachers nested within schools, schools nested within districts, children nested within families, patients nested within hospitals, workers nested within companies, husbands and wives nested within couples (dyads) or even observations across time nested within individuals. 'Once you know that hierarchies exist, you see them everywhere' (Kreft & de Leeuw, 1998, p. 1).

**Multilevel modelling** (MLM) provides a technique for analysing such data. It accounts for the hierarchical structure of the data and the complexity that such structure introduces in terms of correctly modelling variability (Snijders & Bosker, 2012). Multilevel models are often referred to as **hierarchical linear models**, *mixed models*, *mixed-effects models* or *random-effects models*. Researchers often use these terms interchangeably, although there are slight differences in their meanings. For instance, *hierarchical linear model* is a more circumscribed term than the others: it assumes a normally distributed outcome variable. In contrast, mixed-effects or random-effects models are more general than multilevel models: they denote **non-independence** within a data set, but that non-independence does not necessarily need to be hierarchically nested.

In this book, we focus on one particular type of random-effects model: the multi-level model, in which units are hierarchically nested within higher level structures. Other common random-effects models include cross-classified random-effects models, which account for non-independence that is crossed, rather than nested. For example, in longitudinal educational studies, students often change teachers or transfer from one school to another; hence, students experience distinct combinations of teachers or schools. In such scenarios, students are *cross-classified* by two teachers or two schools. Multiple-membership models allow for membership in multiple clusters simultaneously. Although cross-classified and multiple-membership models are random-effects models, they are not purely multilevel models because they do not exhibit clean, hierarchical data structures. This book focuses on **hierarchical linear modelling (HLM)/multilevel modelling (MLM)**. Interested readers should refer to the following resources for more information about cross-classified and multiple-membership models: Airoldi et al. (2015), Beretvas (2008) or Fielding and Goldstein (2006). In addition, this book assumes normally-distributed continuous outcomes. To learn more about using MLM techniques with non-normal (binary, ordinal or count) outcomes, see O'Connell and McCoach (2008) or Raudenbush and Bryk (2002).

In MLM, **organisational models** are models in which people are clustered within hierarchical structures such as companies, schools, hospitals or towns. Multilevel models also prove useful in the analysis of longitudinal data, where observations across time are nested within individuals.

In this chapter, we introduce the MLM framework and discuss two-level multilevel models in which people are clustered within organisations or groups. We begin our introduction to MLM by introducing basic terms and ideas of MLM as well as introducing one of the most fundamental concepts in the analysis of **clustered data**: the **intra-class correlation coefficient** (ICC). Subsequently, Chapter 2 provides an overview of standard two-level multilevel organisational models, and Chapter 3 illustrates fundamental MLM techniques with an applied example. In Chapters 4–6, we turn our attention to structural equation modelling. In Chapters 7 and 8, we return to MLM, demonstrating its application to individual growth models.

## Nested data and non-independence

Most traditional statistical analyses assume that observations are independent of each other. In other words, the assumption of independence means that subjects' responses are not correlated with each other. For example, imagine that a survey company administers a survey to a sample of participants. Under the assumption of independence, one participant's responses do not correlate with the responses of any of the other participants. The assumption of independence might be reasonable when data are randomly sampled from a large population. However, the responses of people clustered within naturally occurring organisational units (e.g. schools, classrooms, hospitals, companies) are likely to exhibit some degree of relatedness, given that they were sampled from the same organisational unit. For instance, students who receive instruction together in the same classroom, delivered by the same teacher, tend to be more similar in their achievement (and other educational outcomes) than students instructed by different teachers.

Observations within a given cluster often exhibit some degree of dependence (or interdependency). In such a scenario, violating the assumption of independence produces incorrect **standard errors** that are smaller than they should be. Therefore, inferential statistical tests that violate the assumption of independence have inflated Type I error rates: they produce statistically significant effects more often than they should. The Type I error rate is the probability of rejecting the null hypothesis when the null hypothesis is correct. Alpha, the desired/assumed Type I error rate, is commonly set at .05. However, alpha may not equal the actual Type I error rate if we fail to meet the assumptions of our statistical test (i.e. normality, independence, homoscedasticity etc.). MLM techniques allow researchers to model the relatedness of observations within clusters explicitly. As a result, the standard errors from multilevel analyses account for the clustered nature of the data, resulting in more accurate Type I error rates.

The advantages of MLM are not purely statistical. Substantively, it may be of great interest to understand the degree to which people from the same cluster are similar to each other and to identify variables that help predict variability both within and across clusters. Multilevel analyses allow us to exploit the information contained in clustered samples and to partition the variance in the outcome variable into **between-cluster variance** and **within-cluster variance**. We can also use predictors at both the **individual level** (level 1) and the group level (level 2) to try to explain this between- and within-cluster variability in the outcome variable.

In general, MLM techniques allow researchers to model multiple levels of a hierarchy simultaneously, partition variance across the levels of analysis and examine relationships and interactions among variables that occur at multiple levels of a hierarchy. In MLM, a level is 'a focal plane in social, psychological, or physical space that exists within a hierarchical structure' (Gully & Phillips, 2019, p. 11). Generally, the levels of interest within an analysis depend on the phenomena and research questions (Gully & Phillips, 2019). For example, in a study of instructional techniques, where students are nested within teachers, students are level-1 units and teachers are level-2 units. In contrast, in a study of teachers' perceptions of their principals' leadership, teachers are nested within principals. In this case, teachers are level-1 units and principals are level-2 units. Often, researchers use the term *organisational model* to refer to cross-sectional MLM where individuals (level-1 units) are clustered within some sort of organisational, administrative, social or political hierarchy (level-2 units).

Traditional correlations and regression-based approaches estimate the relationship between two variables. However, standard single-level analyses (which ignore the clustered/hierarchical nature of the data) assume that the relationship between the variables is constant across the entire sample. MLM allows the relationships among key substantive variables to randomly vary across clusters. For example, the relationship between socio-economic status (SES) and achievement may vary by school. In some schools, student SES may be a strong (positive) predictor of students' subsequent academic achievement; in other schools, SES may be completely unrelated to academic achievement (Raudenbush & Bryk, 2002).

Additionally, in MLM, researchers can study relationships among variables that occur at multiple levels of the data hierarchy as well as potential interactions among variables at multiple levels while allowing relationships among lower-level variables to randomly vary by cluster. How much of the between-cluster variability in these relationships (or in the cluster means) can be explained by cluster-level variables? For instance, imagine we want to study the relationships between student ability, teaching style and academic achievement. The data are clustered: students are nested within teachers (classrooms). For simplicity, assume that each teacher teaches only one class. Therefore, the teacher and classroom levels are synonymous, and student ability varies across different students taught by the same teacher. Consequently, student ability is an individual-level (or level-1) variable. Although teaching style varies across teachers, every student within a given teacher's class is exposed to a single teacher with one individual teaching style. Therefore, teaching style varies across classrooms, but not within classrooms, so teaching style is a classroom/teacher (cluster) level, or level-2 variable.

Of course, the effect of a teacher's teaching style does not necessarily have the same effect on all students. In our current example, we might hypothesise that teaching style moderates the effect of student ability on student achievement. In other words, the relationship between student ability and student achievement varies as a function of teachers' teaching style. For example, some teachers may strive to ensure that all students in the class meet the same set of grade-level standards and are exposed to the same content at the same level, ensuring that all students in the class have the same set of skills and knowledge. In contrast, other teachers may differentiate instruction to meet the needs of individual students. We hypothesise that the relationship between ability and achievement would likely be stronger in the classrooms where teachers differentiate instruction than in the standards-based classrooms. In a standard linear regression model, we can include an interaction between teaching style and student ability. However, the multilevel framework allows the slope for the effect of students' ability on achievement to randomly vary across classrooms, even after controlling for all teacher- and student-level variables in the model. If the ability/achievement slope randomly varies, even after including teaching style in the model, the relationship between ability and achievement does indeed vary across classrooms but the teachers' teaching style does not fully explain the between-class variability in the ability/achievement relationship. Perhaps, other omitted classroom-level variables may explain the variability in the ability/achievement relationship across classes. MLM allows us to ask and answer more nuanced questions than are possible within traditional regression analyses.

As the preceding paragraphs highlight, multilevel models are incredibly useful for studying organisational contexts like schools, companies or families. However, many other types of data exhibit dependence. For instance, multiple observations collected on the same person represent another form of nested data. Growth curve and other longitudinal analyses can be reframed as multilevel models, in which observations across time are nested within individuals. Using the MLM framework, we partition residual or error variance into within-person **residual variance** and between-person residual variance. In such a scenario, between-person residual variance represents between-person variability in any randomly varying level-1 parameters of interest, such as the intercept (which we commonly centre to represent initial status in growth models) and the growth slope. Within-person residual variance represents the variance of time-specific residuals, which is generally referred to as measurement error. We explore multilevel growth models in Chapters 7 and 8 of this book and demonstrate how to reframe the basic MLM framework to analyse longitudinal data. However, for the remainder of Chapters 1 to 3, we focus exclusively on cross-sectional organisational models.

## Intra-class correlation coefficient

This section introduces one of the most fundamental concepts in MLM: the intraclass correlation coefficient (ICC). The ICC measures the proportion of the total variability in the outcome variable that can be explained by cluster membership. The ICC also provides an estimate of the expected correlation between two randomly drawn individuals from the same cluster (Hox et al., 2017).

Most traditional statistical tests (multiple regression, analysis of variance [ANOVA] etc.) assume that observations are independent. The assumption of independence means that cases 'are not paired, dependent, correlated, or associated in any way' (Glass & Hopkins, 1996, p. 295). Nested or clustered data violate this assumption because clustered observations tend to exhibit some degree of interdependence. In other words, observations nested within the same cluster tend to be more similar to each other on a given outcome variable than observations drawn from two different clusters. This interdependence, resulting from the sampling design, affects the variance of the outcome, which in turn affects estimates of the standard errors for model parameters.

Of course, the degree of dependence also varies by outcome variable, and some outcome variables may not exhibit any discernible dependence, even though the observations are clustered. For example, students are clustered within classrooms, so we would expect to see that academic outcomes such as mathematics and reading achievements exhibit some degree of within-class/within-teacher dependence. However, other variables may exhibit little to no dependence, even though they are clustered. Therefore, we compute the ICC separately for each outcome variable of interest.

To better understand this phenomenon, let's imagine that a research assistant named Igor receives the task of surveying 1000 people about how many hours they sleep per night (on average). Instead of randomly sampling 1000 people, he decides that he can accomplish the task much more quickly if he samples 250 households (each of which has four members) and asks all members of each household to respond to the sleep survey. For simplicity, let's assume that each of the 250 households are drawn from different neighbourhoods so that outside noise such as car alarms or sirens that affect one household do not affect any other households. Of course, in households where one member is not sleeping well, other members of the household also tend to sleep less well. (Classic sleep disturbances that affect the sleep patterns of entire households to some degree include crying babies, barking dogs, visitors, late or early household events etc.). However, in this scenario, there is no reason to believe that sleep patterns exhibit any dependencies across households.

So, what happens to variability in the outcome variable, sleep, within households? Because people's sleep hours are more similar within a given household, the expected variability in sleep hours for members of the same household is smaller than the expected variability of members who reside in different households. Therefore, knowing the household in which a person resides can help explain some of the variability in sleep hours. This is *between-cluster variability*. Another way of thinking about between-cluster variability is to imagine computing the mean number of sleep hours for each household in the study. The degree to which those household-aggregated means vary across clusters (households) represents *between-cluster variance*. At first, it may seem counter-intuitive that similarities within clusters actually relate to between-cluster variance. However, a quick thought experiment may help. Imagine that Igor samples households in Stepford, and in these households, every member must sleep exactly as much as the patriarch. In such a scenario, all members of each household have the exact same sleep hours. Therefore, all variance in the outcome variable (sleep hours) must be between-cluster variance; there is no *within-cluster variance*.

Of course, Stepford doesn't actually exist. Even though people within the same household may be more similar to each other than to people from different households, they are not exactly the same. With clustered data, knowing the cluster helps to explain some (but not all) of the variability in the outcome variable of interest.

Instead, we can partition the total variability in sleep time into the portion that is within clusters (i.e. how much do members of the same household differ from their household average in terms of sleep time?) and the portion that is between clusters (i.e. how do households differ from each other in terms of sleep time?). The degree to which people within the same household (or cluster) differ from the household average is *within-cluster variability*. In other words, it is the (pooled) variability across people within the same cluster. Conceptually, *between-cluster variability* represents the variability in the cluster means. Between-cluster variance is analogous to aggregating data to the cluster level and computing cluster means for each cluster and then estimating how much the cluster means vary.

The ICC describes how similar, or homogeneous, individuals are within clusters and how much they vary across clusters: it quantifies the degree of dependence, or the degree of relationship among units from the same cluster (Hox, 2010; Raudenbush & Bryk, 2002; Snijders & Bosker, 2012). The ICC is the proportion of between-cluster variance, or the proportion of the total variability in the outcome variable that can be explained by cluster membership. The calculation of the ICC (often symbolised as $\rho$, 'rho') involves partitioning the total variability in the outcome variable into within-cluster variance ($\sigma^2$) and between-cluster variance ($\tau_{00}$). To compute the ICC, we simply divide the between-cluster variability ($\tau_{00}$) by the total variability ($\tau_{00} + \sigma^2$), as the following equation shows:

$$\rho = \frac{\tau_{00}}{\tau_{00} + \sigma^2} \tag{1.1}$$

A large ICC indicates that there is a large degree of similarity within clusters ($\sigma^2$ is small) and/or a large degree of variability across clusters ($\tau_{00}$ is large). An ICC of 1 indicates that all observations within a cluster are perfect replicates of each other and all variability lies between clusters: the within-cluster variance is 0. An ICC of 0 indicates that observations within a cluster are no more similar to each other than observations from different clusters: the between-cluster variance is 0. We expect a **simple random sample** from a population to have an ICC of 0; the assumption of independence is the assumption that the ICC = 0.

Returning to our sleep example, imagine the ICC (the proportion of between-cluster variability) is .40. This means 40% of the variance in sleep time lies between households and 60% of the variance in sleep time lies within households. It also means that the expected correlation in sleep time for two members of the same household is .40.

To recap, cluster means vary (*between-cluster variance*). People in the same cluster also differ from each other (though two people from a single cluster differ less than two randomly selected people) (*within-cluster variance*). The sum of the within- and between-cluster variances represents total variance in the outcome variable. The ICC indicates the proportion of total variability explained by group membership; an ICC of 1.00 suggests members of each cluster are perfect replicates of one another (so, all variation occurs across clusters), whereas an ICC of .00 implies that cluster members are completely independent of one another (i.e. uncorrelated), akin to a theoretical random sample.

## Effective sample size

A concept related to the ICC is **effective sample size** ($n_{eff}$). Is Igor's sample of 1000 people within 250 households really the same as sampling 1000 people from 1000 different households? No – given the built-in dependence of people within the same household, Igor hasn't really obtained as much information about people's sleep habits as he would have if he had actually sampled 1000 people from 1000 different households.

So, why does ignoring this non-independence (as traditional tests of significance do) increase the possibility of making a Type I error (rejecting the null hypothesis when we should have failed to reject it)? Because people within clusters are more homogeneous than people from different clusters, the variance of the clustered data is smaller than the variance from a truly independent sample of observations. Therefore, treating clustered data as independent underestimates the sampling variance, which in turn produces underestimated standard errors. In other words, given that people in the sample are non-independent (or somewhat redundant with each other), the $n_{eff}$ is smaller than the actual sample size for the study. The $n_{eff}$ for a given sample is a function of the degree of non-independence in the sample and the number of people per cluster. The degree to which the effective sample size and the actual sample size differ determines the degree to which the standard errors from traditional statistical tests are underestimated. To estimate how much the clustered nature of the data impacts the standard errors, we must account for both the homogeneity within clusters (ICC) and the average cluster size ($\bar{n}_j$).

So, what is the effective sample size for Igor's sample? It is certainly less than 1000. One might be tempted to aggregate the data up to the household level and then use the mean sleep score of each household as the outcome variable, resulting in a sample size of 250. However, such an approach is overly conservative. There is still considerable variability in sleep hours within each household. By aggregating to the household level, we would lose all of the information about within-household variability in sleep time. So, aggregating to the cluster level both undersells the amount of information in the sample data and discards substantively interesting information about how different people within the same cluster differ from each other.

So, Igor faces a difficult question: is his sample more like a sample of 250 people, a sample of 1000 people or something in between? The $n_{eff}$ provides the answer. Let's first consider two extremes. When the ICC is .00, then the $n_{eff}$ is equal to the total number of observations. When the ICC is 1.00, observations within a cluster are complete replicates of each other, so sampling more than one unit per cluster is completely unnecessary. Thinking back to our sleep example, an ICC of 1.00 indicates that people within the same household receive identical amounts of sleep. Therefore, sampling more than one person per household would provide no additional information. In such an unlikely situation (at least for those of us who study humans!), the effective sample size would equal the number of clusters. When the ICC is larger than .00 but smaller than 1.00, the effective sample size is somewhere between the total number of people in the sample (as it is when the ICC = .00) and the number of clusters (as it is when the ICC = 1.00).

## Computing the effective sample size n^eff

Computing the effective sample size requires both the ICC ($\rho$) and the average number of observations per cluster ($\bar{n}_j$). Using the effective sample size, we can adjust our standard errors to account for the non-independence.

Using the ICC and the average number of observations per cluster, first we compute the **design effect** (**DEFF**; Kish, 1965). The design effect is a ratio of the **sampling variability** for the study design compared to the sampling variability expected under a simple random sample (SRS). We calculate the DEFF using the following equation:

$$\text{DEFF} = \frac{\text{var}(design)}{\text{var}(SRS)} = 1 + \rho(\bar{n}_j - 1) \tag{1.2}$$

where $\bar{n}_j$ is the average sample size within each cluster and $\rho$ is the ICC. If this ratio equals 1.00 (which only happens when the ICC = .00), then the clustering has no effect. However, DEFF greater than 1.00 indicates some degree of dependence of observations within clusters; this increases the actual Type I error rate above the nominal Type I error rate ($\alpha$). Using the design effect, we can calculate the $n_{eff}$, or the sample size that we should use to more appropriately compute the standard errors for our study. The formula for $n_{eff}$ is simply (Snijders & Bosker, 2012):

$$n_{eff} = \frac{N}{\text{DEFF}} = \frac{N}{1 + \rho(\bar{n}_j - 1)} \tag{1.3}$$

where $\bar{n}_j$ is the average cluster size and $\rho$ is the ICC and $N$ is the total sample size.

Now, let's calculate the DEFF and the $n_{eff}$ for Igor's study, assuming that the ICC = .40. The DEFF for Igor's study is $1 + .4(4 - 1) = 2.2$. This means that the $n_{eff}$ for Igor's study is 1000/2.2, or 454.55, which is about half as large as the actual sample size. So, how do we fix Igor's error? We have two options: (1) we could use MLM techniques or (2) we could adjust the standard errors from a traditional statistical analysis to account for the non-independence in our data. To compute the corrected standard errors, we simply substitute the $n_{eff}$ for $n$. To illustrate, let's correct the standard error of the mean for the degree of clustering in our sample. The standard error of the mean is the square root of the variance ($\sigma^2$) divided by the sample size, $\sqrt{\sigma^2/n}$, which equals the standard deviation divided by the square root of the sample size, $\sigma/\sqrt{n}$. In Igor's sample, the standard deviation in the number of sleep hours per night is 2.00. Therefore, the standard error using simple random sampling is = $2/\sqrt{1000} = 0.063$. However, the effective sample size of Igor's sample is 454.22, which is much smaller than it would have been if he had sampled 1000 people from 1000 different households. We replace $n$ with $n_{eff}$ in the denominator of the standard error formula to correct the standard error of the mean. Replacing $n$ with $n_{eff}$, $2/\sqrt{454.22}$ results in a standard error of 0.094. Thus, the corrected standard error is almost 50% larger than it would have been if we incorrectly assumed our sample of 1000 people were completely independent.

Alternatively, we can also adjust previously computed standard errors using the square root of the DEFF, called the *root design effect* (DEFT; Thomas & Heck, 2001). The DEFT indicates the degree to which the standard errors need to increase to account for the clustering (non-independence). Recall that the DEFF for our study was 2.2. The square root of 2.2 is 1.48 (DEFT). Multiplying 1.48 by the original standard error, 0.063, provides the corrected standard error, 0.094, which matches the standard error computed using $n_{eff}$.

To summarise, two factors influence the design effect: (1) the average cluster size (i.e. the average number of individuals per cluster) and (2) the ICC. Holding average cluster size constant, as the ICC increases, the design effect increases. Similarly, holding ICC constant, as the average cluster size increases, the design effect increases. The effective sample size is the actual sample size divided by the DEFF. In our sleep example, if the average cluster size were 10 (instead of 4) with an ICC of .40, the DEFF would be 4.6, resulting in a DEFT of 2.14 and an effective sample size 217.4. In this scenario, our corrected standard error estimate would be 0.063 * 2.14= 0.135, meaning the corrected standard error is over twice as large as the original standard error.

Figure 1.1 illustrates the effect of increasing average cluster size on the effective sample size. The actual sample size is 1000. This graph presents curves for two common ICC values for school-based research: .15 and .30. The *y*-axis depicts the drop in effective sample size as the average cluster size increases. Holding cluster size constant, $n_{eff}$ is lower when the ICC is higher: $n_{eff}$ is consistently lower when ICC = .30.

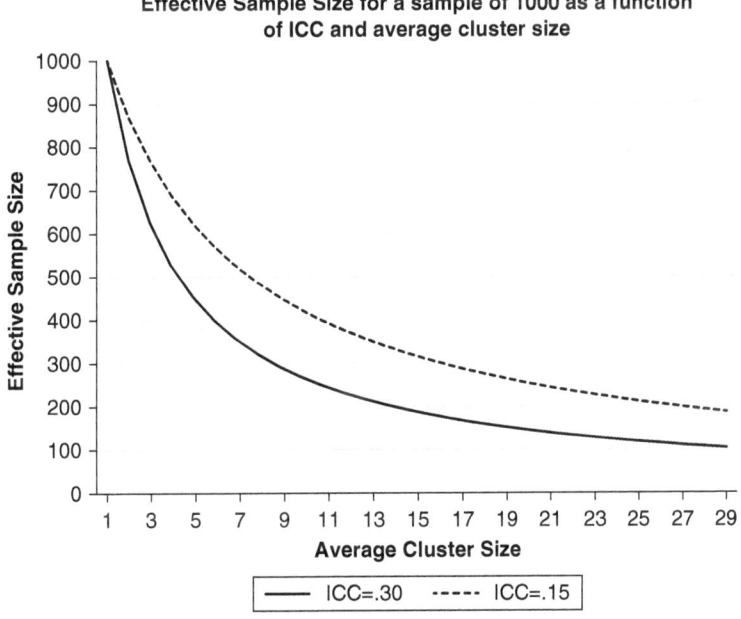

**Figure 1.1** Effective sample size for a sample of 1000 as a function of intra-class correlation coefficient and average cluster size

Some researchers mistakenly believe that they can safely ignore small ICCs. However, small ICCs, coupled with large average cluster sizes, can still result in severely underestimated standard errors. Figure 1.2 illustrates the dangers of ignoring small ICCs when the average cluster size is large. The $n_{eff}$ is a function of ICC and average cluster size for three very small ICC values: .01, .02 and .05, again assuming an original sample size of 1000. For example, with an ICC of .05 and an average of 50 units per cluster, the DEFF is 3.45, so the DEFT is 1.86 ($\sqrt{3.45} = 1.86$) and $n_{eff}$ is 289.86. With an ICC of .01 and 50 people per cluster, $n_{eff}$ is 671.14. With an ICC of .01 and a cluster size of 250 people, $n_{eff}$ is 286.53. In general, if either the ICC or the average cluster size is large, then design effect is non-ignorable. These corrections, which enlarge the standard error and increase the $p$-value, have a major impact on tests of statistical significance. Luckily, when we use MLM, it is not necessary to correct standard errors. MLM produces standard errors that account for the dependency induced by clustering.

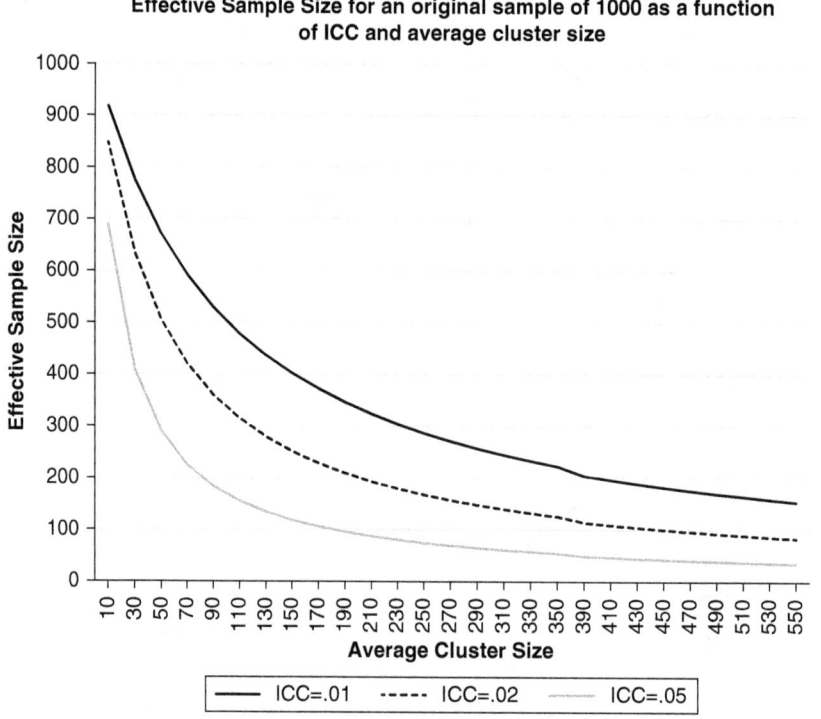

**Figure 1.2** Effective sample size for an original sample of 1000 as a function of intra-class correlation coefficient and average cluster size. This figure demonstrates that even a seemingly-small intra-class correlation coefficient can have a large effect on standard errors and tests of statistical significance if the average cluster size is very large

**Chapter Summary**

- Observations within a given cluster often exhibit some degree of dependence (or interdependency). In such a scenario, violating the assumption of independence produces incorrect standard errors that are smaller than they should be. Therefore, inferential statistical tests that violate the assumption of independence have inflated Type I error rates: they produce statistically significant effects more often than they should.
- Multilevel modelling (MLM) allows researchers to model the relatedness of observations within clusters explicitly. The standard errors from multilevel analyses account for the clustered nature of the data, resulting in more accurate Type I error rates.
- Multilevel analyses can partition the variance in the outcome variable into between-cluster variance and within-cluster variance.
- Predictors at both the individual level (level 1) and the group level (level 2) may explain this between-cluster variability in the outcome variable. Level-1 predictors may explain within-cluster variability in the outcome variable.
- The intra-class correlation coefficient (ICC) measures the proportion of the total variability in the outcome variable that is explained by cluster membership. The ICC is also the expected correlation between two randomly drawn individuals from the same cluster (Hox et al., 2017).
- The effective sample size ($n_{eff}$) for a given sample is a function of the degree of non-independence in the sample and the number of people per cluster. The degree to which the effective sample size and the actual sample size differ indicates the degree to which the standard errors from traditional statistical tests are underestimated.

## Further Reading

McCoach, D. B., & Adelson, J. L. (2010). Dealing with dependence: Part I.
    Understanding the effects of clustered data. *Gifted Child Quarterly, 54*(2), 152–155.
This article is a good supplement to the material covered in this chapter. It provides a conceptual introduction to the issue of clustering and dependence as well as an illustration of the effect of non-independence on the standard error.

Kreft, I. G., & de Leeuw, J. (1998). *Introducing multilevel modeling*. Sage.
This book provides a non-technical, accessible and practical introduction to multilevel modelling. The book also provides a broad overview of multilevel modelling, applications of multilevel modelling and its historical development.

# 2

# MULTILEVEL MODELLING

## A CONCEPTUAL INTRODUCTION

## Chapter Overview

Review of single-level regression .............................................. 16

Random effects and variance components.............................................. 19

Intercepts as outcomes models .............................................. 20

Full contextual (slopes-as-outcomes) model.............................................. 23

Variance–covariance components.............................................. 23

Advice on modelling randomly varying slopes .......................................... 24

Centring level-1 predictors ......................................... 26

Important guidance on group mean centring........................................... 27

Estimation........................................................................................... 29

Further Reading ................................................................................. 44

Having discussed the importance of taking the nested nature of data into account when conducting statistical analyses, we now introduce the multilevel model. This chapter provides a conceptual introduction to multilevel modelling (MLM). Chapter 3 provides additional guidance on building models in MLM and presents an applied example using real data.

## Review of single-level regression

A multilevel regression model is still a regression model, so let's briefly review the standard regression equation before extending to the multilevel case. Typically, we represent a regression model with one predictor as $Y_i = \beta_0 + \beta_1(X_i) + r_i$, where $Y_i$ = person $i$'s score on the outcome variable, $Y$. The intercept, $\beta_0$, represents the expected value of $Y$ when $X$ (the independent variable) is equal to 0. Generally, the intercept receives relatively little attention in multiple regression. However, in multilevel models, the intercept is often the star of the show! (We will have much more to say about this later.) The error term or residual, $r_i$, represents individual $i$'s actual score on the outcome variable ($Y$) minus their model-predicted score on the outcome variable ($\hat{Y}$), which is $\beta_0 + \beta_1 (X_i)$. A positive residual indicates that the person's actual score is higher than their predicted score, whereas a negative residual indicates that a person's actual score is lower than their predicted score. In multiple regression, we assume that these errors are normally distributed with a mean of 0 and a constant variance $\sigma^2$.

## Regression model with no predictors

Let's begin with the simplest possible regression model: a standard regression model with no predictors: $Y_i = \beta_0 + r_i$. In this case, the intercept $\beta_0$ represents the expected value on the outcome $Y$; absent any other information, $\beta_0$ is the mean of $Y$. The error term $r_i$ denotes the difference between person $i$'s actual score ($Y_i$) and his/her predicted score on $Y$; in this model, the person's predicted score is simply $\beta_0$, the mean of $Y$. Furthermore, with standard regression approaches, we make the assumption of independence, which means that we assume the $r_i$'s are uncorrelated with each other.

## Multilevel model with no predictors

Recall from Chapter 1, in a clustered sample (like Igor's), people within a given cluster are more similar to each other than to individuals from other clusters. Therefore, in clustered samples, we expect the $r_i$'s to be correlated within clusters, but independent across clusters. So, how does a multilevel model with no predictors differ from a multiple regression model with no predictors? Given that the residuals for observations within clusters co-vary, some of the variance in the dependent variable can be explained by cluster membership. Therefore, we introduce an additional error term, $u_{0j}$, to capture the portion of the residual that is explained by membership in cluster $j$. The residual for the intercept for each cluster ($u_{0j}$) represents the deviation of a cluster's intercept from the overall intercept. The term $u_{0j}$ allows us to model the dependence of observations from the same cluster because $u_{0j}$ is the same for every person within cluster $j$ (Raudenbush & Bryk, 2002).

Returning to our sleep example, some of the variability in sleep time is between households, meaning that households differ in terms of their expected/mean sleep time. Households do vary in terms of their average sleep time, so we allow the intercept $\beta_0$, which is the mean of the outcome variable (sleep time) to vary across clusters. Conceptually, allowing the intercept to randomly vary across clusters is analogous to allowing separate intercepts for each cluster. Therefore, our level-1 equation is now $Y_{ij} = \beta_{0j} + r_{ij}$, where $j$ indexes the cluster and $i$ indexes the person. Instead of having only one intercept (as we did in the multiple regression equation), we now have $j$ intercepts, one for each cluster. This means that person $i$ in cluster $j$'s score on the outcome variable $Y$ equals the expected **cluster mean** for cluster $j$, $\beta_{0j}$, plus person $i$'s deviation from this expected cluster mean, $r_{ij}$. So in Igor's sample, each household has its own intercept, $\beta_{0j}$, which is the predicted household (cluster) mean. Given that there are 250 clusters (households), there are also 250 $\beta_{0j}$'s or intercepts.

For simplicity, let's assume we have no predictors at level 2. The level-2 equation is then $\beta_{0j} = \gamma_{00} + u_{0j}$. In MLM, we refer to these $\beta_{0j}$'s as *randomly varying* intercepts. The randomly varying intercept, which was on the right-hand side of the **level-1 equation** (acting as a predictor of $Y$) is now on the left-hand side of the **level-2 equation** (acting as an outcome variable). The 250 intercepts ($\beta_{0j}$'s) are predicted by an overall intercept, $\gamma_{00}$, and a level-2 residual (error), $u_{0j}$, which captures the deviation of cluster $j$'s predicted intercept, $\beta_{0j}$, from the overall intercept, $\gamma_{00}$. Each of the $j$ clusters has its own level-2 residual, $u_{0j}$, which allows each cluster to have its own intercept ($\beta_{0j}$). Rearranging the level-2 equation so that $u_{0j} = \beta_{0j} - \gamma_{00}$, it becomes clear that the level-2 residual, $u_{0j}$, is the difference between $\beta_{0j}$ (the expected cluster mean for the outcome variable) and $\gamma_{00}$ (the overall expected value on the outcome variable). Thus, our set of multilevel equations for a completely unconditional model is

$$Y_{ij} = \beta_{0j} + r_{ij}$$
$$\beta_{0j} = \gamma_{00} + u_{0j}$$

(2.1)

The subscript for $\gamma_{00}$ contains no $i$'s or $j$'s, meaning that $\gamma_{00}$ is *fixed*: there is only one value of $\gamma_{00}$, the overall intercept. Because we have no predictors, $\gamma_{00}$ is also the predicted mean (average) on the outcome variable, $Y$. Notice that $\beta_{0j}$ occurs in both equations. Therefore, we can substitute $\gamma_{00} + u_{0j}$ for $\beta_{0j}$ to obtain one combined (or mixed) equation,

$$Y_{ij} = \gamma_{00} + u_{0j} + r_{ij} \tag{2.2}$$

What does this mean? Person $i$ in cluster $j$'s score on $Y$ $(Y_{ij})$ is equal to the overall expected/mean $Y$ score $(\gamma_{00})$ plus the amount by which his/her cluster (cluster $j$) deviates from that overall mean $Y$ score $(u_{0j})$ plus the amount by which he/she (person $i$ in cluster $j$) deviates from his/her cluster mean $(r_{ij})$. So each person's score $(Y_{ij})$ equals the expected (predicted) mean $(\gamma_{00})$ plus their cluster's deviation from the overall mean $(u_{0j})$ plus their deviation from their own cluster's mean $(r_{ij})$.

---

**Box 2.1**

**A note on multilevel versus combined equations**

Conceptually, separating the equations from a multilevel model into multiple levels is often more intuitive than the combined equation. In reality, the multilevel model that is estimated is the **combined model**, and different statistical software packages require users to convey models in different formats. For example, users of SAS, Stata, R and SPSS must specify the combined model, whereas users of the software packages HLM, Mplus and MLWin can use the multiple-equation notation to estimate multilevel models. Thus, some prefer to use the combined notation, while others prefer the multiple-equation notation. Either convention is acceptable, as both sets of equations are equivalent and contain the same information. In this book, we tend to favour the multilevel equations, but we sometimes present the combined form as well.

---

For a more concrete example, let's return to Igor's sample. Imagine that, on average, people report sleeping 8 hours per night $(\gamma_{00} = 8)$. Suzie lives in a house where the average number of sleep hours per night is 6 $(\beta_{0j} = 6)$, but Suzie herself sleeps 7 hours per night $(Y_{ij} = 7)$. Conceptually, $Y_{ij} = \gamma_{00} + u_{0j} + e_{ij}$ for Suzie would be $7 = 8 + (-2) + 1$. In a non-multilevel framework (with a non-clustered, simple random sample), the prediction equation for Suzie would simply be $Y_i = \beta_0 + e_i$, or $7 = 8 + (-1)$. The single-level regression equation contains only one error term, Suzie's deviation from the overall average (or predicted) score. The multilevel regression equation contains two residuals: (1) the deviation of Suzie's household from the overall mean (which in this case is $-2$) and (2) Suzie's deviation from her household mean (which in this case is $+1$). So, the overall mean (the overall intercept or predicted score) is the same in the multilevel and single-level frameworks above. What differs is our treatment of the residual(s).

## Random effects and variance components

Without predictors, each person's score on the dependent variable is composed of three elements: (1) the expected mean ($\gamma_{00}$), (2) the deviation of the cluster mean from the overall mean ($u_{0j}$) and (3) the deviation of the person's score from his/her cluster mean ($r_{ij}$). In this equation, $\gamma_{00}$ is a **fixed effect**: $\gamma_{00}$ is the same for everyone. The $u_{0j}$ term is called a **random effect** for the intercept because $u_{0j}$ randomly varies across the level-2 units (clusters). In MLM, *fixed effects* are parameters that are fixed to the same value across all clusters (or individuals), whereas *random effects* differ (vary) across clusters (or individuals) (West et al., 2015).

## Residual variances in MLM

Multilevel models and standard regression models do not differ in terms of their fixed effects. However, they differ in terms of the complexity of their residual variance/covariance structures. This more complex residual variance/covariance structure is at the heart of MLM. Therefore, understanding the meaning and utility of the included random effects is essential.

To account for the dependence/clustering, we break the residual into two pieces, $u_{0j}$ and $e_{ij}$: $u_{0j}$ captures the deviation of the cluster mean (intercept) from the overall mean (intercept), and $r_{ij}$ captures the deviation of the individual's score from the mean for that individual's cluster. We can then compute variances for each of these residuals. (You may have noticed that we spend a lot more time thinking about intercepts and residuals in MLM than we ever did in ordinary least squares [OLS] regression!) The variance of $r_{ij}$, $\sigma^2$, represents the within-cluster residual variance in the outcome variable, and the variance of $u_{0j}$, $\tau_{00}$, represents the between-cluster residual variance in the outcome.

We also make several important assumptions related to our model's residual variance terms: (a) the set of $u$'s is normally distributed with a mean of 0 and a variance of $\tau_{00}$, (b) the set of $r$'s is normally distributed with a mean of 0 and a variance of $\sigma^2$ and (c) the within-cluster residuals ($r_{ij}$'s) and between-cluster residuals ($u_j$'s) are uncorrelated. This last assumption allows us to cleanly partition the variance in the outcome variable into within- and between-cluster **variance components**. Therefore, in the simplest unconditional model with no predictors, the total variance in the outcome variable (var($Y_{ij}$)) equals the sum of the between-cluster variance, $\tau_{00}$, and the within-cluster variance, $\sigma^2$. The ability to partition variance into within-cluster variance and between-cluster variance is one of MLM's greatest assets.

## Intercepts as outcomes models

Because the intercepts vary across clusters, we can build a regression equation at level 2 to try to explain the variation in these randomly varying intercepts. For instance, in our sleep example, we could include household-level covariates to predict between-cluster variance in households' sleep time. For example, the number of dogs in the house or the average age in the household are potential level-2 covariates. Raudenbush and Bryk (2002) refer to these models as *means as outcomes models* because the level-2 model predicts differences in the intercepts across clusters (level-2 units).

The level-2 covariates may help to explain why some households sleep more than others. However, level-2 variables can never explain within-cluster variance (i.e. household-level variables cannot explain why certain members of the family sleep more or less than other family members). To explain within-cluster (level-1) variance, we need to include within-cluster (level-1) covariates.

## Adding level-1 predictors

Now, let's consider a model in which there is one predictor at the lowest level (level 1). Imagine that we want to predict sleep hours using age. (The age–sleep relationship might actually be non-linear: people in middle adulthood might sleep less than children and older adults. However, for simplicity, let's assume a linear relationship between age and sleep time.) We regress sleep hours ($Y_{ij}$) on age ($X_{ij}$). Now our level-1 model is as follows:

$$Y_{ij} = \beta_{0j} + \beta_{1j}(X_{ij}) + r_{ij} \tag{2.3}$$

Remember that in standard linear regression, the intercept is the predicted value on $Y$ when all predictors are held constant at 0. Similarly, we interpret the intercept ($\beta_{0j}$) as the predicted mean sleep hours in cluster $j$ when $X_{ij}$ (age) is equal to 0. Because age is equal to 0 at birth, the intercept is the expected amount of sleep time for a new-born infant. The slope $\beta_{1j}$ (the effect of age on sleep time) can vary by cluster, just like $\beta_{0j}$ does. If we allow $\beta_{1j}$ to randomly vary by cluster, $\beta_{1j}$ becomes an outcome variable in a level-2 equation and has its own residual term, $u_{1j}$. Equation (2.4) contains the multilevel model with a randomly varying intercept and a randomly varying slope.

$$\begin{aligned} Y_{ij} &= \beta_{0j} + \beta_{1j}\left(X_{ij}\right) + r_{ij} \\ \beta_{0j} &= \gamma_{00} + u_{0j} \\ \beta_{1j} &= \gamma_{10} + u_{1j} \end{aligned} \tag{2.4}$$

In Equation (2.4), $\gamma_{00}$ represents expected (predicted) number of sleep hours when age = 0, and $\gamma_{10}$ represents the average effect of age on sleep time across the entire sample. So, if age is measured in years, we expect a $\gamma_{10}$-hour change in sleep for every additional year. The error term, $u_{1j}$, represents the difference between the average slope and cluster $j$'s slope. In our example, $u_{1j}$ is the difference between cluster $j$'s age–sleep slope and the overall age–sleep slope. If the 'effect' of age on sleep time does not vary across clusters, then all clusters should have the same (or very similar) age–sleep slopes. In such a scenario, the value of $u_{1j}$ for each cluster would be 0 (or near zero), and the variance of $u_{1j}$ would also be approximately 0. If the slope is the same across all clusters (i.e. the slope does not vary across clusters), it is not necessary to estimate a randomly varying slope. Instead, we could estimate a model in which the intercept for sleep time randomly varies across clusters, but the age–sleep slope (the effect of age on sleep) remains constant across clusters. In that scenario, our multilevel model equations would be

$$Y_{ij} = \beta_{0j} + \beta_{1j}(X_{ij}) + r_{ij}$$
$$\beta_{0j} = \gamma_{00} + u_{0j} \tag{2.5}$$
$$\beta_{1j} = \gamma_{10}$$

Again, using substitution to combine these level-specific equations into one mixed-format model produces the combined model for each set of multilevel equations above. If the age–sleep slope does not randomly vary across clusters, the combined model is simple. Substituting $\gamma_{00} + u_{0j}$ for $\beta_{0j}$ and $\gamma_{10}$ for $\beta_{1j}$, the mixed-format equation is

$$Y_{ij} = \gamma_{00} + \gamma_{10}(X_{ij}) + u_{0j} + r_{ij} \tag{2.6}$$

such that person $ij$'s score on $Y$ is a function of $\gamma_{00}$, the overall intercept (the predicted score when $X_{ij} = 0$, which in this case is when age = 0), $\gamma_{10}$ (the slope of age on sleep hours) multiplied by person $ij$'s age ($X_{ij}$), the deviation of his/her household's intercept (the predicted number of sleep hours at age = 0) from the overall intercept ($u_{0j}$), and $r_{ij}$, the deviation of person $ij$'s score from his/her model-predicted score.

If the age–sleep slope does randomly vary by cluster, then substituting $\gamma_{10} + u_{1j}$ for $\beta_{1j}$ results in the following combined equation:

$$Y_{ij} = \gamma_{00} + \gamma_{10}(X_{ij}) + u_{0j} + u_{1j}(X_{ij}) + r_{ij} \tag{2.7}$$

Now person $ij$'s score is a function of $\gamma_{00}$, the overall intercept (the predicted score when $X_{ij} = 0$), $\gamma_{10}$, the slope of age on sleep hours, multiplied by person $ij$'s age ($X_{ij}$), the deviation of his/her household's intercept from the overall intercept ($u_{0j}$), the deviation of his/her household's slope from the overall slope ($u_{1j}$) multiplied by person $ij$'s age ($X_{ij}$) and the deviation of person $ij$'s score from their model-predicted score, $r_{ij}$.

Allowing the age–sleep slope to randomly vary across households by including a *random effect* for the slope ($u_{1j}$) specifies a model in which the age–sleep slope is different for different households. Therefore, in some households, there could be no relationship between age and sleep time, resulting in an age–sleep slope of 0; in other households, the sleep slope could be negative, indicating that older members of the household tend to sleep less than younger members of the household. Finally, the age–sleep slope could be positive, indicating that older members of the household tend to sleep more than younger members of the household. The fixed effect, $\gamma_{10}$, indicates the expected (average) value of the age–sleep slope across the entire sample. The variance in the age–sleep slope, $var(u_{1j})$, indicates how much households vary from the overall average. If the variance of $u_{1j}$ is large, there is a lot of between-household variability in the age–sleep slope. In contrast, if the variance of $u_{1j}$ is 0, then there is no variability across households in terms of their age–sleep slopes: in this case, we would want to fix $u_{1j}$ to 0, as that would greatly simplify our model.

Figure 2.1 illustrates the concept of randomly varying intercepts and randomly varying slopes by graphing the relationship between a hypothetical independent variable (*X*), such as age, and a hypothetical dependent variable (*Y*), such as hours of

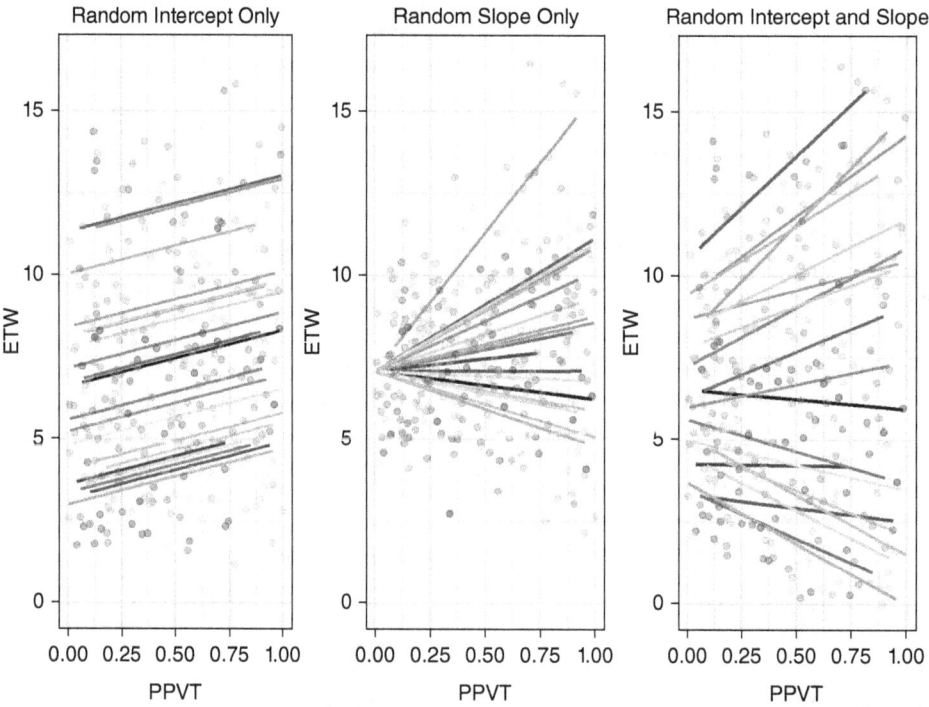

**Figure 2.1** (1) Randomly varying intercepts, (2) randomly varying slopes and (3) randomly varying intercepts and slopes

*Note.* ETW = expressive target word assessment; PPVT = Peabody Picture Vocabulary Test.

sleep, under three conditions: (1) when the intercept randomly varies but the slope is fixed, (2) when the slope randomly varies but the intercept is fixed and (3) when both the slope and the intercept randomly vary. (Note each line in Figure 2.1 represents the regression line for a specific cluster.) When only the intercepts randomly vary, all clusters have equal slopes, but they differ in their intercepts, producing parallel regression lines. Likewise, when only the slopes randomly vary, all clusters have equal intercepts, but they differ in their slopes. Therefore, all regression lines appear to originate in the same location (at $X = 0$) but then diverge. Lastly, when the slopes and intercepts randomly vary, each cluster has its own unique intercept and slope.

## Full contextual (slopes-as-outcomes) model

The **full contextual/theoretical model** contains both level-1 and level-2 predictors. Level-2 predictors may help to explain between-cluster differences in the intercept (the expected value of the outcome variable when all level-1 variables are held constant at 0). Level-2 predictors may also help explain between-cluster differences in level-1 slopes. In other words, the level-2 variable helps to predict how the relationship between the level-1 predictor and the outcome variable differs across clusters. Returning to our example, age is a level-1 predictor of sleep hours. We could include a household-level variable, such as the average noise level in the home, to predict the average number of sleep hours within the household (the intercept). A level-2 variable (i.e. the noise level in the house) could also predict a level-1 slope (i.e. the age–sleep hours slope). We refer to a level-2 predictor of a level-1 slope as a **cross-level interaction** because it represents an interaction between a level-2 variable and a level-1 variable. In this example, the cross-level interaction indicates whether the noise level in the house moderates the relationship between age and sleep hours.

## Variance–covariance components

The $\gamma$ terms are the *fixed effects* and the $u$ terms are the *random effects*. All of the $\gamma$ terms could be estimated using multiple regression models with interaction terms. However, the $u$ terms, the random effects, are unique to mixed/multilevel models. Multilevel techniques allow us to model, estimate and test the variances (and covariances) of these random effects (also known as *variance components* and denoted by the symbol, $\tau_{qq}$). Specifically, $\tau_{00}$ represents the variance of the randomly varying intercepts ($u_{0j}$), $\tau_{11}$ signifies the variance of the first randomly varying slope ($u_{1j}$) and so on. In addition, we generally allow the random effects (within a given level) to co-vary with each other.

Therefore, in our simple example above, $\tau_{01}$ represents the covariance between residuals for the randomly-varying intercept and the randomly varying slope.

$$\mathrm{var}\begin{bmatrix} u_{0j} \\ u_{1j} \end{bmatrix} = \begin{bmatrix} \tau_{00} & \tau_{01} \\ \tau_{01} & \tau_{11} \end{bmatrix} \qquad (2.8)$$

If we standardise $\tau_{01}$, it represents the correlation between the residuals of the intercept and slope. If $\tau_{01}$ is positive, then clusters with more positive intercepts also tend to have more positive/less negative slopes. If $\tau_{01}$ is negative, then clusters with more positive intercepts tend to have less positive/more negative slopes. Although we do allow random effects to co-vary within a given level, we assume that the residuals are uncorrelated across the levels of our analysis. As a result, although $\tau_{00}$ and $\tau_{11}$ are allowed to co-vary, we assume that both $\tau_{00}$ and $\tau_{11}$ are uncorrelated with $\sigma^2$.

## Advice on modelling randomly varying slopes

Depending on the researcher's theoretical framework and the sample size at level 1, the slopes for some of the level-1 predictors may be estimated as randomly varying across level-2 units, or they can be estimated as fixed across all level-2 units. A random-coefficients model is a model in which one or more level-1 slopes randomly vary (Raudenbush & Bryk, 2002). Although our simple example contains only one level-1 variable (age), often multilevel models contain several level-1 variables. For example, many multilevel educational studies in which students are nested within schools include a variety of level-1 control variables, such as gender, race/ethnicity (which is often a set of four to six dummy-coded variables), free lunch status, English learner status and special education status. A set of control variables could easily include 10 or more level-1 variables. In such a situation, the researcher must decide which level-1 slopes to allow to randomly vary across schools and which level-1 slopes to fix to a single value across all schools. Why not allow all 10 level-1 covariates to randomly vary across schools? First, remember the structure of the residual covariance matrix. The unstructured tau ($\tau$) matrix contains a variance for the randomly varying intercept and every randomly varying slope as well as all possible covariances among the slopes and intercepts. Therefore, the number of unique variance–covariance components in the tau matrix is equal to $r(r + 1)/2$, where $r$ equals the number of random effects. As we saw earlier, with a **random intercept** and a **random slope**, the Tau matrix contains $(2 * 3)/2 = 3$ parameters (two variances and a covariance). However, in a model that contains five randomly varying slopes and a randomly varying intercept, the tau matrix contains $(6 * 7)/2 = 21$ unique parameters, and the tau matrix for a model with 10 randomly varying slopes and a randomly varying intercept contains

(11 * 12)/2 = 66 unique parameters. In other words, in a model with 10 randomly varying slopes, we need to estimate a total of 67 different residual parameters: $\sigma^2$ and an 11 * 11 tau matrix, containing 66 unique level-2 variance components. Partitioning the residual variance in a model into 67 separate pieces feels like a Sisyphean task. (Remember, in multiple regression, we estimate just one residual variance parameter.)

Raudenbush and Bryk (2002) cautioned against succumbing to the 'natural temptation to estimate a "saturated" level-1 model' in which all level-1 predictors are specified to have randomly varying slopes (p. 256). We cannot overstate the importance of this advice. It is essential to be parsimonious when specifying randomly varying slopes for several reasons:

1   First, as demonstrated above, adding random slopes radically increases the complexity of the model.
2   There is an upper limit on the number of random slopes based on the sample size at level-1. Minimally, cluster size must be larger than the number of variance components for the model to be identified. Therefore, in dyadic data, it is only possible to estimate one random effect. Our sleep example had a cluster size of four, so the maximum number of potential random effects is three (presumably a randomly varying intercept and two randomly varying slopes). This does not mean that it is a good idea to estimate such nearly saturated models; we generally prefer to fit level-1 models in which the number of random effects is comfortably less than the cluster size.
3   It is common to experience convergence problems when trying to estimate randomly varying slopes that are unnecessary. Multilevel models that contain a random slope that has no between-cluster variance often fail to converge (or require thousands of iterations to converge). Because variances cannot be less than 0, trying to estimate randomly varying slopes that are actually 0 in the population often leads to boundary issues, resulting in models that fail to converge. Unfortunately, such results may not provide guidance about which of the random effects to eliminate (McCoach et al., 2018).

Of course, it is easy to think of at least one logical reason that each slope in a multilevel model might randomly vary, and it is tempting to allow most or all of them to do so 'just to see' what happens. However, we implore you – don't do it! In our experience, some people who learn about randomly varying slopes become 'greedy' and want to be able to allow every slope parameter in large regression models to randomly vary across clusters. Attempting to estimate and interpret dozens of residual variance–covariance components is unrealistic and unreasonable under most circumstances.

We recommend including random slope effect only if the randomly varying slope is central to your research question or if you have compelling evidence from prior research that the slopes are likely to randomly vary. Use randomly varying slopes carefully, sparingly and cautiously: be judicious and parsimonious about which random slopes to include in multilevel models. Also, eliminate any unnecessary random effects for level-1 coefficients that do not vary across level-2 units (McCoach et al., 2018).

## Centring level-1 predictors

In regression models, we often *centre* covariates for both substantive and analytic reasons. As mentioned earlier, the intercept is the predicted value of the outcome variable when all of the predictor variables are held constant at 0. So, in our sleep example, the intercept for sleep was the predicted sleep hours at age 0. In single-level regression, one common strategy is to *centre* continuous predictor variables by subtracting the mean of the variable ($\bar{X}$) from each person's score ($X_i$). This transforms person $i$'s score on $X$ into a deviation score, which indicates how far above or below the mean person $i$ scored. Therefore, the mean of a centred variable is 0 and the variance is the same as the variance of the score in its original metric (because all scores change only by a single constant value, the mean). In single-level regression, centring continuous covariates is especially important when including interaction terms. The choice of centring influences the main effects for the predictor variables included in the interaction term: the regression coefficient is the predicted effect of $X$ on $Y$ when the other predictor variable in the interaction term equals 0 (Aiken & West, 1991).

In MLM, we may centre continuous predictor variables for substantive and/or analytic reasons. First, centring continuous covariates allows for a more substantively useful and interpretable intercept. Second, the magnitude of the between-person (residual) variance in the intercept, $\tau_{00}$, and the correlation between the intercept and any randomly varying slopes is dependent on the location of the intercept.

In organisational applications of MLM, the two main centring techniques for lower-level covariates are **grand mean centring** and **group mean centring**. Grand mean centring subtracts the overall mean of the variable from all scores. Therefore, the grand mean–centred score captures a person's standing relative to the full sample. Group mean centring subtracts the cluster's mean from each score in the cluster. As such, the transformed score captures a person's standing relative to their own cluster.

As an example, let us grand mean and group mean centre age ($X_{ij}$) for person $i$ in cluster $j$. In our example, the grand mean represents the mean age across all individuals $i$ and all clusters $j$ ($\bar{X}_{..}$), and the cluster mean represents the mean age of all individuals $i$ in a household $j$ ($\bar{X}_{.j}$). To grand mean centre age, we subtract the mean age in the entire sample from each person $ij$'s age ($X_{ij} - \bar{X}_{..}$), so under grand mean centring $X_{ij}$ is person $ij$'s deviation from the average age in the entire sample ($\bar{X}_{..}$). To group mean centre age, we subtract the average household age from each person's age ($X_{ij} - \bar{X}_{.j}$); so, under group mean centring, $X_{ij}$ is person $ij$'s deviation from his/her household's average age ($\bar{X}_{.j}$).

Obviously, the decision about how to centre independent variables has major implications for the interpretation of the intercept. Grand mean centring age sets the intercept at the overall mean. This holds age constant at the overall mean, thereby controlling for age. When grand mean centring age, the randomly varying intercept, $\beta_{0j}$, denotes the predicted number of sleep hours for household $j$ assuming that this household's average age is the same as the overall average age. The intercept is the predicted number of sleep hours in household $j$, holding age constant at the overall mean ($\bar{X}_{..}$). Person $i$ in cluster $j$'s score on the grand mean–centred $X$ variable represents the deviation of that person's score from the overall average. In this case, grand mean–centred age represents each person's deviation from the average age across the entire sample. Grand mean centring represents a simple linear transformation of the original variable.

One problem with grand mean centring arises when no one in a given cluster has scores near the overall mean. In such cases, the intercept for that cluster is extrapolated outside the range of data for the cluster. For example, if the average age across households is 40, but in household $j$, the four members are 55, 55, 65 and 65 years old, then the grand mean–centred scores are 15, 15, 25 and 25, respectively. No one in the household has a centred score near 0. Thus, the intercept in household $j$ is the predicted sleep score for a 40-year-old, even though there are no 40-year-olds in that household. For a detailed discussion of the statistical and interpretational issues that such extrapolation can cause, see Raudenbush and Bryk (2002).

On the other hand, if we group mean centre age, then the randomly varying intercept ($\beta_{0j}$) is the mean number of sleep hours in household $j$. Having subtracted each cluster's own mean ($\bar{X}_{.j}$) from each score, the mean of the cluster-mean centred age variable is 0 in every cluster. Therefore, the randomly varying intercept for each cluster is the mean (expected/predicted) number of sleep hours in that household ($j$). Person $i$ in cluster $j$'s score on the group mean–centred $X$ variable represents the deviation of their score from their cluster's average score. In this case, group mean–centred age represents each person's deviation from their household's average age. So, in a household where the ages are 55, 55, 65 and 65 years, the household's mean age is 60 years. To group mean centre, we subtract 60 from each score, producing group mean–centred scores of –5, –5, 5 and 5, respectively. The mean of the group mean–centred variable is 0 in every cluster, so the overall intercept is the mean of cluster means: it is the overall average household sleep time.

## Important guidance on group mean centring

A group mean–centred score provides information about individuals' relative standing as compared to their cluster, but it provides no information about the individual or the group's relative standing as compared to the overall sample. So, for example, grand mean–centred scores of 15, 15, 25 and 25 indicate that two of the members of the household are 15 years older than the sample average and two of the members of the household are 25 years older than the sample average. In contrast, the group mean–centred scores of –5, –5, 5 and 5 tell us nothing about how the ages in this household compare to ages in the other households in the sample.

Group mean centring removes between-cluster variation from the level-1 covariate, so the variance of a group mean–centred variable provides an estimate of the pooled within-cluster variance (Enders & Tofighi, 2007). Using group mean centring, we can partition the variance in the predictor, the outcome and the relationship between the predictor and the outcome into within-cluster and between-cluster components.

However, using group mean centring does not preserve information about between-cluster differences on the $X$ variable. Using a different cluster mean to centre each cluster results in a centred $X$ variable that contains information about how much a person deviates from his/her group, but it contains no information about how much the person deviates from the overall mean on $X$. Therefore, when using group mean centring, be sure to introduce the aggregate of the group mean–centred variable (or a higher-level variable that measures the same construct) into the analysis. Without an aggregate or **contextual variable** at level 2, all of the information about between-cluster variability in the $X$ variable would be lost – a considerable but avoidable drawback. In our age example, the grand mean–centred mean age for our cluster, +20, provides information indicating that the average age in this cluster is 20 years older than the average age in the overall sample. In contrast, the group mean–centred age for our cluster (and every other cluster in the sample) is 0. However, adding the cluster mean into the model as a level-2 predictor preserves between-cluster component of the age variable. Finally, in group mean centring, a different cluster mean is subtracted in every cluster. Therefore, group mean centring is not a simple linear transformation, and it does not produce results that are statistically equivalent to the uncentred and grand mean–centred results.

Within the multilevel literature, some debate exists about whether to use grand mean centring or group mean centring. Because centring decisions affect the interpretations of important model parameters involving the intercept, it is important to carefully and thoughtfully decide if and how to centre covariates. The decision to use grand mean or group mean centring may vary depending on the context of the study, the research questions asked and the nature of the variables in question. For instance, if the primary research question involves understanding the impact of a level-2 variable on the dependent variable and the level-1 variables serve as control variables, grand mean centring may be an appropriate choice. On the other hand, when level-1 variables are of primary research interest or for research on contextual and compositional effects, group mean centring may be more appropriate. In addition, group mean centring aids in the computation of variance explained (**R-squared**) measures, a point we discuss more fully in Chapter 3. To preserve between-cluster information from the covariate, we recommend including the aggregates of group mean–centred variables at level 2.

What about centring level-2 variables? Grand mean centring is the only available option at level 2. As a general rule, it is advisable to grand mean centre all level-2 continuous variables. When using level-2 variables as part of a cross-level interaction, grand mean centring is especially important. However, even for level-2 variables that predict only randomly varying intercepts (not randomly varying slopes), grand mean centring the level-2 variable usually facilitates interpretation of the intercept. When reporting MLM results, it is important to explain centring decisions and procedures and to interpret the parameter estimates accordingly. See Enders and Tofighi (2007) for an excellent discussion of centring in organisational multilevel models.

## Estimation

This book does not delve into the computational details required to actually estimate multilevel models. However, it is helpful to conceptually understand the analytic challenges of multilevel data and the estimation strategies that MLM employs.

MLM does not require balanced data: the number of units per cluster can vary across clusters. In fact, there is no minimum or maximum number of units per cluster, and multilevel models can easily accommodate data sets that include some clusters with very few level-1 units and other clusters with very large numbers of level-1 units. MLM employs a variety of estimation strategies to handle unbalanced data.

To keep things simple, we contextualise this discussion in the context of the unconditional random effects model: $Y_{ij} = \gamma_{00} + u_{0j} + e_{ij}$. Let's start by identifying a simple but important issue that arises when estimating parameters from unbalanced data: how to determine the expected value of the outcome variable ($\gamma_{00}$) from multiple clusters of multiple sizes. In non-clustered data, the sample mean provides our 'best guess' about the population mean (the *expected value*). In clustered data, what is the expected value of the outcome variable, $\gamma_{00}$? Let's imagine that we have randomly sampled 100 schools, and the sample sizes within the schools vary widely: the smallest cluster size is two students and the largest cluster size is 1000 students. How should we determine expected achievement?

One option would be to ignore clustering and compute the sample mean. In such a scenario, every person is weighted equally; however, the schools with larger numbers of sampled students have a much larger influence on the expected mean than the small schools do. On the other hand, we could compute the mean of school means. However, in that case the schools with very few students have an outsized influence on the expected mean. (In addition, the school mean that is computed from a school with 1000 students is likely to be a much better estimate of the school's performance than a school mean that is computed from only two students, a point to which we return when we discuss empirical Bayes estimates.)

In MLM, larger clusters do have a larger influence on the expected mean. However, the ICC tempers that influence. Remember, the ICC provides valuable information about the proportion of between-school variance in the outcome, which indicates the degree of dependence (redundancy) within a cluster. If the ICC is 0, then students within a given school are no more similar to each other than students from different schools. In such a situation, taking the sample mean ignoring clustering seems reasonable. On the other hand, if the ICC is 1.0, all students in a school are complete replicates of each other. In such a scenario, the mean of school means might be of greater interest, given the deterministic nature of within-school performance. When the ICC is 0, the influence of each cluster on $\gamma_{00}$ is proportional to its cluster size. When the ICC is 1.0, each cluster has an equal influence on $\gamma_{00}$, regardless of its size (Snijders & Bosker, 2012). In reality, the ICC lies between 0 and 1.0. So $\gamma_{00}$ is a compromise between a proportional weighted average (as it would be when ICC = 0) and a mean of cluster means (as it would be when ICC = 1). The higher the ICC, the more $\gamma_{00}$ approaches a mean of cluster means; the lower the ICC, the more $\gamma_{00}$ approaches a proportional weighted average.

## Conceptual introduction to maximum likelihood

Both MLM and structural equation modelling (SEM) use maximum likelihood (ML) estimation techniques. In ML estimation, we estimate parameters that maximise the probability of observing our data. This section provides a very rudimentary, conceptual introduction to ML. The probability of observing an event implicitly assumes a model. We make statements about the probability of observing some event, based on the model parameters. For example, take the case of a coin toss. Everyone knows that the probability of tossing a head is .5. Let's formalise this notion. Our model contains one parameter, $p$, the probability of tossing a head, and that parameter $p$ is equal to .5. In probability, we know the value of the parameter, and we try to predict future outcomes based on that known parameter. The likelihood, on the other hand, turns probability on its head. With likelihood, we already have the data, and we try to determine the most likely value for a parameter, given the data. The goal of ML estimation is to find the set of parameter values that makes the actual data most likely to have been observed. So imagine we know nothing about the probability of tossing heads, but we want to use data empirically to determine the value for that parameter, so we flip a coin 100 times. The coin lands on heads 55 times and on tails 45 times. Then we can ask – what is the most likely parameter value for $p$, the probability that I will flip a head, given the data that I have collected? The answer in this case would be .55 (not .50): the parameter value of $p$ = .55 maximises the probability of observing our results. For a much more detailed and nuanced discussion of likelihood estimation, see Myung (2003).

## Maximum likelihood estimation in MLM

The most common estimation techniques for estimating variance components for multilevel models with normal response variables are full information maximum likelihood (FIML) and restricted maximum likelihood (REML).

In FIML, the estimates of the variance and covariance components are conditional upon the point estimates of the fixed effects (Raudenbush & Bryk, 2002). FIML chooses estimates of the fixed effects, the level-2 variance–covariance components (T) and the level-1 residual variance ($\sigma^2$) 'that maximise the joint likelihood of these parameters for a fixed value of the sample data, $Y$' (Raudenbush & Bryk, 2002, p. 52). Thus, the number of parameters in the model includes both the fixed effects and the variance–covariance components. In contrast, REML maximises the joint likelihood of the level-2 variance–covariance components (T) and the level-1 residual variance ($\sigma^2$) given the observed sample data, $Y$. Thus, when estimating the variance components, REML takes the uncertainty due to loss of degrees of freedom from estimating fixed parameters into account, while FIML does not (Goldstein, 2011; Raudenbush & Bryk, 2002; Snijders & Bosker, 2012).

When the number of clusters (level-2 units) is large, REML and FIML results produce similar estimates of the variance components. However, when there are small numbers of clusters, the FIML estimates of the variance components ($\tau_{qq}$) are smaller than those produced by REML. With few clusters, FIML tends to underestimate variance components, and the REML results may be more realistic (Raudenbush & Bryk, 2002). A simple formula, $(J - F)/J$, where $J$ is the number of clusters and $F$ is the number of fixed effects in the model, provides a rough approximation of the degree of underestimation of the FIML estimates (Raudenbush & Bryk, 2002). For example, when estimating a model with three fixed effects using a sample containing observations from 20 clusters, we estimate that the level-2 variance components are .85 = ((20 − 3)/20) as large in FIML as they are in REML; this means the FIML variance components are underestimated by 15%. However, there is an advantage to using FIML: it allows us to compare the fit of two different models, as we explain in the section 'Deviance and Model Fit'.

## Reliability and estimation of randomly varying level-1 coefficients

The randomly varying level-1 coefficients ($\beta_{0j}$, $\beta_{1j}$,..., $\beta_{qj}$) are not parameters in the model; they are a function of the fixed effects and the cluster-level residuals. Generally, standard MLM uses empirical Bayes estimation to generate the 'best estimates' of $\beta_{qj}$ (Raudenbush & Bryk, 2002). The computation of empirical Bayes residuals follows a different logic and process than the computation of OLS residuals. Below, we present a very simple, conceptual introduction to empirical Bayes estimation of the randomly varying level-1 intercept for an unconditional random effects model. For a more detailed description, we recommend Goldstein (2011) and Raudenbush and Bryk (2002).

For simplicity, consider the estimation of the randomly varying intercept, $\beta_{0j}$, the 'true cluster mean' under the simplest model, the random effects ANOVA model, $\beta_{0j} = \gamma_{00} + u_{0j}$. We do not know the true cluster mean for cluster $j$; cluster $j$ has sample size $n_j$. MLM allows for unbalanced data: the within-cluster sample sizes ($n_j$) can vary greatly across clusters. Some clusters could have many observations (i.e. – $n_j$ is large); other clusters could have small $n_j$.

How could we estimate the *true* mean for cluster $j$? The sample mean, $\bar{Y}_{.j}$, provides an estimate of the *true* mean. However, the smaller the cluster size ($n_j$), the less confidence we should have in using $\bar{Y}_{.j}$ (the cluster's observed mean) as an estimate of the cluster's *true* mean. In the most extreme situation, imagine we had no observations from cluster $j$ with which to estimate the *true* cluster mean. With a sample size of 0, what is our best guess about the true mean of cluster $j$? It is the overall mean, $\gamma_{00}$. Why? We know nothing about this cluster, but we have information about lots of other clusters. Our best guess for the mean of this cluster is the overall mean (expected value). So, there are two potential competing estimates for the true mean for a cluster: the overall mean (expected value) across the entire sample, which allows us to 'borrow' information from other clusters to estimate the true mean of cluster $j$, and the observed sample mean of cluster $j$, which contains some degree of error or imprecision. As the sample size in cluster $j$ increases, the precision with which we can estimate the true mean from the sample mean increases; there is less error in our measurement of the true cluster mean based on the sample mean. Of course, another factor influences our ability to estimate the true school mean from the sample mean: the ICC. Again, imagine an extreme example: if the ICC were 1.0, every observation within a cluster is a replicate of every other observation. When there is very little or no within-cluster variance, $\bar{Y}_{.j}$ is an especially good estimate of the true mean of cluster $j$. When there is a great deal of within-cluster variance, $\bar{Y}_{.j}$ is a poor estimate of the true cluster mean (especially with small sample sizes; Raudenbush & Bryk, 2002).

## Reliability of Cluster $j$

Our two potential estimates of the *true* cluster mean are $\gamma_{00}$ and $\bar{Y}_{.j}$. Empirical Bayes estimation combines these estimates of the true cluster mean, based on the reliability of cluster $j$. The reliability of cluster $j$ incorporates three pieces of information: the within-cluster variability ($\sigma^2$), the between-cluster variability ($\tau_{00}$) and the number of observations per cluster, $n_j$ (Raudenbush & Bryk, 2002):

$$\text{Reliability of } \hat{\beta}_{0j} = \frac{\tau_{00}}{\tau_{00} + \sigma^2 / n_j} \tag{2.9}$$

When the reliability in cluster $j$ is higher, more weight is placed on the sample mean as the estimate of the true school mean. When the reliability of cluster $j$ is lower, more weight is placed on $\gamma_{00}$ as an estimate of the true school mean. Holding between- and within-school variance constant, larger cluster sizes ($n_j$'s) result in higher reliability. Each cluster has its own estimate of reliability; however, variance estimates $\tau_{00}$ and $\sigma^2$ remain constant across clusters. Therefore, larger clusters have larger reliability estimates. Nevertheless, larger between-cluster variance (relative to within-cluster variance) also increases reliability. In other words, reliability is higher when the group means vary substantially across level-2 units (holding constant the sample size per group). So, increasing group size, increasing homogeneity within clusters and increasing heterogeneity between clusters all increase reliability. The formula for the ICC, $\tau_{00} / (\tau_{00} + \sigma^2)$, features prominently in the reliability formula above. With a bit of algebra, we can re-express the reliability formula in terms of ICCs (Raudenbush & Bryk, 2002). Larger ICCs, which indicate that within-cluster group variance is small relative to between-cluster variance, result in higher reliability. Although reliability can range from 0 to 1, the lower bound for the reliability in any given sample is the ICC, and that occurs when $n_j = 1$.

## Empirical Bayes estimates of randomly varying parameters

Imagine we need to estimate the true political attitudes for a set of counties, and we have an incomplete set of information. In most counties, pollsters randomly sampled 1000 or more respondents. However, in one county, pollsters randomly sampled only two respondents. In the counties where the pollsters sampled 1000 respondents, the best guess about the true political attitudes would be near the sample mean for the 1000 respondents. In the county where the pollsters sampled only two of the respondents, we can compute the sample mean. But how confident would we be that the mean of the two respondents accurately reflects the true political attitudes in that county?

If one of the two people in the sample is extreme, our sample mean could actually be a very poor estimate of the true political attitudes in the county. Imagine an even more extreme example: what if the pollsters missed one county entirely? What would be our best estimate of the political attitudes in that county? There are two logical possibilities for estimating the true county political attitudes. One is the sample mean in the county, and this seems like a good estimate of the true mean in counties where we have many observations (more information). However, in the counties without much information, what is our best guess about the county's political attitudes? We could use the overall mean across all of the counties as an estimate of the county's political attitudes. If we know nothing else about the county and we have no information from the county, using the overall mean provides our best estimate. What do we do for the county with only two respondents? The small sample of respondents does give us some information about the political attitudes in the county, but we cannot completely trust that the sample mean of those two respondents provides a good estimate of the true county mean. In such a situation, we could use a combination of the overall mean and the sample county mean to derive an estimate of the true mean for the county. To do so, we would want to give more weight to the cluster (county) mean when we have more information, and we would want to place more weight on the overall mean (expected value) when we have less information from the cluster (county). Conceptually, this is the essence of empirical Bayes estimates of the randomly varying parameters (intercepts and slopes).

Again, assuming an unconditional random effects ANOVA model, the empirical Bayes estimate of the true cluster mean ($\beta_{0j}^*$) weights the two potential estimates for each cluster as a function of the reliability for that cluster.

$$\beta_{0j}^* = \lambda_j \bar{Y}_{.j} + (1 - \lambda_j)\hat{\gamma}_{00} \qquad (2.10)$$

where $\lambda_j$ is the reliability of the sample mean, $\bar{Y}_{.j}$ (Raudenbush & Bryk, 2002) and $\hat{\gamma}_{00}$ is the expected value of the intercept (which is the expected value of the outcome variable in the unconditional random effects ANOVA model).

The sample mean ($\bar{Y}_{.j}$) is weighted by the reliability for that cluster ($\lambda_j$); the model-based mean (expected value, $\hat{\gamma}_{00}$) is weighted by 1 minus the reliability ($1 - \lambda_j$). Thus, the empirical Bayes estimate of the true cluster mean is a compromise between the sample mean and the model-based mean, and the degree to which we trust the sample-based mean ($\bar{Y}_{.j}$) determines the weight ($\lambda_j$) that we place on the sample-based mean ($\lambda_j \bar{Y}_{.j}$). Our lack of trust in the sample mean ($1 - \lambda_j$) determines the weight that we place on the overall expected value. Thus, the higher the reliability of the estimate for cluster $j$, the more weight is placed on the sample mean ($\bar{Y}_{.j}$) as the estimate of the true cluster mean, $\beta_{0j}$. In contrast, the lower the reliability of cluster $j$ estimate, the more weight is placed on the model-based estimate $\hat{\gamma}_{00}$ as the estimate of the true cluster mean. In the extremes, if the reliability were 1, the sample mean would be the estimate of the true cluster mean. If the reliability were 0, $\hat{\gamma}_{00}$ would serve as the estimate of the true cluster mean. Sometimes the empirical Bayes estimators are referred to as shrinkage estimators because the $\overline{Y}_j$ estimate is 'shrunken' towards the model-based estimate; empirical Bayes residuals are like OLS estimates of the residuals which are 'shrunken' towards 0 (Raudenbush & Bryk, 2002).

## Deviance and model fit

### Deviance

Using ML to estimate the parameters of the model also provides the likelihood, which easily can be transformed into a deviance statistic (Snijders & Bosker, 2012). The **deviance** is –2 multiplied by difference of the log likelihood of the specified model and the log likelihood of a saturated model that fits the sample data perfectly. Therefore, deviance is actually a measure of the *badness* of fit of a given model: higher deviances are indicative of greater model misfit (Singer & Willett, 2003). Although lower deviances indicate better **model fit**, we cannot interpret deviance in isolation, and it is a function of sample size as well as model fit. However, we can interpret differences in deviance for competing models as long as the models (a) are hierarchically nested, (b) use same observations and (c) use FIML to estimate the parameters (if we wish to compare two models that differ in terms of their fixed effects).

### Likelihood ratio (deviance difference) test

When one model is a subset of the other, the two models are said to be hierarchically nested (e.g. Kline, 2015), such that 'the more complex model includes all of the parameters of the simpler model plus one or more additional parameters' (Raudenbush et al., 2004, pp. 80–81). In sufficiently large samples, under standard normal theory assumptions and using the same set of observations, the difference between the deviances of two hierarchically nested models follows an approximate chi-square distribution with degrees of freedom equal to the difference in the number of parameters being estimated between the two models (de Leeuw, 2004; Raudenbush & Bryk, 2002; Singer & Willett, 2003). Using the likelihood ratio test (LRT), we can compare two hierarchically nested models. The simpler model (the model with fewer parameters) is the null model ($M_0$); the more parameterised model is the alternative model ($M_1$). The deviance of the simpler model ($D_0$) has $p_0$ parameters; the deviance of the more parameterised model ($D_1$) has $p_1$ parameters. The simpler model must have fewer parameters ($p_0 < p_1$), and the deviance of the simpler model must be at least as large as the deviance of the more parameterised model ($D_0 \geq D_1$). We compare the difference in deviance ($\Delta D = D_0 - D_1$) to the critical value of chi-square with degrees of freedom equal to the difference in the number of estimated parameters ($\Delta p = p_1 - p_0$). Using the LRT, we prefer the more **parsimonious model**, as long as it does not result in (statistically significantly) worse fit. Put another way, if the model with the larger number of parameters fails to reduce the deviance by a substantial amount, we retain the simpler model ($M_0$). However, when the change in deviance ($\Delta D$) exceeds the critical value of chi-square with $p_2 - p_1$ df (degrees of freedom), then the additional parameters result in statistically significantly improved model fit. In this scenario, we favour the more-complex model (i.e. Model $M_1$, with $p_1$ df).

Having described the LRT, we must now attend to a few subtle but important details about using the LRT to compare two nested models within multilevel modelling.

First, comparing two nested models that differ in their fixed effects ($\gamma$) requires FIML, not REML. In FIML, the number of reported parameters includes the fixed effects (the $\gamma$ terms) as well as the variance–covariance components. In REML, the number of reported parameters includes only the variance and covariance components. REML allows for comparison of models that differ in terms of their random effects, but both models must have the same fixed effects structure. Therefore, comparisons of models with differing fixed and random effects should utilise the deviance provided by FIML (Goldstein, 2011; McCoach & Black, 2008; McCoach et al., 2018; Snijders & Bosker, 2012). The major advantage of using FIML over REML is the ability to compare the deviances of models that differ in terms of their fixed and/or random effects. Most statistical programs use REML as the default method of estimation, so remember to select FIML estimation to use the deviance estimates to compare two nested models with differing fixed effects (McCoach & Black, 2008).

Second, when comparing the fit of two models that differ in terms of their variance components, we sometimes encounter a boundary issue that affects the way in which we must conduct such model comparisons. Variances cannot be negative. Therefore, if the variance of the random effect is 0 in the population, then the estimation of this variance component hits a lower boundary (variance = 0). (Similar issues can arise when testing correlation coefficients, which are bounded by ±1.00; however, covariance values are generally not bounded is this way.) Given the lower bound of 0, the sampling distribution for a variance with a population value of 0 is not normally distributed. Instead, it has a median and mode of 0 and is leptokurtic and positively skewed. Therefore, for multilevel models with random effects, the standard LRT is too conservative (Self & Liang, 1987; Stram & Lee, 1994).

To adjust for this issue, if the model has only one random effect, and therefore only one $\tau$, we can use the chi-square value for $p = .10$ to test for statistical significance when we set the Type I error rate (alpha) at .05 (Snijders & Bosker, 2012). The critical value of chi-square with 1 *df* is 3.841 for $p < .05$ and 2.706 for $p < .10$. To compare two models that differ in terms of one variance component, we should use the critical value of 2.706. In contrast, to compare two models that differ in terms of one fixed effect (and no random effects), the critical value is 3.841.

Comparing a model with one random effect to a model with two random effects is a bit more complex. The simpler model has two fewer parameters: it eliminates both a variance and a covariance. Although the variance has boundary issues (the variance cannot be less than 0), the covariance does not. Therefore, the correct critical value of $\chi^2$ comes from the $\bar{\chi}^2$ distribution, which is actually a mixture of $\chi^2$ distributions (Snijders & Bosker, 2012). Technically, the correct critical value of $\chi^2$ for a model that eliminates one random slope variance ($\tau_{11}$) and one covariance ($\tau_{01}$) is 5.14, rather than 5.99, as would normally be the case for a model that differs by two parameters (Snijders & Bosker, 2012, p. 99). The rejection regions for LRT that include variance components are 2.706 for a single variance parameter, 5.14 for a variance and a covariance, 7.05 for a variance and two covariances and 8.76 for a variance and three covariances (Snijders & Bosker, 2012). Snijders and Bosker (2012) present a more detailed discussion of this issue, as well as a table with the correct critical values to compare nested models that differ in terms of one or more randomly varying slopes.

## Akaike information criterion and Bayesian information criterion

Information criteria such as *Akaike information criterion* (AIC) and the *Bayesian information criterion* (BIC) also provide a method to compare the fit of competing model. There is an advantage to using *information criteria* for model comparison: they allow for comparison of non-nested models. Using AIC and BIC, we can compare competing models fit using the same sample, whether or not they are hierarchically nested. Lower information criteria (ICs) are indicative of better fitting models; therefore, the model with the lowest IC is considered the best fitting model (McCoach & Black, 2008). For additional details regarding the conceptual and methodological underpinnings of the AIC and the BIC, see Bozdogan (1987), Burnham and Anderson (2004), Raftery (1995), Schwarz (1978), Wagenmakers and Farrell (2004), Weaklim (2004, 2016) and Zucchini (2000).

## The Akaike information criterion (AIC)

To compute the AIC, simply multiply the number of parameters by 2 and add this product to the deviance statistic. The formula for the AIC is

$$AIC = D + 2p \qquad (2.11)$$

where $D$ is the deviance and $p$ = the number of estimated parameters in the model. The model with the lowest AIC value is considered the best model.

The deviance (or –2 log likelihood [–2LL]) represents the degree 'of inaccuracy, badness of fit, or bias when the maximum likelihood estimators of the parameters of a model are used' (Bozdogan, 1987, p. 356). The second term, $2p$, imposes a penalty based on the complexity of the model. This penalty indicates that the deviance must decrease by more than two points per additional parameter to favour the more parameterised model.

Compare this to the LRT for model selection. The critical value of $\chi^2$ with 1 $df$ at $\alpha$ = .05 is 3.84 (or 2.706 for a 1 $df$ change involving a variance). Therefore, when comparing two models that differ by 1 $df$, the LRT imposes a more stringent criterion for rejecting the simpler model. In fact, for comparisons of models that differ by seven or fewer parameters,[i] using the LRT results in an equivalent or more parsimonious model than the AIC. Conversely, when comparing models that differ by more than seven parameters, the AIC favours more parsimonious models than the LRT.

## The Bayesian information criterion (BIC)

The BIC equals the sum of the deviance and the product of the natural log of the sample size and the number of parameters. The formula for the BIC is

$$BIC = D + \ln(n) * p \tag{2.12}$$

where $D$ is deviance (–2LL), $p$ is the number of parameters estimated in the model and $n$ is the sample size. As with the AIC, the model with the lowest BIC is considered the best fitting model.

Therefore, the penalty the BIC imposes for each additional parameter is a function of the sample size ($n$). However, in MLM, it is not entirely clear which sample size should be used with the BIC: the total number of observations, the number of clusters at the highest level or some weighted average of the two. Furthermore, different software packages compute the BIC differently. Some (e.g. SPSS) use the overall sample size, whereas others (e.g. SAS PROC MIXED) use the number of clusters (level-2 units). Therefore, even when different statistical packages produce identical –2LL and AIC values, the BIC value may differ. Hence, the choice of sample size to compute the BIC could potentially change the outcome(s) of the model selection process.

Regardless of the choice of sample size for BIC, the per parameter penalty for the BIC is higher than the per parameter penalty for the AIC. Generally, multilevel models

---

[i]The number 7 assumes that we are using the standard critical values for chi-square with $\alpha$ = .05, not critical values that have been adjusted for boundary issues in the variances.

have at least 10 clusters, and for a sample size of 10, the penalty for the BIC is 2.3 times the number of parameters. (In fact, the sample size must be less than eight for the per parameter penalty for the BIC to drop below 2.) In contrast, the penalty for the AIC is 2 times the number of parameters. Therefore, whenever the AIC favours the simpler (less parameterised model), the BIC also favours the simpler (less parameterised) model. Whenever the BIC favours the more complex (more parameterised model), the AIC also favours the more parameterised model.

Not all software programs provide AIC and BIC measures in their output. However, it is easy to compute the AIC and the BIC from the deviance statistic. FIML is the most appropriate estimation method to use when computing information criteria (Verbeke & Molenberghs, 2000) compare two models that differ in terms of their fixed effects.

## Using model fit criteria for model selection

Unfortunately, the AIC, BIC and LRT may differ in terms of which model they favour. Honestly, the differences in the model fit criteria can be a bit overwhelming, and the various criteria do not always favour the same model. Table 2.1 displays the total penalty imposed by each of the model fit criteria for models that differ by 1, 2, 3 and 4 parameters. For example, imagine that we want to compare two models that differ by one fixed effect. Our total sample size is 1000 people, nested within 50 clusters. The deviance of the model that includes the parameter is 3.9 points lower than the deviance of the model that did not. In such a scenario, the LRT and AIC would favour the model that includes the fixed effect parameter; both the $BIC_2$ and the $BIC_1$ would favour the model that eliminates the fixed effect parameter. In this situation, two different researchers, faced with the same results, might make different decisions about which model to favour.

**Table 2.1** The total penalty imposed by each of the model fit criteria for models that differ by 1, 2, 3 and 4 parameters

|  | 1 Parameter | 2 Parameters | 3 Parameters | 4 Parameters |
|---|---|---|---|---|
| AIC | 2 | 4 | 6 | 8 |
| BIC (n = 10) | 2.3 | 4.6 | 6.9 | 9.2 |
| LRT (var) | 2.7 | 5.14 | 7.05 | 8.76 |
| LRT (trad) | 3.84 | 5.99 | 7.82 | 9.49 |
| BIC (n = 50) | 3.91 | 7.82 | 11.73 | 15.64 |
| BIC (n = 100) | 4.61 | 9.22 | 13.83 | 18.44 |
| BIC (n = 1000) | 6.91 | 13.82 | 20.73 | 27.64 |
| BIC (10,000) | 9.21 | 18.42 | 27.63 | 36.84 |

*Note.* AIC = Akaike information criterion; BIC = Bayesian information criterion; LRT = likelihood ratio test.

We suggest some simple heuristics to help navigate the morass of model fit criteria. These are meant to provide guidance in using the various model fit criteria; they are not meant to supersede them. Because AIC and BIC do not require nested models, we can apply this strategy to both nested and non-nested models. Table 2.2 displays the *per-parameter* penalty imposed by each of the model fit criteria for models that differ by 1, 2, 3 and 4 parameters. Across all of the model fit criteria, the per-parameter penalty is always at least 2. Therefore, when comparing two models that differ by one or more parameters (in terms of the number of parameters estimated by the model), first compute the difference in the deviances of the two models. Then divide that number by the difference in the number of estimated parameters ($\Delta/p$). If the ratio of the deviance difference to the number of parameters ($\Delta/p$) is less than 2, the fit criteria favour the more parsimonious (less parameterised) model. This is the simplest scenario, given that all criteria always favour the more parsimonious model when $\Delta/p$ is less than 2. When this ratio is above 10, we recommend favouring the more parameterised model. Again, this is a staightforward decision, as all criteria suggest favouring the more complex model (except perhaps the $BIC_1$, but the total sample size needs to be almost 25,000 people for the $BIC_1$ to favour the simpler model, and even then, all other criteria favour the more parameterised model). In small- to moderate-sized samples, we tend to favour the more parameterised model when the ratio is above 4 because the AIC and LRT always favour the more parameterised model when the ratio is above 4. In very large sample sizes, this heuristic may not be appropriate, given that deviance is a function of sample size. Therefore, in very large samples, using the $BIC_1$ may be advisable.

**Table 2.2** The per-parameter penalty imposed by each of the model fit criteria for models that differ by 1, 2, 3 and 4 parameters

|  | 1 Parameter | 2 Parameters | 3 Parameters | 4 Parameters |
|---|---|---|---|---|
| AIC | 2 | 2 | 2 | 2 |
| BIC ($n = 10$) | 2.3 | 2.3 | 2.3 | 2.3 |
| LRT (vdi) | 2.7 | 2.57 | 2.35 | 2.19 |
| LRT (trad) | 3.84 | 3 | 2.61 | 2.37 |
| BIC ($n = 50$) | 3.91 | 3.91 | 3.91 | 3.91 |
| BIC ($n = 100$) | 4.61 | 4.61 | 4.61 | 4.61 |
| BIC ($n = 1000$) | 6.91 | 6.91 | 6.91 | 6.91 |
| BIC ($n = 10,000$) | 9.21 | 9.21 | 9.21 | 9.21 |

*Note.* AIC = Akaike information criterion; BIC = Bayesian information criterion; LRT = likelihood ratio test.

$\Delta/p$ ratios between 2 and 4 represent the 'grey zone', where some model fit criteria favour the simpler model and other model fit criteria favour the more parameterised model.

When $\Delta/p$ is between 2 and 4, compute the various model fit criteria. Then supplement the model fit criteria with Rights and Sterba's (2019) variance-explained measures (discussed in the next section). What proportion of within, between and/or total variance does this parameter explain? Does removing the parameter substantially reduce the predictive ability of the model? Also, given the purpose of the model, decide which would be a more grievous error: an error or omission or the inclusion of an unnecessary parameter. If omitting a potentially important parameter is more problematic, then we recommend favouring the more parameterised model. If including an unnecessary parameter is more problematic, then we recommend favouring the simpler model. Generally speaking, in the grey zone, we favour retaining potentially important *fixed effects* but eliminating unnecessary *random effects* (variance components).

If the difference in the number of parameters is 0 (which can happen when comparing two non-nested models), then we favour the model with the lower deviance. Why? We cannot use the LRT for non-nested models. If two models have the same number of parameters, then the model with the lower deviance always has the lower AIC, $BIC_1$ and $BIC_2$ (and in fact has the lower IC, regardless of which is IC chosen).

Model selection decisions should consider both the fit and the predictive ability of the multilevel model. Next, we turn our attention to quantifying explained variance in multilevel models.

## Proportion of variance explained

In single-level regression models, an important determinant of the utility of a model is the proportion of variance explained by the model, $R^2$. In MLM, computation of the proportion of variance explained becomes far more complex. Variance components exist at each level of the multilevel model. In addition, in random coefficients models, the relation between an independent variable at level 1 and the dependent variable varies as a function of the level-2 unit or cluster. Given that variance in the outcome variable is decomposed into multiple components, quantifying the variance explained by a set of predictors becomes more complicated than in the single-level case.

Conceptually, we could be interested in measuring variance explained within clusters, variance explained between clusters and/or total variance explained (both within and between clusters). For example, adding a cluster-level (level-2) variable to a multilevel model cannot possibly explain within-cluster variance. Similarly, cluster mean centred level-1 (within-cluster) variables cannot explain between-cluster variance. However, a cluster-level variable can explain between-cluster variance; and because it can explain between-cluster variance, it can also explain some of the total variance. Imagine a situation in which 5% of the variance in the outcome variable lies between clusters and 95% of the variance lies within clusters. Suppose we find a variable that explains most (80%) of the between-cluster variance. This variable is a powerful predictor of the between-cluster variance, but it only explains 4% (80% * 5%) of the total variance. Therefore, deciding how to compute and report variance-explained measures in multilevel modelling requires explicit consideration of the context and goals of the research.

To this end, Rights and Sterba (2019) developed 'an integrative framework of $R^2$ measures for multilevel models with random intercepts and/or slopes based on a completely full decomposition of variance' (p. 309). To use Rights and Sterba's variance-explained measures to partition outcome variance into between-cluster outcome variance and within-cluster outcome variance, we group mean centre all level-1 predictor variables and add the aggregate level-1 variables into the level-2 model. (There is one exception: if the level-1 variable has only within-cluster variance, and has no between-cluster variance, then this is not necessary.) Then we can decompose the model-implied total outcome variance into five specific sources of variance: (1) variance attributable to level-1 predictors via fixed slopes ($f_1$), (2) variance attributable to level-2 predictors via fixed slopes ($f_2$), (3) variance attributable to level-1 predictors via random slope variation and covariation ($v$), (4) variance attributable to cluster-specific outcome means via random intercept variation ($m$)[ii] and (5) variance attributable to level-1 residuals ($\sigma^2$) (Rights & Sterba, 2019).

Assuming all level-1 variables have been group mean centred, three of the sources contain only within-cluster variance: (1) variance attributable to level-1 predictors via fixed slopes ($f_1$), (2) variance attributable to level-1 predictors via random slope variation and covariation ($v$) and (3) variance attributable to level-1 residuals ($\sigma^2$). Therefore, we can evaluate the proportion of within-cluster variance explained by level-1 predictors ($f_1$), and we can determine what proportion of within-cluster variance is accounted for by the variances and covariances of the randomly varying slopes ($v$) (Rights & Sterba, 2019).

Two of these sources contain only between-cluster variance: (1) variance attributable to level-2 predictors via fixed slopes ($f_2$) and (2) variance attributable to cluster-specific outcome means via random intercept variation ($m$). Therefore, we can determine the proportion of between-cluster variance that is explained by our level-2 predictors ($f_2$) and the proportion of between-cluster variance that is random intercept variance not explained by the level-2 predictors in our model ($m$) (Rights & Sterba, 2019).

In every model, it is possible to decompose the model-implied total variance into these five sources. Then, using Rights and Sterba's (2019) **integrative framework of $R^2$** measures in multilevel models, researchers can compute a variety of variance-explained

---

[ii]When all level-1 variables are cluster mean centred (and the aggregate is included at level-2), $m = \tau_{00}$.

measures, each of which provides potential insights into the model's predictive capabilities. Rights and Sterba (2019) show the correspondence between their integrative framework and other $R^2$ measures that have been used in MLM. Their integrative framework allows for easy computation of previously used variance-explained measures without needing to estimate multiple multilevel models. In addition, Shaw et al. (2020) developed an R package, *r2mlm*, that computes all measures in Rights and Sterba's integrative framework and provides graphical representations of the various measures. In Chapter 3, we return to this topic in greater detail, when we describe the process of fitting and evaluating multilevel models. There, we provide more details on Rights and Sterba's $R^2$ measures and provide concrete recommendations for using Rights and Sterba's integrative framework within the model building process.

## Effect size

An *effect size* is a practical, interpretable, quantitative measure of the magnitude of an effect. As with any statistical analyses, it is important to report effect size measures for multilevel models. The $R^2$ measures described above can help researchers and readers to determine the impact that a variable or a set of variables has on a model, with respect to variance explained. In addition, researchers can compute Cohen's $d$-type effect sizes to describe the mean differences among groups. To calculate the equivalent of Cohen's $d$ for a group-randomised study (where the treatment variable occurs at level 2), use the following formula (Spybrook et al., 2011):

$$\delta = \frac{\hat{\gamma}_{01}}{\sqrt{\hat{\sigma}^2 + \hat{\tau}_{00}}} \qquad (2.13)$$

Assuming two groups have been coded as 0/1 or −.5/+.5, the numerator of the formula represents the difference between the treatment and control groups. The denominator utilises the $\sigma^2$ and $\tau_{00}$ from the unconditional model, where the total variance in the dependent variable is divided into two components: (1) the between-cluster variance, $\tau_{00}$, and (2) the within-cluster variance, $\sigma^2$. There are numerous ways to compute effect sizes in MLM (or any analysis), and not all effect sizes need to be standardised, especially when unstandardised metrics are commonly used and easily understood. We encourage you to present the results of your MLM as clearly as possible and to contextualise the parameters in practically meaningful and easily interpretable ways.

Now that we have introduced most of the fundamental concepts in MLM, let's turn our attention to an applied example so that we can provide concrete guidance on how to build and interpret multilevel models. Chapter 3 focuses on building, evaluating and interpreting multilevel models.

**Chapter Summary**

- In a multilevel model without predictors, each person's score on the dependent variable is composed of three elements: (1) the expected mean ($\gamma_{00}$), (2) the deviation of the cluster mean from the overall mean ($u_{0j}$) and (3) the deviation of the person's score from his/her cluster mean ($r_{ij}$). In this equation, $\gamma_{00}$ is a *fixed effect*: $\gamma_{00}$ is the same for everyone. The $u_{0j}$ term is called a *random effect* for the intercept because $u_{0j}$ randomly varies across the level-2 units (clusters).
- In multilevel models, *fixed effects* are parameters that are fixed to the same value across all clusters (or individuals), whereas *random effects* differ (vary) across clusters (or individuals).
- The variance in the outcome variable can be partitioned into within and between-cluster variance components. The ability to partition variance into within-cluster variance and between-cluster variance is one of MLM's greatest assets.
- Intercepts and slopes can vary across clusters in a multilevel model. We can build a regression equation at level 2 to try to explain the variation in these randomly varying intercepts and slopes.
- Model selection decisions should consider both the fit and the predictive ability of the multilevel model.
- Variance components exist at each level of the multilevel model. In addition, in random coefficients models, the relation between an independent variable and the dependent variable at level 1 varies as a function of the level-2 unit or cluster.
- Rights and Sterba (2019) developed 'an integrative framework of $R^2$ measures for MLM with random intercepts and/or slopes based on a completely full decomposition of the outcome variance'.

# Further Reading

Raudenbush, S. W., & Bryk, A. S. (2002). *Hierarchical linear models: Applications and data analysis methods* (2nd ed.). Sage.

This is a classic book on the theory and use of hierarchical linear modelling and multilevel modelling. The book is a must read for researchers interested in diving further into the mathematical details of hierarchical linear models (e.g. estimation theory and multivariate growth models).

Enders, C. K., & Tofighi, D. (2007). Centering predictor variables in cross-sectional multilevel models: A new look at an old issue. *Psychological Methods, 12*(2), 121–138.

This article provides a detailed overview of grand mean centring and group mean centring in the context of two-level multilevel models. In addition to the expansive discussion of centring in multilevel models, it provides illustrative examples that should provide readers with a foundation to answering questions with their data.

# 3

# MULTILEVEL MODEL BUILDING STEPS AND EXAMPLE

## Chapter Overview

A review of traditional model building steps in MLM ................................ 47

Model building steps for confirmatory/predictive models ........................ 50

An example of a two-level multilevel model ............................................. 54

Further Reading ....................................................................................... 65

Building a multilevel model is not nearly as straightforward as running an ANOVA or a single-level regression model. Decisions about whether to include random effects can influence tests of the fixed effects, and inclusion of fixed effects can impact the magnitudes of the random effects. In addition, multilevel models are large, often containing many parameters at multiple levels, so it is easy to become lost in a forest of fixed and random effects or to inadvertently engage in $p$-hacking during the modelling process. Therefore, it is essential to approach model building with some sort of system or guiding framework.

Several authors (e.g. Hox, 2010; Raudenbush & Bryk, 2002; Snijders & Bosker, 2012; West et al., 2015) have provided suggestions for how to approach the task of model building. These prescriptions are principled and rational, and they play an important role in the tradition of MLM. However, in our experience, some researchers view model building methods as a licence to engage in exploratory analyses. In an exploratory framework, tests of statistical significance lose their original meaning. After all, if there is no hypothesis, how can hypothesis testing be sensible? Researchers who use multilevel models in an exploratory capacity must present their methods and results in a manner consistent with their approach. Exploratory studies are important sources for new theories and hypotheses; however, descriptions of such research should clarify its role as hypothesis generating rather than hypothesis confirming. Exploratory studies must clearly describe the model building and modification process in enough detail so that other researchers can reproduce the entire series of model building steps and models. Researchers should also describe how much the focal parameters of interest varied across the range of estimated models. Well-articulated, honest, exploratory models can be quite informative and can provide guidance, especially for newer, uncharted areas of research.

Unfortunately, some exploratory research masquerades as confirmatory research. In these instances, researchers run an endless array of level-1 models and various combinations of full contextual models before settling on a model that maximises the number of statistically significant parameters and tells an interesting story, whether or not that story was part of the original hypothesis. In this era of open science, such research antics are finally being scrutinised and criticised. 'Lack of clarity between post-diction and prediction provides the opportunity to select, rationalize, and report tests that maximize reward over accuracy' (Nosek et al., 2018, p. 2601). For this reason, **preregistration** has become an increasingly popular practice in psychology, education and the social sciences. Preregistration involves (a) committing to a series of analytic steps prior to engaging with the data, (b) providing detailed descriptions of the proposed analytic techniques and (c) specifying exactly how the authors will conduct all data analyses, without advance knowledge of the research outcomes (Nosek et al., 2018). In addition, preregistration reports describe which variables will be included, how those variables will be recoded or transformed and how decisions about modelling and statistical tests will occur during the analysis. Preregistration achieves the scientific goal of prediction because the observed data does not influence selection of tests and because all conducted tests are knowable (Nosek et al., 2018).

Historically, modelling techniques (MLM/SEM) have involved a series of model building steps and model comparisons (Raudenbush & Bryk, 2002). However, multistep model building procedures have the potential to increase researcher degrees of freedom. Given the movement towards preregistration, it is important to more carefully consider how to conduct reproducible and defensible *confirmatory* analyses within a modelling framework. Therefore, after reviewing traditional multilevel model building procedures, we recommend a modified set of steps to build and test confirmatory multilevel models. Our approach incorporates Rights and Sterba's (2019) unified framework for computing the proportions of variance explained in MLM, which we introduced in Chapter 2.

## A review of traditional model building steps in MLM

Traditionally, many MLM textbooks and instructors have suggested fitting a series of models. Although the exact steps vary from book to book, one of the most common approaches has been to build the multilevel model up, starting with a completely unconditional model, then moving to the level-1 model, and before finally moving onto the full contextual model. In fact, until recently, we advocated the traditional model building approach outlined below:

1  *Model 1:* The unconditional, random-effects model allows us to compute the **unconditional ICC**.
2  *Model 2:* Estimate the level-1 model. Deciding which level-1 coefficients to allow to randomly vary across clusters is often the most difficult decision when building a multilevel model. Before adding variables to the model, consider theory, prior research and the substantive focus of the study to decide which level-1 variables should be allowed to randomly vary by cluster.
3  *Model 3:* Add level-2 variables as predictors of the randomly varying slopes and intercepts. If theory specifies a cross-level interaction between a level-2 variable and a level-1 variable, add that interaction to the model, even if the level-1 slope is not randomly varying.
4  *Model 4 (Optional):* After running the full contextual model, if any random effects are unnecessary, eliminate them. Then, rerun the model and compare the fit of the simpler model (Model 4) to that of the full contextual model (Model 3).

One major reason for fitting this series of increasingly parameterised models was to allow for the computation of pseudo-$R^2$ measures, such as Raudenbush and Bryk's (2002) *proportional reduction in variance.* In MLM, it is useful and informative to evaluate the degree to which the predictors in the multilevel model can help to explain between- and/ or within-cluster variability in the outcome variable of interest. Computing Raudenbush and Bryk's *proportional reduction in variance* requires estimating multiple models. However, using Rights and Sterba's (2019) framework, we can estimate variance-explained measures within a single multilevel model. For context and comprehensiveness, we briefly review Raudenbush and Bryk's (2002) *proportional reduction in variance.* Then we outline model building steps for confirmatory MLM using the Rights and Sterba (2019) integrated framework for computing variance explained in MLM.

## Raudenbush and Bryk's proportional reduction in variance

The most common traditionally reported multilevel pseudo-$R^2$ statistics are Raudenbush and Bryk's (2002) *proportional reduction in variance statistics*, available for each of the variance components. In models that do not include randomly varying slopes, we can estimate the proportional reduction in level-1 residual variance and the proportional reduction in variance for the level-2 intercept (residual) variance. We can also compute the proportional reduction in variance for the level-2 slope (residual) variance for slopes that randomly vary across clusters.

The *proportional reduction in variance* compares the residual variance from the full (more parameterised) model to the residual variance from a simpler 'base' model. If the full model explains additional variance, then the residual variance should decrease. In such a scenario, the residual variance for the baseline model should be greater than the residual variance for the full model. The proportional reduction in variance divides the difference in the residual variance across the two models (in the numerator) by the baseline residual variance (in the denominator).

To compute the proportional reduction in level-1 variance, subtract the remaining variance in the more parameterised, level-1 model (i.e. Model 2) from the variance of the simpler, unconditional model (i.e. Model 1). Then, divide this difference by the variance from the simpler model (Model 1). The formula is

$$\frac{\hat{\sigma}_b^2 - \hat{\sigma}_f^2}{\hat{\sigma}_b^2} \tag{3.1}$$

where $\hat{\sigma}_b^2$ is the estimated level-1 variance for the simpler 'base' model and $\hat{\sigma}_f^2$ is the estimated level-1 variance for the fitted model (Raudenbush & Bryk, 2002).

At level-2, $\hat{\tau}_{qq}$ are the estimated variance components for the intercept ($\beta_{0j}$) and each slope estimate ($\beta_{1j}, \beta_{2j}, ..., \beta_{qj}$) that is allowed to randomly vary across clusters. Therefore, the proportional reduction in the variance of a given slope, $\beta_{qj}$, is

$$\frac{\hat{\tau}_{qq_b} - \hat{\tau}_{qq_f}}{\hat{\tau}_{qq_b}} \tag{3.2}$$

where $\hat{\tau}_{qq_b}$ is the estimated variance of slope $q$ in the simpler base model (e.g. Model 2) and $\hat{\tau}_{qq_f}$ is the estimated variance of slope $q$ in the fitted model (e.g. $\tau_{11}, \tau_{22}, ..., \tau_{qq}$). The level-2 proportion reduction in variance statistics must utilise two models with identical level-1 models (Raudenbush & Bryk, 2002, p. 150).

However, Raudenbush and Bryk's (2002) proportion reduction in variance statistic does not behave like the familiar $R^2$. It is a simple comparison of the residual variances from two models, so it cannot be interpreted as an explanation of the absolute amount of variance in the dependent variable, and it cannot be computed without running two separate models. In addition, the proportion reduction in variance statistic can be negative. This frequently happens when comparing the level-2 intercept variance from a completely null model (an unconditional, random-effects ANOVA model with no predictors at level 1 or level 2) to the level-2 intercept variance from a model with a group mean–centred predictor at level 1. Finally, we cannot compute the proportion reduction in variance for two models that differ in terms of the number of random slopes being estimated.

More recent developments in the multilevel modelling literature offer improvements to this traditional framework. After outlining several shortcomings of traditional approaches for calculating $R^2$ measures, Rights and Sterba (2019) detailed a new, integrative framework for computing the proportion of total and level-specific variance explained that only requires estimates from a *single* fitted model. As mentioned in Chapter 2, in every two-level model, it is possible to decompose the model-implied total variance into five sources: (1) variance attributable to level-1 predictors via fixed slopes ($f_1$), (2) variance attributable to level-2 predictors via fixed slopes ($f_2$), (3) variance attributable to level-1 predictors via random-slope variation and covariation ($v$), (4) variance attributable to cluster-specific outcome means via random intercept variation ($m$) and (5) variance attributable to level-1 residuals ($\sigma^2$) (Rights & Sterba, 2019). Using these five sources of variance, we can compute a variety of variance-explained measures (Rights & Sterba, 2019).

Rights and Sterba's (2019) integrated framework for computing explained variance arrived during a period in which expectations for research transparency, reproducibility and replicability have transformed the research landscape. Their framework provides an essential new tool for building and evaluating confirmatory multilevel models in fewer steps, as well as a simpler method for engaging in confirmatory/predictive modelling using MLM. Thinking more broadly, Rights and Sterba's (2019) approach both supports and facilitates the movement towards open science and research reproducibility. Given the current zeitgeist of open science, we recommend revising and substantially simplifying the traditional multilevel model building approach for non-exploratory models. Below, we outline our suggested approach for fitting confirmatory multilevel models.

## Model building steps for confirmatory/predictive models

*Step 1:* First, fit the **unconditional random-effects ANOVA model** (Model 1). This is the model with no predictors. The main goal of fitting this model is to compute and report the unconditional ICC for the outcome variable. This unconditional model has only three parameters: (1) the within-cluster variance, (2) the between-cluster variance and (3) the parameter estimate for the intercept, which represents the overall expected value on the outcome variable.

*Step 2:* Fit the full theoretical model, including all within- and between-cluster variables, all hypothesised cross-level (and same-level) interactions and all theoretically relevant random effects (Model 2). Using Rights and Sterba's framework to partition the variances into the five components mentioned above requires *group mean centring* all level-1 variables and including the aggregates of those variables at level 2. Carefully consider which level-1 slopes are allowed to randomly vary at level 2. (See the sections above and below for additional guidance on random slopes.) If your theory specifies a cross-level interaction between a level-2 variable and a level-1 variable, add that interaction to the model, even if the level-1 slope is not randomly varying. *Therefore, the decision to allow for a random slope should be based on your hypothesis about whether you think that the slopes will randomly vary AFTER accounting for slope variability attributable to cluster-level variables* (LaHuis & Ferguson, 2009).

Rights and Sterba (2019) describe 12 different multilevel model $R^2$ measures; we introduce a few of these measures below. The equations below were first introduced by, and follow the labelling conventions of, Rights and Sterba. Compute, interpret and report the following proportions of variance explained for the full contextual model:

A   Compute the proportion of within-cluster outcome variance explained by the level-1 predictors via fixed slopes ($R_w^{2(f1)}$):

$$R_w^{2(f1)} = \frac{f_1}{f_1 + v + \sigma^2} \tag{3.3}$$

$R_w^{2(f1)}$ indicates how well the *fixed effects* for the set of level-1 predictors (set of level-1 slopes) explain the within-cluster variance in the outcome variable.

B   In addition, consider computing the total proportion of within-cluster outcome variance explained by level 1 predictors (via fixed-slopes and random-slope variation/covariation). This is the sum of the variance attributable to level-1 predictors via fixed slopes $(f_1)$ and the variance attributable to level-1 predictors via random-slope variation and covariation $(v)$ divided by the total level-1 variance, which includes both terms from the numerator plus the variance attributable to the level-1 residuals, $\sigma^2$:

$$R_w^{2(f_1 v)} = \frac{f_1 + v}{f_1 + v + \sigma^2} = 1 - \frac{\sigma^2}{f_1 + v + \sigma^2} \tag{3.4}$$

This is analogous to computing Raudenbush and Bryk's proportion reduction in level-1 residual variance (Rights & Sterba, 2019). One minus $R_w^{2(f_1 v)}$ is the proportion of within-cluster residual variance (variance that is not explained by either the fixed-effects or the randomly varying slopes for the set of level-1 predictors.)

$$1 - R_w^{2(f_1 v)} = \frac{\sigma^2}{f_1 + v + \sigma^2} \tag{3.5}$$

C   (If no slopes randomly vary, then skip C.) If any level-1 slopes randomly vary across clusters, compute the proportion of within-cluster variance explained by level-1 predictors via random-slope variation/covariation:

$$R_w^{2(v)} = \frac{v}{f_1 + v + \sigma^2} \tag{3.6}$$

$R_w^{2(v)}$ indicates how much within-cluster variance is explained by allowing the slopes to randomly vary across clusters. Adding cross-level interactions that explain the between-cluster variability in the slopes, decreases $v$ and increases $f_2$. Therefore, $R_w^{2(v)}$ is the slope variance that is unexplained by level-2 (between-cluster) variables. Inspecting $R_w^{2(v)}$ can also provide insight into the necessity for allowing the slopes to randomly vary. If the model contains multiple randomly varying slopes, it can be useful to assess how much within-cluster variance each of the randomly varying slopes explains. In such a situation, you may wish to examine the $R_w^{2(v)}$ separately for each randomly-varying slope. To do so, fit a model in which only one slope randomly varies. The $R_w^{2(v)}$ for the model indicates the proportion of within-cluster variance attributable to that randomly varying slope.

D   Compute the proportion of between-cluster outcome variance explained by level-2 predictors via fixed slopes:

$$R_b^{2(f_2)} = \frac{f_2}{f_2 + m} = \frac{f_2}{f_2 + \tau_{00}} \tag{3.7}$$

This provides information about the degree to which the set of level-2 predictors can explain the between-cluster variance in the outcome variable. This is equivalent to Raudenbush and Bryk's proportion reduction in level-2 variance.

One minus the proportion of between-cluster outcome variance explained by level-2 predictors via fixed slopes is the proportion of the between-cluster variance that remains unexplained by the model. This is the proportion of the residual variance in the intercepts:

$$1 - \frac{f_2}{f_2 + m} = \frac{m}{f_2 + m} = \frac{\tau_{00}}{f_2 + \tau_{00}} \tag{3.8}$$

E   Compute the proportion of total (outcome) variance explained by cluster-specific outcome means via random-intercept variation:

$$R_t^{2(m)} = \frac{m}{m + f_2 + f_1 + v + \sigma^2} = \frac{\tau_{00}}{\tau_{00} + f_2 + f_1 + v + \sigma^2} \tag{3.9}$$

This is the **conditional ICC**, the proportion of (residual) between-cluster variance in the intercepts after accounting for all the variables in the model. It indicates how much variance is explained by allowing the intercept to randomly vary across clusters (Rights & Sterba, 2019). As a side note, if both $v$ and $m$ were 0, there would be no need for a multilevel model because there would be only one residual component, $\sigma^2$.

F   What about an overall $R^2$ type measure? You may wish to report the proportion of variance in the outcome variable that is explained by the model.

The proportion of total variance in the outcome that is explained by the level-1 and level-2 fixed effects ($R_t^{2(f)}$) includes variance explained by all level-1 and level-2 variables but does not include unexplained variance in the randomly varying level-1 slopes or in the randomly varying intercepts.

$$R_t^{2(f)} = \frac{f_1 + f_2}{f_1 + f_2 + v + m + \sigma^2} \tag{3.10}$$

$R_t^{2(f)}$ can be separated into the proportion of total variance that is explained by the level-1 slopes and the cross-level interactions (given that all level-1 predictors are group (cluster) mean centred), $R_t^{2(f_1)}$ :

$$R_t^{2(f_1)} = \frac{f_1}{f_1 + f_2 + v + m + \sigma^2} \tag{3.11}$$

And the proportion of total variance that is explained by level-2 predictors, $R_t^{2(f_2)}$:

$$R_t^{2(f_2)} = \frac{f_2}{f_1 + f_2 + v + m + \sigma^2} \tag{3.12}$$

G   As always, the reporting of results should include effect sizes and descriptions of the practical importance of the findings. These effect size measures could include some of the $R^2$ measures described above, Cohen's $d$ measures of effect size for dichotomous variables and/or other readily interpretable metrics that contextualise the results in a way that highlights their substantive meaningfulness and practical importance.

*Step 3:* (Optional/Not always necessary). After running the full contextual model, if any random effects prove to be unnecessary, eliminate them. Re-estimate the model and compare the model fit and proportions of variance explained within clusters, between clusters and overall for the simpler model to the full contextual model (Model 2). Determining whether or not to eliminate random effects for one or more slopes is not completely straightforward. We recommend examining the following four pieces of information to determine whether any of the randomly varying slopes are unnecessary:

1   Model convergence issues are often a sign that your multilevel model contains an unnecessary random slope. If you experience model convergence problems, or if it takes thousands of iterations for the model to converge, you may need to eliminate one or more random effects.
2   Some software packages provide tests of the statistical significance for the variance components. You can consult these tests for guidance; however, statistical tests of variance components are somewhat controversial and should be treated as approximations or heuristics to provide guidance, not exact and infallible tests. Also, in programs that usually report standard errors, their absence for some or all of the variance components is often a sign that one or more of the random effects in the model is unnecessary.
3   Use Rights and Sterba's framework to compute the proportion of within-cluster (outcome) variance explained by the level-1 predictors via random-slope variation/ covariation in the model that includes the randomly varying slope(s). If the model contains multiple randomly varying slopes, to determine whether to eliminate one of the slopes, compare the $R^2_{vw}$ from the model that includes all of the slopes as randomly varying to the $R^2_{vw}$ from the model that constrains one of the slopes to be non-random. If the change in $R^2$ is near .00, allowing the slope to randomly vary explains very little of the within-cluster variance, suggesting that the random slope could potentially be eliminated.
4   Compare the model fit (deviance, AIC, BIC) of the model that includes the random slope to the model that does not. The AIC, which tends to be more liberal than either the BIC or the chi-square difference test (with a small number of degrees of freedom), suggests a penalty of 2.00 points per parameter. So, deviance changes of less than 2 points per eliminated parameter suggest that model fit is very similar across the two models – the fit of the more parsimonious model and the fit of the more complex model (the one that includes the random slope) are nearly the same. In such cases, we favour the simpler model. Generally, deviance differences of less than 2 points per parameter suggest eliminating the random slope does not adversely impact model fit.

But, be careful! Remember – eliminating a random slope does not necessarily reduce the model by one parameter. Fitting an unstructured variance/covariance matrix for the level-2 variance components allows covariances among all of the level-2 variance components. Therefore, a model with one random slope and one random intercept has three variance/covariance parameters, whereas a model with only a random intercept has one variance/covariance parameter. Thus, the model with the random slope contains two additional parameters; and, according to the AIC, the deviance should drop by at least 4 points to justify including the additional parameters. Furthermore, when comparing a model with two random slopes (and a randomly varying intercept) to a model with one random slope (and a randomly varying intercept), the model with two random slopes has three more parameters than the model with two random slopes (one additional variance and two additional covariances). In this scenario, deviance increases of less than 6 points when eliminating one of the random effects from the model would favour the simpler model according to the AIC (or 7.05 points when using the adjusted LRT for the variance components).

*Step 4:* (Optional). You may wish to compare a model that eliminates one or more fixed effects from the model to the full contextual model. The most common rationale for fitting the reduced model and comparing it to the full model is to compute a change in $R^2$ measure, which provides a method for determining how much variance is uniquely explained by the variable eliminated from the reduced model. This change in $R^2$ provides an indication of the predictive utility of the predictor. Rights and Sterba (2019) provide a detailed demonstration of using their framework to compare their $R^2$ measures across models.

## An example of a two-level multilevel model

To illustrate how to fit and interpret a multilevel model, we present a simple example, using data gathered from a multisite cluster-randomised trial evaluating the impact of a high-quality, evidence-based vocabulary instruction intervention in kindergarten (Coyne et al., 2019). We illustrate the model building steps outlined above by first fitting the unconditional model, followed by the full model. Then, for pedagogical purposes, we compare the full model to (a) a more complex model that includes an additional randomly varying slope and (b) a more parsimonious model that eliminates the fixed effect for a same-level interaction.

The data consists of 1,428 students, nested within 233 different clusters/classrooms, with roughly six at-risk students per classroom. (Although these are actually three-level data, students are nested within classes, which are nested within schools; however, for illustrative purposes, we fit a simple two-level multilevel model, with students nested within classes.) To compare two models in terms of fit, variance explained or even parameter estimates, the two models must use the exact same sample (Singer & Willett, 2003). Therefore, we deleted any cases with missing data on any of the variables used in the analyses (Peabody Picture Vocabulary Test scores [PPVT], expressive target word [ETW] and treatment [TRT]).

At the start of the kindergarten school year, researchers randomly assigned half of the at-risk students within a given class to the treatment group and half to the control group. The treatment group received supplemental small-group vocabulary instruction in addition to the whole-group instruction; the control group received only whole-group vocabulary instruction. (The observations were not independent because students were nested within classrooms.) The multilevel analyses presented below explore the question: *Does supplemental small-group kindergarten vocabulary instruction intervention increase student's knowledge of the vocabulary words taught during the intervention?* Researchers measured vocabulary knowledge using an ETW task, which assessed the student's ability to explain the meaning of a given word. The researcher administered the ETW assessment after the intervention concluded, in the spring of kindergarten. In the sample, ETW scores ranged from 0 to 52 (the maximum score), with a mean of 13.65 and standard deviation of 11.10. For pedagogical purposes, we fit a series of 4 two-level multilevel models:

1   A completely unconditional random intercept-only model, which allowed us to estimate the (unconditional) ICC.
2   A random intercept and slope model with two group mean-centred level-1 variables: treatment and PPVT (Dunn & Dunn, 2007) and two level-2 variables (the cluster means for PPVT and TRT). TRT was allowed to randomly vary across clusters.
3   A random intercept and slope model with two group mean-centred level-1 variables, treatment and PPVT, their interaction (PPVT × TRT) (at level 1), and one level-2 variable (the cluster mean for PPVT). TRT was allowed to randomly vary across clusters.
4   A random intercept and slope model with two group mean-centred level-1 variables, treatment and PPVT, their interaction (PPVT × TRT) (at level 1), and one level-2 variable (the cluster mean for PPVT). Here, *both* level-1 variables (TRT and PPVT) were allowed to randomly vary across clusters.

Our conceptual model of interest is Model 3. Model 3 allows the treatment effect to vary across classrooms; however, it assumes that the effect of PPVT on ETW and the moderating effect of TRT on the PPVT–ETW slope do not vary across clusters. To illustrate many of the concepts and techniques that we have presented in this chapter, we compare Model 3 to both a simpler model (Model 2, which does not include the cross-level interaction effect of TRT on the PPVT–ETW slope) and a more complex model (Model 4, which allows both the TRT slope and the PPVT–ETW slope to randomly vary across classes). Comparing Models 2 and 3, we can determine how much within-cluster outcome variance the PPVT-TRT interaction explains. Comparing Models 3 and 4 indicates how much between-cluster variance exists in the PPVT × TRT interaction.

Treatment was a dichotomous effect-coded variable (treatment = 1/2, control = −1/2). To partition variance into five separate within- and between-variance components, we group mean centred treatment, and we included the cluster mean of the treatment variable at level 2.

## The random intercept-only, one-way ANOVA or unconditional model (Model 1)

Recall, the level-1 model notation for the unconditional random-effects ANOVA (random intercept-only) model is

$$Y_{ij} = \beta_{0j} + r_{ij} \tag{3.13}$$

The outcome, $Y_{ij}$ measures students' expressive word knowledge. Students' expressive word knowledge scores ($Y_{ij}$) vary randomly around their classroom means ($\beta_{0j}$). The level-1 error, $r_{ij}$ is assumed to be normally distributed with a mean of 0 and a constant variance of $\sigma^2$.

The level-2 intercept-only model is

$$\beta_{0j} = \gamma_{00} + u_{0j} \tag{3.14}$$

In the level-2 model, $\beta_{0j}$ represents the expected (mean) expressive word knowledge in the $j$th classroom, $\gamma_{00}$ represents the overall expected value of expressive word knowledge in the population and $u_{0j}$ is a random effect for the $j$th classroom. The random effect, $u_{0j}$ is assumed to have a mean of 0 and variance of $\tau_{00}$. Merging the level-1 and level-2 equations yields the combined intercept-only model:

$$Y_{ij} = \gamma_{00} + u_{0j} + r_{ij} \tag{3.15}$$

The multilevel model decomposes the total variance of expressive word knowledge ($Y_{ij}$) into two orthogonal components, the within-classroom variance ($\sigma^2 = 104.38$) and the between-classroom variance ($\tau_{00} = 19.70$):

$$\mathrm{var}(Y_{ij}) = \mathrm{var}(u_{0j} + r_{ij}) = \tau_{00} + \sigma^2 \tag{3.16}$$

Here, $\sigma^2$ represents within-classroom variability and $\tau_{00}$ represents between-classroom variability in students' expressive word knowledge. Having partitioned the total variance into unconditional within- and between-level sources of variance, we can calculate the ICC ($\rho$):

$$\rho = \tau_{00} / (\tau_{00} + \sigma^2) \tag{3.17}$$

The parameter estimates for $\rho$, $\tau_{00}$, $\sigma^2$ appear in Table 3.1 (Model 1). The ICC is $\rho = \hat{\tau}_{00} / (\hat{\tau}_{00} + \hat{\sigma}^2) = 19.70/(19.70 + 104.38) = 0.158$. The ICC indicates that 15.8% of the variance in expressive word knowledge is between classrooms; 84.2% of the variance in expressive word knowledge is within classrooms. Put differently, we expect two students from the same classroom to have expressive word knowledge scores that correlate at .158.

**Table 3.1** Results of two-level analysis evaluating the effectiveness of a supplemental vocabulary instruction intervention with kindergarten students

|  | Model 1 | Model 2 | Model 3 | Model 4 |
|---|---|---|---|---|
| Fixed effects |  |  |  |  |
| Intercept ($\gamma_{00}$) | 13.72** (0.41) | 13.81** (0.39) | 13.83** (0.39) | 13.83** (0.39) |
| Mean TRT ($\gamma_{01}$) |  | 8.40** (2.41) | 8.46** (2.41) | 7.75** (2.40) |
| Mean PPVT ($\gamma_{02}$) |  | 0.45** (0.12) | 0.45** (0.12) | 0.42** (0.12) |
| TRT ($\gamma_{10}$) |  | 10.53** (0.63) | 10.53** (0.63) | 10.54** (0.63) |
| PPVT ($\gamma_{20}$) |  | 0.43** (0.05) | 0.44** (0.05) | 0.43** (0.05) |
| TRT × PPVT ($\gamma_{30}$) |  |  | 0.23* (0.09) | 0.23* (0.09) |
| Intercept variance ($\tau_{00}$) | 19.73 | 24.19 | 24.12 | 24.57 |
| Slope variance ($\tau_{11}$) |  | 46.94 | 46.90 | 47.83 |
| Slope variance ($\tau_{22}$) |  |  |  | 0.02 |
| Residual variance ($\sigma^2$) | 104.44 | 57.50 | 57.20 | 56.29 |
| Covariance ($\tau_{00}, \tau_{11}$) |  | 31.14 | 30.94 | 31.10 |
| Covariance ($\tau_{00}, \tau_{22}$) |  |  |  | 0.53 |
| Covariance ($\tau_{11}, \tau_{22}$) |  |  |  | 0.34 |
| Students | 1427 | 1427 | 1427 | 1427 |
| Classrooms | 222 | 222 | 222 | 222 |
| −2 Log likelihood | 10855.9 | 10196.6 | 10190.6 | 10185.7 |
| AIC | 10861.9 | 10214.6 | 10210.6 | 10211.7 |
| BIC$_1$ | 10877.7 | 10262.0 | 10263.2 | 10280.2 |
| BIC$_2$ | 10872.1 | 10245.2 | 10244.6 | 10255.9 |

*Note.* Model 1 – random intercept only; Model 2 – model includes one random effect (for TRT), but excludes the PPVT × TRT interaction; Model 3 – model includes one random effect (for TRT) and the PPVT × TRT interaction; Model 4 – model includes two random effects (for TRT and PPVT) and the PPVT × TRT interaction. Sample size for all four models is 1427 students nested within 222 classrooms/clusters. TRT = treatment; PPVT = Peabody Picture Vocabulary Test; AIC = Akaike information criterion; BIC = Bayesian information criterion.

*p < .05. **p < .01.

*Fixed effects.* From Table 3.1, the Model 1 intercept (i.e. the overall average expressive word knowledge) is $\hat{\gamma}_{00} = 13.73$ and has a standard error (*SE*) of 0.41. We can construct the 95% confidence interval around the parameter estimate $\hat{\gamma}_{00} \pm 1.96(SE) = 13.73 \pm 1.96(.41) = [12.93, 14.53]$. Given our sample estimate, we are 95% confident that expressive word knowledge lies somewhere between 12.93 and 14.53.

*Plausible range of classroom means.* The expected expressive word knowledge score (ETW) in the spring of kindergarten is 13.73, but mean ETW varies considerably across classrooms. How do we know? The between-class variance, $\tau_{00} = 19.70$, tells us how much the expected ETW varies across classrooms. Taking the square root of the between-class variance ($\sqrt{19.70} = 4.44$) produces the between-class standard deviation. As we discussed in Chapter 2, assuming a normal distribution, 68% of classroom means should lie within 1 *SD* of $\gamma_{00}$ and 95% of scores should lie within 1.96 *SD* of $\gamma_{00}$. Therefore, assuming ETW is normally distributed, we expect 95% of the class (cluster) means on ETW to fall within the range $\hat{\gamma}_{00} \pm 1.96(\sqrt{\hat{\tau}_{00}}) = 13.73 \pm 1.96(4.44) = [5.03, 22.43]$. This interval of plausible values describes the degree of variability in the parameter across clusters. Even though the expected ETW across all classrooms is 13.73, there is a fair amount of between-class variability in ETW scores: 95% of the classrooms have average expressive word knowledge scores between 5.03 and 22.43.

## Random coefficients models (random intercept and slope models)

Our conceptual model (Model 3) includes a treatment indicator variable, coded −1/2 for control students and 1/2 for treatment students to evaluate the impact of the vocabulary intervention. To account for the potential impact of prior vocabulary knowledge on ETW, we also included the students' pre-intervention PPVT score as a covariate. To partition the variance in PPVT into within- and between-cluster variance, we group mean centred both PPVT and TRT and added the cluster means into the model as level-2 variables. Furthermore, we included the (same level) interaction between pretest PPVT score and treatment assignment (PPVT × TRT). We can interpret the same-level interaction from two different perspectives: (1) the relationship between PPVT and post-test expressive vocabulary depends on whether the student was assigned to the vocabulary intervention or (2) the treatment effect (the effect of TRT on ETW) varies as a function of students' initial vocabulary knowledge (PPVT scores); that is, the effectiveness of the treatment varies as a function pre-intervention PPVT score.

Therefore, the level-1 model is

$$Y_{ij} = \beta_{0j} + \beta_{1j}\left(TRT_{ij} - TRT_{.j}\right) + \beta_{2j}\left(PPVT_{ij} - PPVT_{.j}\right) + \beta_{3j}\left(TRT_{ij} - TRT_{.j}\right)^{*}\left(PPVT_{ij} - PPVT_{.j}\right) + r_{ij} \tag{3.18}$$

The set of level-2 equations is

$$\beta_{0j} = \gamma_{00} + \gamma_{01}\left(MeanTRT_j - \overline{MeanTRT}\right) + \gamma_{02}\left(MeanPPVT_j - \overline{MeanPPVT}\right) + u_{0j}$$
$$\beta_{1j} = \gamma_{10} + u_{1j}$$
$$\beta_{2j} = \gamma_{20} \qquad\qquad\qquad\qquad\qquad\qquad\qquad\qquad\qquad\qquad\quad (3.19)$$
$$\beta_{3j} = \gamma_{30}$$

The randomly varying intercepts $\beta_{0j}$, are predicted by $\gamma_{00}$, the expected value of the intercept when both TRT and PPVT are at the overall (grand) mean. Because both level-2 variables are grand mean centred, $\gamma_{00}$ is the expected ETW score across both treatment and control conditions: $\gamma_{00}$ is essentially the overall mean ETW score. The interpretation of $\gamma_{00}$ depends on the coding of the predictors. For example, if we had dummy coded TRT at level 1 and had not added the aggregate back in at level 2, as is common, then the intercept and its interpretation would dramatically change. In that case, the intercept would be the expected ETW score in the control group (assuming TRT = 0 in the control group.)

$\gamma_{01}$ is the effect of the proportion of students in the TRT group on the intercept. $\gamma_{02}$ is the effect of class mean PPVT on the intercept. The level-2 model contains a random component, $u_{0j}$, for the intercept ($\gamma_{00}$), so the intercept randomly varies across clusters.

The level-2 equations also contain the fixed-effect coefficients for each of the level-1 slopes (i.e. the within-cluster effect of treatment [$\gamma_{10}$], the within-cluster effect of PPVT [$\gamma_{20}$], and the treatment by PPVT interaction effect [$\gamma_{30}$]). $\gamma_{10}$ is the most substantively important parameter: it represents the average treatment effect, holding class mean PPVT constant at its overall mean. The residual, $u_{1j}$ for the treatment slope, $\beta_{1j}$, allows the treatment slope to randomly vary across clusters. The variance of $u_{1j}$, $\tau_{11}$, captures the between-classroom variability in treatment effects. In other words, a randomly varying TRT slope allows the treatment to be differentially effective across classrooms. (Each classroom can have its own treatment effect, $\beta_{1j}$.)

There is no $u_{2j}$ term for the PPVT slope ($\beta_{2j}$), nor is there a $u_{3j}$ term for the TRT × PPVT slope ($\beta_{3j}$). These two slopes do not vary across clusters, so the effect of PPVT on ETW is assumed to be constant across clusters, as is the effect of the PPVT × TRT interaction on ETW.

Substituting the level-2 equations into the level-1 equation, the combined model for the **random coefficients model** (random intercept and slope model) is

$$Y_{ij} = \gamma_{00} + \gamma_{01}MeanPPVT_j + u_{0j} + \left(\gamma_{10} + u_{1j}\right)TRT_{ij} +$$
$$\gamma_{20}\left(PPVT_{ij} - PPVT_{.j}\right) + \gamma_{30}TRT_{ij} \times \left(PPVT_{ij} - PPVT_{.j}\right) + r_{ij} \qquad (3.20)$$

Rearranging this combined equation produces the standard expression of the mixed model by (a) using the distributive property to multiply TRT × ($\gamma_{10}+u_{1j}$), (b) grouping the fixed effects together and (c) grouping the random effects together. The reorganised model is

$$
\begin{aligned}
Y_{ij} = {}& \gamma_{00} + \gamma_{01} MeanPPVT_j + \gamma_{02} MeanTRT_j + \\
& \gamma_{10}\left( TRT_{ij} - TRT_{\cdot j}\right) + \gamma_{20}\left( PPVT_{ij} - PPVT_{\cdot j}\right) + \\
& \gamma_{30}\left( TRT_{ij} - TRT_{\cdot j}\right) x \left( PPVT_{ij} - PPVT_{\cdot j}\right) + u_{0j} + u_{1j} TRT_{ij} + r_{ij}
\end{aligned}
\tag{3.21}
$$

Table 3.1 contains the parameter estimates for this model in the column labelled Model 3.

We assume that $u_{0j}$ and $u_{1j}$ are multivariate normally distributed with means of 0 and variances of $\mathrm{var}(u_{0j}) = \tau_{00}$ ; and $\mathrm{var}(u_{1j}) = \tau_{11}$ the covariance between $u_{0j}$ and $u_{1j}$ $\mathrm{cov}(u_{0j}, u_{1j}) = \tau_{01}$ . In Model 3, the residual covariance between the intercept and the treatment slope is 31.39; standardising that covariance produces a correlation of .92 ($\tau_{01}$=.92). The residuals for the treatment slope tend to be more positive in clusters where the intercept residuals are higher. In other words, holding class PPVT scores and class proportion of TRT students constant at the mean, classes with higher ETW scores also tend to have higher treatment slopes (i.e. the treatment effect is stronger in classes with higher mean ETW scores. The confidence interval for this correlation ranges from .69 to .98. In our experience, very noisy, uninterpretable estimates of these covariance parameters are not uncommon. Therefore, always examine the standard errors and/or confidence intervals for the covariance parameters (or correlations) before interpreting them.

*Fixed effects.* Model 3 includes four predictors: PPVT (group mean centred), mean PPVT (grand mean centred), treatment (group mean centred) and mean treatment (grand mean centred). The intercept for Model 3 is: $\gamma_{00}$ = 13.81. Because we group mean centred both level-1 variables and added the aggregates back in at level 2, the intercept is virtually identical to the intercept from the unconditional model. It is still the overall expected ETW score (across conditions). The TRT slope ($\gamma_{10}$ = 10.53) is the treatment effect: it represents the expected *difference* between the treatment group and the control group when PPVT is at the mean (when PPVT = 0). We can use this fixed-effect estimate to compute an effect size for the treatment. The difference between the treatment and control groups (i.e. the TRT slope; $\gamma_{10}$ = 10.53) appears in the numerator of the effect size equation. The denominator is the standard deviation of the outcome variable, computed by taking the square root of the sum of the within- and between-cluster variance components from the unconditional model (model 1): $\sqrt{\sigma^2 + \tau_{00}} == \sqrt{104.44 + 19.73} = 11.14$ . Therefore, the effect size is 10.53/11.14, or $d$ = 0.95, indicating an average treatment effect of 0.95 *SD* units when PPVT is at the mean.

However, there is an interaction between treatment and PPVT scores: the treatment slope varies as a function of students' relative PPVT scores. The interaction provides evidence of moderation: the effect of the treatment varies as a function of initial PPVT score (or equivalently, the effect of PPVT varies as a function of treatment group). For every point higher a student scores on the PPVT (relative to his/her cluster), their expected ETW score increases by $\gamma_{20}$ = .44 points in the control group and by .67 ($\gamma_{20}$ = .44 + $\gamma_{30}$ = .23) points in the treatment group. Therefore, relative PPVT score has a greater influence on ETW scores for treatment students (or the difference between the treatment and control students is larger for students with higher PPVT scores).

As the cluster mean of PPVT increases by 1 unit, the cluster's predicted mean ETW increases by $\gamma_{01}$ = .45 points, indicating that classes with higher mean PPVT scores also tend to have higher mean ETW scores. Finally, as the proportion of students in the treatment group increases by 1 point, the expected ETW score increases by 8.46 points. This effect seems enormous, but the class treatment variable is akin to the grand mean centered proportion of students in the treatment group. Therefore, a 1 point change in the class treatment variable is only possible when moving from a class where every student is in the control group to a class in which every student is in the treatment group, which is not even possible, given the study design.

## Evaluating variance components/random effects

Using Rights and Sterba's integrative $R^2$ framework, in Model 3, the level-1 fixed effects (TRT, PPVT and PPVT × TRT) account for 30.2% of the within-cluster variance ($R_w^{2(f_1)}$ = .302) and 23.8% of the total variance ($R_t^{2(f_1)}$ = .238 ). The random-slope variation and covariation explain 11.2% of the within-cluster variance ($R_w^{2(v)}$ = .112) and 8.8% of the total variance ($R_t^{2(v)}$ = .088). Together, the level-1 predictors via both fixed-effects and random-slope variation/covariation explain 41.5% of the within-cluster variance ($R_w^{2(f_1 v)}$ = .415) and 32.6% of the total variance ($R_w^{2(f_1 v)}$ =.326). This means 58.5% of the within-cluster variance ($1 - R_w^{2(f_1 v)}$) and almost 46% of the total cluster variance ($1 - R_t^{2(f_1 m)}$) is unexplained (residual) within-cluster variance. In other words, almost 46% of the total variance in ETW is variability among students within classrooms that is not explained by either their PPVT scores or their treatment condition (i.e. assignment to TRT or control).

The two between-cluster variables, class mean PPVT score and class mean TRT explain 9.4% of the between-cluster variance ($R_b^{2(f_2)}$) and 2% of the total variance ($R_t^{2(f_2)}$) in ETW. The conditional ICC is .19 ($R_t^{2(m)}$ = .19). In other words, after controlling for class mean PPVT and the class proportion of students in the treatment condition, 19% of the total variance is random intercept variance, which represents unexplained mean differences across classrooms in ETW. Overall, the fixed effects at level-1 and level-2 explain 25.8% of the total variance in ETW ($R_t^{2(f)}$). As mentioned above, the level-1 slopes explain 23.8% of the total variance, whereas the level-2 predictors explain only 2% of the total variance.

To illustrate what happens if we add a random effect with very little between-cluster variance, we fit Model 4, which also allows the slope for PPVT to randomly vary across clusters. Models 3 and 4 are nested: Model 4 contains all of the parameters from Model 3 and adds a random effect for the PPVT slope. Model 4 estimates three additional parameters: (1) a variance for the PPVT slope ($\tau_{22}$), (2) the covariance between the PPVT slope and the intercept ($\tau_{02}$) and (3) the covariance between the PPVT slope and the TRT slope ($\tau_{12}$).

As seen in Table 3.1, the fixed-effects estimates are virtually identical for Models 3 and 4. Allowing PPVT to randomly vary by cluster increases the within-cluster variance explained by the level-1 predictors via random-slope variation and covariation from 11.2% to 11.9% and increases the total variance explained by the level-1 predictors via random-slope variation and covariation from 8.8% to 9.3%. Thus, adding a random effect for the PPVT slope explains less than 1% of the within-cluster variance (and only 0.5% of the total variance) in ETW. The small amounts of within-cluster and overall variance explained provide one indication that it may be unnecessary to allow the PPVT slope to randomly vary across clusters.

As mentioned above, Model 4 contains three additional covariance parameters – one slope variance for PPVT [$\tau_{22}$] and two covariances ($\tau_{02}$ and $\tau_{12}$). Adding those three parameters decreases the deviance by only 3.9 points, or approximately 1.3 points per parameter. For the AIC to be lower in the more parameterised model, deviance must decrease by at least 2 points per parameter. Therefore, the AIC is lower for Model 3. As mentioned in Chapter 2, the per-parameter penalty for the BIC is natural log of the sample size. Additionally, the BIC penalty depends upon whether we consider $n$ the number of observations at level-1 (BIC$_1$) or the number of clusters (BIC$_2$). However, $n$ must be less than 8 for the BIC penalty to be less than the AIC penalty of 2 points per parameter. Therefore, regardless of which sample size we select to compute the BIC, if the AIC favours the simpler model, the BIC$_1$ and BIC$_2$ also favour the simpler model. In this example, we have 222 classrooms; therefore, the per-parameter penalty is 5.40 for the BIC$_2$ and 7.26 for the BIC$_1$. The deviance would need to decrease by 16.21 points for the BIC$_2$ to favour Model 4 and by 21.78 points for the BIC$_1$ to favour Model 4. The drop of 3.9 points in deviance falls far short of these criteria. Therefore, all three information criteria (AIC, BIC$_1$ and BIC$_2$) favour Model 3 over Model 4.

We can also compare the difference in the deviances using the LRT. As we discussed in Chapter 2, the LRT compares the difference in the deviances of the two nested models to the critical value of $\chi^2$ with degrees of freedom equal to the difference in the number of estimated parameters between the two models. When the difference in deviance is less than the critical value of $\chi^2$, we favour the more parsimonious model. Conversely, when the deviance difference is larger than the critical value, we favour the more complex (more parameterised) model. However, when comparing models that differ in terms of their random effects, we should obtain the critical value from the $\chi^2_{p+1}$ distribution (Snijders & Bosker, 2012). The critical value for the $\chi^2_{p+1}$ distribution in a model with two slopes is 7.05. (*Note:* The critical value of $\chi^2$ with 3 *df* is 7.81. Using the $\chi^2_{p+1}$ is more liberal. It sets a lower bar for rejecting the null hypothesis, so it more frequently favours the more parameterised model.) Even so, 3.90 is lower than 7.05; therefore, we favour the simpler model (which excludes the random PPVT slope). In this case, all three information criteria and the LRT all indicate that including the random PPVT slope does not substantially improve the fit of the model, suggesting that we should favour the more parsimonious model. As a side note, given that the PPVT slope does not vary randomly across classes, we should not allow the PPVT × TRT slope to randomly vary across classes. In general, if either one of the two variables in the interaction does not randomly vary across clusters, it is unnecessary (and inadvisable) to allow the interaction to randomly vary across clusters. In summary, allowing the PPVT slope to randomly vary explained very little (0.7%) of the within-cluster variance and did not improve model fit; therefore, we favour Model 3 (which does not allow the PPVT slope to randomly vary across classrooms.

## Do we really need the fixed effect for the PPVT × TRT interaction?

In Model 3, approximately 30.3% of the within-cluster variance ($R_w^{2(f_1)} = .303$) and 23.8% of the overall variance is explained by students' PPVT scores ($R_t^{2(f_1)} = .238$), their treatment assignment and the interaction between treatment and PPVT. Let's compare Model 3 to Model 2, which eliminates the fixed effect for the PPVT × TRT interaction term. To determine whether we prefer Model 2 or Model 3, we again examine both measures of predictive utility (e.g. variance explained) and measures of model 'fit' (e.g. AIC, BIC and the LRT).

We can compute the additional variance explained by the interaction effect by comparing the model that includes the interaction effect (Model 3) to model that excludes that fixed effect (Model 2). When we remove the PPVT × TRT interaction term, the model still explains virtually the same proportion of within-cluster variance ($R_w^{2(f_1)}{}_{mod2}.299$ vs. $R_w^{2(f_1)}{}_{mod3} = .303$) and the same proportion of total outcome variance ($R_t^{2(f_1)}{}_{mod2}.235$ vs. $R_t^{2(f_1)}{}_{mod3} = .238$).

Note that we are unable to explain 58.5% of the within-cluster variance and 46% of the total variance in the model when we include the interaction effect, and we are unable to explain 58.8% of the within-cluster variance and 46.3% of the total variance when we exclude the interaction effect. What does all this tell us? Even though the PPVT × TRT interaction is statistically significant, adding the interaction effect to the model does not actually explain much additional variance in ETW, over and above the model that contains only the main effects for treatment and PPVT scores: the interaction effect explains less than 0.4% of the within-cluster variance.

So, which model fits better? Because Model 3 estimates one more fixed effect but has the same variance/covariance structure as Model 2, the two models differ by one parameter. The change in deviance is 6.0, which we compare to the critical value of chi-square with 1 *df* (3.84). The LRT is statistically significant (because 6.0 is greater than 3.84). Therefore, we favour the more complex model, which includes PPVT × TRT interaction. The AIC and $BIC_2$ are lower for Model 3 than for Model 2, but the $BIC_1$ is lower for Model 2. Hence, according to the AIC, $BIC_2$ and the LRT, the model with the interaction effect (Model 3) fits better than the model that excludes it (Model 2). In addition, the interaction effect is statistically significant, so favouring the model that includes the interaction effect certainly seems reasonable. However, we should be realistic about the relatively negligible role of the interaction effect in explaining additional variance in ETW scores, over and above the effects of TRT and PPVT. These results could be summarised as follows: *Although there is evidence of PPVT × TRT interaction, the magnitude of the interaction effect is quite small, explaining approximately 0.4% of the within-cluster variance (and 0.3% of the total variance) in ETW scores.*

Although we fit a series of four models for pedagogical reasons, if we were conducting these analyses as part of a research project, we would not have estimated Model 4 if we did not have strong theoretical reasons to hypothesise that the effect of PPVT on ETW would vary across classrooms. In contrast, fitting Model 2 provides useful information about the proportion of variance in ETW that is explained by interaction between PPVT and TRT. Thus, Model 2 serves as a supplementary model to compute the incremental variance that was explained by one of our substantive variables of interest. However, Model 3 is our hypothesised and focal model.

This concludes our whirlwind tour of MLM. We have provided a solid conceptual overview of the technique, but there are several important areas that we have not addressed. These include residual analysis, power and sample size issues, three-level models and modelling heterogeneity of the level-1 residual variances. If you are planning to conduct multilevel analyses, we encourage you to delve into each of these topics in greater depth. We also recommend continuing your journey with a full-length book on MLM. Our personal favourite books are Raudenbush and Bryk (2002), Hox et al. (2017), Goldstein (2011) and Snijders and Bosker (2012).

In Chapters 4, 5 and 6, we turn our attention to SEM. Then in Chapter 7, we return to multilevel modelling as a method to fit longitudinal growth models.

## Chapter Summary

- A set of steps to build and test confirmatory multilevel models is recommended. The steps incorporate Rights and Sterba's (2019) integrative framework of $R^2$ for computing the proportions of variance explained in multilevel models.

- A simple example illustrates how to fit and interpret a multilevel model using data gathered from a multisite cluster-randomised trial evaluating the impact of a high-quality, evidence-based vocabulary instruction intervention in kindergarten. The proposed model building steps are outlined by first fitting the unconditional model, followed by the full model. Then, the full model is compared to (a) a more complex model that includes an additional randomly-varying slope and (b) a more parsimonious model that eliminates the fixed effect for a same-level interaction.

## Further Reading

Hox, J. J., Moerbeek, M., & Van de Schoot, R. (2017). *Multilevel analysis: Techniques and applications* (3rd ed.). Routledge.

This book provides an extensive and technical overview of multilevel modelling as well as example applications. Furthermore, the book provides extensions of multilevel modelling including multilevel factor models, multilevel path models and latent curve models.

Rights, J. D., & Sterba, S. K. (2019). Quantifying explained variance in multilevel models: An integrative framework for defining R-squared measures. *Psychological Methods, 24*, 309–338.

This article provides an elegant solution to the issue of computing variance explained ($R^2$) measures in multilevel modelling.

# 4

# INTRODUCTION TO STRUCTURAL EQUATION MODELLING

## Chapter Overview

Advantages of SEM ................................................................. 68

Assumptions and requirements of SEM ..................................... 70

Understanding SEM ................................................................ 71

What is path analysis? ............................................................. 72

Estimating direct, indirect and total effects in path
models with mediation ............................................................. 75

What are latent variables? ........................................................ 78

Factor analysis ....................................................................... 80

Measurement models versus structural models ........................... 81

Estimation ............................................................................. 84

Model fit and hypothesis testing in SEM ................................... 84

Model comparison .................................................................. 87

Model modification and exploratory analyses ............................. 88

Model fit versus model prediction ............................................. 89

When SEM gives you inadmissible results ................................... 90

Alternative models and equivalent models ................................. 90

A word of caution ................................................................... 91

Further Reading ..................................................................... 93

**Structural equation modelling (SEM)** refers to a family of techniques, including (but not limited to) path analysis, **confirmatory factor analysis (CFA)**, structural regression models, autoregressive models and latent change models (Marcoulides & Schumacker, 2001; Raykov & Marcoulides, 2000). SEM utilises the analysis of covariances to examine the complex interrelationships within a system of variables. SEM builds upon, integrates and greatly expands the capabilities of more traditional statistical techniques such as multiple linear regression and ANOVA (Hoyle, 2012).

This chapter provides a conceptual introduction to SEM. We describe the advantages of SEM, outline the assumptions and requirements for SEM, define key terms and concepts and provide brief, non-technical introductions to path analysis and **latent variables**. In Chapter 5, we delve into more detail, and we discuss specification, **identification** and estimation in SEM. In addition, we discuss Wright's rules and systems of equations. In Chapter 6, we discuss model building and provide an example where we fit and estimate a **hybrid model**.

## Advantages of SEM

SEM is extremely versatile; it places very few restrictions on the kinds of models that can be tested (Hoyle, 1995, 2012). Consequently, SEM allows researchers to test a wider variety of hypotheses than would be possible with most traditional statistical techniques (Kline, 2015). Using SEM to specify a given model based on theory, researchers can examine the degree to which the model can reproduce the relationships among **observed variables** (and patterns of means). Alternatively, researchers can test competing theories by fitting alternative models to determine which of the competing models appears to best fit the observed data (Kline, 2015).

SEM allows researchers to distinguish between observed and *latent* variables and to explicitly model both types of variables. Latent variable models have both conceptual and statistical advantages over traditional observed variable techniques. Using latent variables, researchers can include **latent constructs** in their analyses. A **construct** is a concept, model or schematic idea (Bollen, 2002). Thus, *latent constructs* are non-observable concepts, such as cognitive ability, self-concept or optimism. Although latent constructs themselves are not directly observed, their presence and influence can be inferred based on variables that are directly observed. For example, educators use observable indicators such as test scores, self-reports, teacher ratings and/ or behavioural observations to infer latent constructs such as students' academic engagement. Latent variable models permit a level of methodological and theoretical freedom that is nearly impossible in most other statistical analyses.

Furthermore, using latent variables in SEM accounts for potential errors of measurement, allowing researchers to explicitly account for (and model) **measurement error**

(Raykov & Marcoulides, 2000). The ability to separate measurement error or 'error variance' from 'true variance' is one of the reasons that SEM provides such powerful analyses. In multiple regression, measurement error within a predictor variable attenuates the regression weight from the predictor variable to the dependent variable, downwardly biasing the parameter estimates (Baron & Kenny, 1986; Campbell & Kenny, 1999; Cole & Preacher, 2014). Structural equation models that include latent variables use multiple indicators to estimate the effects of latent variables. This approach corrects for the unreliability within the measured predictor variables, providing more accurate estimates of the effects of the predictor on the criterion.

Measurement error in the **mediator** (or other variables in the mediational model) can also produce biased estimates of direct, indirect and **total effects** (Baron & Kenny, 1986; Cole & Preacher, 2014). Cole and Preacher (2014) outline four serious consequences of ignoring measurement error in mediational analyses: (1) measurement error can cause path coefficients to be either over- or underestimated (and predicting the direction of bias becomes quite difficult as the complexity of the model increases), (2) measurement error can decrease the power to detect incorrect models, (3) even seemingly small amounts of measurement error can make valid models appear invalid and (4) differential measurement error across the model can actually change substantive conclusions. Generally, these four issues become increasingly problematic as model complexity increases (p. 300). Fortunately, using latent variables as the structural variables in mediational models eliminates these issues: using latent variables in SEM accounts for the measurement error and produces unbiased estimates of the direct, indirect and total effects.

Finally, SEM allows researchers to specify a priori models and to assess the degree to which the data fits the specified model. SEM provides a comprehensive statistical approach to test existing hypotheses about relations among observed and latent variables (Hoyle, 1995). In this way, SEM forces the researcher to think critically about the relationships among the variables of interest and the hypotheses being tested. Further, SEM allows researchers to test competing theoretical models to determine which model best reproduces the observed variance–covariance matrix.

SEM analyses can include means as well as covariances. In fact, researchers can also use SEM techniques to model *latent* means. Thus, SEM provides a framework for examining between-group differences: **multiple group SEM (MG-SEM)** enables between-group comparisons of any model parameters, including latent means. Therefore, MG-SEM facilitates the examination of both differences in patterns of interrelationships among variables across groups and differences in means and variances across groups. Latent growth curve models also incorporate means into the structural equation model (Bollen & Curran, 2004; Duncan et al., 1999; Grimm et al., 2016). For the remainder of this section (Chapters 4–6), we confine our discussion to modelling covariance structures without means. Chapter 7 introduces means structure analysis in the context of latent growth curve models to study change across time.

## Assumptions and requirements of SEM

SEM is a regression-based technique. As such, it rests upon four key assumptions and requirements necessary for any regression-based analyses: normality, linearity, independence and adequate variability. We briefly discuss each of these requirements below.

*Normality.* Many of the assumptions of SEM parallel those of multiple linear regression. Namely, SEM assumes that the variables of interest are drawn from a multivariate normal population (Hancock & Mueller, 2006; Kaplan, 2000). ML estimation performs optimally when the data is continuous and normally distributed (Finney & DiStefano, 2006; Kaplan, 2000). Generally, SEM is fairly robust to small violations of the normality assumption; however, extreme non-normality can cause problems. (For more information about dealing with non-normal data, see Curran et al., 1996; Finney & DiStefano, 2006; Hayakawa, 2019; S. G. West et al., 1995.)

*Linearity.* As in multiple regression, SEM assumes that the variables of interest are linearly related to each other. In addition, we can examine non-linear and interaction effects using SEM. Interested readers should consult Maslowsky et al. (2015) for a comprehensible introduction to modelling interaction effects in SEM.

*Sampling.* ML estimation assumes that the data represents a simple random sample from the population. Chapter 1 of this book discussed the pitfalls of assuming that data are independent. Multilevel modelling provides a solution to issue of non-independence. Multilevel SEM combines multilevel modelling and SEM techniques to analyse data that have been collected using multistage or cluster sampling techniques (Kaplan, 2000). However, we need to walk before we run, so our introduction to SEM assumes that data are independent. Space precludes us from discussing multilevel SEM in this book. For more information about multilevel SEM, we recommend Bauer et al. (2006), Heck and Thomas (2000, 2015), Hox et al. (2017), Muthén and Muthén (1998–2017), Preacher, Zhang et al. (2011, 2016) and Preacher, Zyphur et al. (2010).

*Sample Size.* Because standard SEM uses ML estimation to minimise the discrepancy between the observed covariance matrix and the model-implied covariance matrix, SEM is a large sample technique. (For information about alternative estimation methods, see Kaplan, 2000, or Hoyle, 1995.) There are no definitive rules for a minimum sample size, but the literature provides general rules of thumb. Under most circumstances, sample sizes below 100 are too small to use SEM techniques (Kline, 2015; Schumacker & Lomax, 1996). Latent variable models generally require larger sample sizes than comparable path models (that include only observed variables). In general, models with larger numbers of freely estimated parameters require larger sample sizes. As the ratio of the number of cases to the number of parameters declines, the estimates generated by SEM become more unstable (Kline, 2015). Kline (2015) recommends maintaining a ratio of at least 10 cases per freely estimated parameter. In many instances, sample sizes of 200 or more are sufficient for estimating structural equation models, especially if the variables are normally distributed and obtained from a random sample of subjects. Sample sizes of 250 to 500 are typical in SEM studies (Kline, 2015). In summary, SEM generally requires large samples; however, the minimum sample size ultimately depends on the size and complexity of the structural equation model.

*Range of values.* Because SEM is essentially a correlational technique, anything that affects the magnitudes of the covariances among the variables in the model impacts the SEM analysis. For example, restriction of range in one variable attenuates the covariance between that variable and any other variables in the model. Similarly, a small number of influential outliers can have a large impact on the variance–covariance matrix. Such issues are likely to lead to biased parameter estimates.

## Understanding SEM

The basic building block of any structural equation model is the variance–covariance matrix. (Means are also required for some analyses such as growth curve analysis and MG-SEM, but to lay the groundwork of SEM, we begin with traditional covariance structure analysis.) SEM provides a way to use the covariance matrix to explain complex patterns of interrelationships among variables. In fact, the covariance matrix is the **sufficient statistic** for standard structural equation model: it contains all the information needed to fit a standard SEM analysis. We can compute the covariance matrix using the correlation matrix and the standard deviations for each variable. Because it is possible to create and analyse a structural equation model without the raw data file, interested researchers can conduct SEM using a published covariance or correlation matrix.[1] The covariance matrix is the unstandardised version of the correlation matrix; therefore, it is simple to create the covariance matrix using standard deviations and correlations. As we delve into what might seem to be a complex barrage of symbols, jargon and numbers, remember that the covariance matrix is at the heart of SEM.

SEM's versatility stems from the fact that it incorporates path analysis (or simultaneous equation modelling) and factor analysis (or latent variable modelling) into one modelling framework. We begin our tour of SEM by providing a brief definition and overview of path analysis and factor analysis. In the chapters that follow, we demonstrate how path analysis and factor analysis can be combined into a single latent variable modelling framework.

---

[1] Technically, it is considered proper form to analyse a covariance matrix, but under a variety of conditions, analysing a correlation matrix will produce the same results, as a correlation matrix is simply a standardised version of a covariance matrix.

## What is path analysis?

More than 100 years ago, Sewall Wright (1920, 1923) developed path analysis, a technique that allows for the estimation of a system of simultaneous equations (Bollen, 1989). In traditional statistical analyses such as regression, a variable can serve as either an independent variable or a dependent variable; however, the same variable cannot serve as both a predictor and an outcome simultaneously (Hoyle, 2012). Mediators are a classic example of variables that serve as both independent and dependent variables, and mediational models cannot be estimated in a single multiple regression analysis. In contrast, in path analysis, a single variable can act as both an independent variable and a dependent variable. Therefore, path models can specify a complex system of predictive pathways among a large number of variables.

### Path diagrams

**Path diagrams**, visual displays of structural equations, are perhaps the most intuitive way to conceptualise the process of developing and testing a specified model. Most predictive models can be represented as path models. Exhibit 4.1 provides a summary of the typical symbols in a path diagram.

**Exhibit 4.1** Symbols in a path diagram

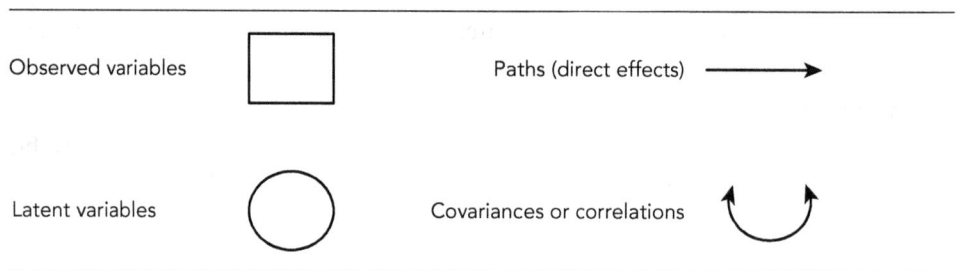

*Note.* The double-headed arrow represents a covariance in the unstandardised solution and a correlation in the standardised solution.

An observed (or manifest) variable is a variable that is actually measured. For example, a student's score on a test or a subscale is an observed variable. In a path diagram, rectangles indicate observed or measured variables. In contrast, circles or ellipses represent latent variables, which are not directly observed in the sample (data). Straight single-headed arrows represent paths. Just as in multiple regression, these paths represent the degree to which the predictor variable predicts a given outcome variable *after controlling for* (or holding constant) the other variables that also contain direct paths to (arrows pointing to) the dependent variable. In fact, we can construct a path diagram to display any multiple regression model. Double-headed arrows, which are generally curved, indicate a simple bivariate correlation between two variables.

Figure 4.1 illustrates a simple three-variable mediation model as a path diagram. The three observed variables are growth mindset, academic persistence and academic achievement. Straight single-headed arrows, called paths, connect growth mindset → academic persistence, academic persistence → academic achievement and growth mindset → academic achievement. The direction of the arrowhead is important: the arrow points from the predictor variable to the outcome variable.

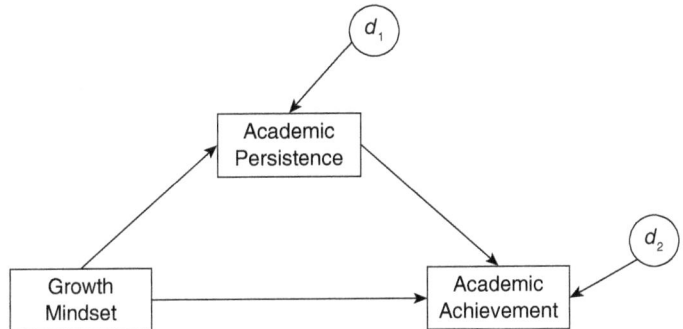

**Figure 4.1** A path model in which growth mindset predicts academic persistence, which in turn predicts academic achievement

Just as in regression, path coefficients in SEM can be unstandardised or standardised. **Unstandardised path coefficients** depict the expected unit of change in the dependent variable given a one-unit change in the predictor variable, holding the other variables in the model constant. Unstandardised path coefficients or parameters reflect the scales of measurement of both the independent and dependent variables. Therefore, the interpretation of unstandardised path coefficients depends on the scales of both the predictor and criterion variables. In contrast, **standardised path coefficients** are analogous to beta coefficients in regression; conceptually they represent path coefficients in a model where all variables are standardised (i.e. $z$ scores with mean = 0 and standard deviation/variance = 1).

Each of the parameters has an associated standard error, and we can test the statistical significance of every parameter that we estimate. As in multiple regression, each unstandardised path coefficient is divided by its standard error to compute a critical ratio. If the absolute value of this ratio is greater than or equal to 1.96, the path is statistically significant (at $\alpha$ = .05). If the ratio of the unstandardised path coefficient to its standard error is less than 1.96, the path is considered non-statistically significant.

For example, the path from growth mindset → academic achievement indicates that growth mindset predicts academic achievement. Because both growth mindset and academic persistence predict academic achievement, the growth mindset → academic achievement path represents the direct *effect* of growth mindset → academic achievement, after controlling for academic persistence. Similarly, the academic persistence → academic achievement path represents the direct effect of academic persistence → academic achievement, after controlling for growth mindset. Only growth mindset predicts academic persistence; therefore, the path from growth mindset → academic persistence does not control for any other variables.

The latent variables, $d_1$ and $d_2$ in Figure 4.1 represent *disturbance variances*. We predict the outcome variable, $Y$, with one or more predictors, $X$. The total variance in the outcome variable can be partitioned into the variance in $Y$ that is explained by the predictor(s) and the variance in $Y$ that is unexplained by the predictors. In multiple regression, we often refer to the unexplained variance as residual or error variance. In path analyses, this residual is called the **disturbance**, and the disturbance variance is the variance in the outcome variable that is unexplained by the model.

## Exogenous and endogenous variables

SEM makes a key distinction between **exogenous variables** and **endogenous variables**. *Exogenous* variables predict other variables, but they are not predicted by any other variables in the model. In our simple example, growth mindset is an exogenous variable: it is purely a predictor variable. Exogenous variables may be (and generally are) correlated with any other exogenous variables, and they predict one or more variables in the model. However, we assume the causes of exogenous variables (or variables that explain the variance in the exogenous variables) lay outside the model.

In contrast, *endogenous* variables are predicted by one or more variables in the model. Just as in regression, every endogenous variable in the model contains a residual (called a disturbance), representing the unexplained variance in the vari able. Therefore, the total variance in academic persistence equals the variance that is explained by growth mindset plus the disturbance (unexplained) variance. In path analysis, endogenous variables can also predict other endogenous variables. In other words, a variable can be a predictor only (exogenous), a predictor and an outcome (endogenous) or an outcome only (endogenous). In our simple model, both academic persistence (which is a mediator) and academic achievement (which is an outcome only) are endogenous variables.

Walking through our simple conceptual example in Figure 4.1 illustrates how the same variable can be both a predictor and an outcome. (This example assumes all variables are observed.) Academic persistence predicts subsequent academic achievement.

Students who are more persistent tend to have higher academic achievement. Academic persistence is the predictor and academic achievement is the outcome. However, growth mindset predicts academic persistence (Dweck et al., 2014). Here, growth mindset is the predictor and academic persistence is the outcome variable. Growth mindset predicts academic persistence, which in turn predicts academic achievement. Thus, academic persistence is both an outcome variable and a predictor: academic persistence is predicted by growth mindset and a predictor of academic achievement.

## Estimating direct, indirect and total effects in path models with mediation

Path analysis provides a method for estimating the *direct*, *indirect* and *total effects* of a system of variables in which there are mediator (intermediate) variables (Bollen, 1989). A mediator is a 'middle man', an intervening variable that explains the relationship between a predictor variable and a dependent variable (Baron & Kenny, 1986).

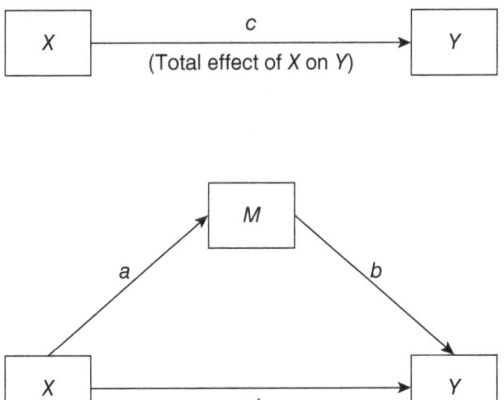

**Figure 4.2** A simple mediational model

## Direct effects

A **direct effect** represents the independent contribution of a predictor variable ($X$) on an outcome variable ($Y$), after controlling for all of the other variables that also predict $Y$ (and share variance with $X$). In our simple example, growth mindset is $X$, academic persistence (which is both a predictor and an outcome) is $M$ and academic achievement (which is only an outcome) is $Y$. In multiple regression, the **partial regression coefficient** is a direct effect: it is the effect of $X$ on $Y$, after controlling for all the predictor variables in the model. In our model above, the direct effect of growth mindset on academic achievement is the effect of growth mindset on academic achievement, after controlling for academic persistence. If this direct effect is 0, then growth mindset does not predict academic achievement after controlling for the effects of academic persistence on academic achievement. In other words, any variance in academic achievement that is explained by growth mindset is also explained by academic persistence. Therefore, once we control for academic persistence, growth mindset does not predict any additional variance in academic achievement. If there is a direct effect of growth mindset on academic achievement, then growth mindset explains additional variance in academic achievement, over and above the amount that is explained by academic persistence. This direct effect is just like a partial regression coefficient in a multiple regression equation. In fact, if we ran a multiple regression with academic persistence and growth mindset as predictors of academic achievement, the partial regression coefficient for growth mindset would be identical to the direct effect from the path analysis.

## Indirect effects

An **indirect effect** refers to the effect of a predictor variable ($X$) on an outcome variable ($Y$) that is mediated by one or more intervening variables ($M$) (Raykov & Marcoulides, 2000). In other words, the indirect effect is the effect of the predictor variable on the outcome variable that 'passes through' one or more intervening variables. Growth mindset ($X$) predicts academic persistence ($M$), which in turn predicts academic achievement ($Y$). The indirect effect of growth mindset on academic achievement is the effect of growth mindset on academic achievement that is also shared with academic persistence. Models with indirect effects are often referred to as mediational models. In our example, academic persistence is a *mediator* variable: it is an intermediate variable that explains how growth mindset influences academic achievement (Baron & Kenny, 1986). Figure 4.2 illustrates a simple mediational model with an indirect effect of $X$ on $Y$ via $M$. The coefficient for path from $X$ to $M$ is $a$ and the coefficient for the path from $M$ to $Y$ is $b$. The product of the two paths ($a * b$) provides an estimate of the indirect effect.

Indirect effects do not exist in *standard* multiple regression models (in which variables are either predictors or outcomes, but not both), but they do exist in path analysis (and SEM). Recursive path models can be estimated using multiple regression analyses in a traditional OLS framework. However, using multiple regression to estimate the indirect effect requires estimating two separate regression models. The first model regresses $Y$ on $X$. The second model regresses $Y$ on $M$ and $X$. The indirect effect of $X$ on $Y$ (via $M$) is the effect of $X$ on $Y$ (the total effect) – the effect of $X$ on $Y$ after controlling for $M$ (the direct effect). Because the Total effect = Direct effect + Indirect effect, the indirect effect of $X$ on $Y$ via $M$ is the difference in those two coefficients. When we refer to path analysis and SEM, we are referring to single-step methods for estimating these models. When we refer to 'standard multiple regression analysis', we are referring to the process of running a single multiple regression model.

## Total effect

The total effect of a predictor variable ($X$) on an outcome variable ($Y$) is the effect of $X$ on $Y$, whether or not it is mediated by a third variable, $M$. There are two ways to compute the total effect. The first is quite simple: the simple linear regression of $Y$ on $X$ produces the total effect of $Y$ on $X$. The second method to compute the total effect is to sum the direct and indirect effects. In other words, the total effect of $X$ on $Y$ is the sum of the direct effect of $X$ on $Y$ and the indirect effect of $X$ on $Y$ that is mediated by the intermediate variable, $M$. In the top panel of Figure 4.2, the total effect is $c$, the regression coefficient for the model that regresses $Y$ on $X$ but does not include $M$. Alternatively, in the bottom panel of Figure 4.2, we can compute the total effect by summing the direct effect ($c'$) and the indirect effect ($a * b$).

## Mediation and bootstrapping

Although the parameter estimates for direct, indirect and total effects are easy to estimate for mediational models, correctly determining whether the *indirect effect* is statistically significantly different from 0 requires additional analytic attention. The indirect effect ($a * b$) is multiplicative. Even if the sampling distribution of both $a$ and $b$ are normally distributed, the sampling distribution of the $a * b$ product is not necessarily normally distributed. Therefore, using the analytic standard error to determine the statistical significance of the $a * b$ path may result in incorrect statistical inferences. Instead of trying to derive the standard error analytically, it is easy to *bootstrap* the sampling distribution around the $a * b$ path. Bootstrapping is a resampling technique used to empirically derive the sampling distribution when an analytic solution is not feasible. Treating the sample (of size $n$) as the population, bootstrapping involves drawing repeated samples with replacement. The parameter estimates vary across samples. The variance of the parameter estimates provides an empirical estimate of the sampling variance; the standard deviation of the parameter estimates is an empirically derived standard error. However, because we believe that the sampling distribution of the indirect effect is not likely to be normal, we eschew standard errors and $p$-values (which assume that the distribution is normally distributed) in favour of empirically derived confidence intervals (CIs). To determine the 90% CI, we locate the 5th and 95th percentiles of the sampling distribution: those values become the upper and lower limits of our CI. For additional information about bootstrapping in SEM, we recommend Shrout and Bolger (2002), Preacher and Hays (2008) and MacKinnon and Fairchild (2009). Interestingly, recent research suggests that many of the bootstrap approaches have inflated Type I error rates (Yzerbyt et al., 2018). Therefore, Yzerbyt et al. (2018) suggest first examining the statistical significance of each of the paths separately. If both paths are statistically significant, then examine the magnitude and CI of the indirect effect using bootstrapping (Yzerbyt et al., 2018). Thus, in their approach, the tests of the individual components evaluate the statistical significance of the indirect effect whereas 'the confidence interval reveals its magnitude' (Yzerbyt et al., 2018, p. 942).

## Mediation and causality

Mediation implies the existence of an underlying causal mechanism: the effect of a putative cause is transmitted through the mediator to the outcome variable (Mayer et al., 2014). However, since the advent of path analysis, controversy has surrounded the technique's causal aspirations (Wright, 1923). Recently, a great deal of methodological work has focused on whether and how researchers can make strong causal inferences from mediational models (Preacher, 2015; VanderWeele, 2015). Mediation analysis requires several fairly strong assumptions to attribute a causal interpretation to the indirect effect: (a) there are no omitted variables (confounders), (b) there is no measurement error in the predictor variable or mediators, (c) the functional form of the model is correct and (4) we have correctly modelled temporal precedence and the timing of measurement allows us to capture the mediation process (MacKinnon, 2008; MacKinnon et al., 2020). Because the term *mediation* implies an underlying causal mechanism, some researchers avoid using the term entirely, and instead refer only to the direct, indirect and total effects within path analytic (or structural equation) models with intermediate variables. To interpret estimates of direct and indirect effects causally does require fairly strong assumptions (VanderWeele, 2015). However, we choose to use the term *mediation* to describe models in which the effect of one variable is presumed to be transmitted through an intermediate variable to an outcome variable of interest, even when we fail to meet the strict assumptions of causal inference. We encourage our readers to read Volume 10 of this series, which is devoted to the topic of causal inference.

## What are latent variables?

SEM is often referred to as a latent variable modelling technique (Hoyle, 2012). What are *latent variables*? The term *latent* means 'not directly observable'. Latent variables appear in a model but are not directly measured. We often teach our students a crude (but effective) rule of thumb for identifying whether a variable is latent or not. If the variable appears in the data file, it is observed. If the variable does not, it is latent. Bollen (2002) defines latent variables as variables for which there are no values (for at least some observations) in a given sample. Often, we use latent variables to model the hypothetically existing *constructs* of interest in a study that we cannot directly measure, such as peace, intelligence and apathy. However, not all latent variables are latent constructs. Bollen's definition of latent variables is broader and more inclusive than the definition of a latent construct. According to Bollen (2002), residuals (e.g. errors or measurement and disturbances) are also technically latent variables: they are not directly observed in a given sample. However, they are generally not latent constructs of substantive interest. We use the term *latent construct* to indicate a latent variable of substantive interest that assumes theoretical importance in a latent variable model. We use the term *latent variable* more broadly: a latent variable may be a latent construct of substantive interest, but it need not be. See Bollen (2002) for a far more nuanced discussion of the ways to define latent variables.

## Measuring/modelling latent constructs

So, how can we model constructs that we cannot directly measure? In SEM, the existence of latent constructs is inferred or derived from the relationships among observed variables that measure the latent construct. To model latent constructs using reflective[ii] latent variables, we make two philosophical assumptions. First, we assume that the constructs are real, even if they cannot be directly measured (Borsboom, 2005; Borsboom et al., 2003; Cook & Campbell, 1979; Edwards & Bagozzi, 2000; Nunnally & Bernstein, 1994). Second, we assume that a latent construct has a causal relationship with its indicators (the observed variables that measure the construct of interest; Borsboom et al., 2003). In other words, the latent construct influences people's responses to the observed variables (or indicators; McCoach, Gable et al., 2013). Figure 4.3 illustrates this assumption. The circle represents the latent construct, the squares represent the observed variables that serve as indicators of the latent construct and the single-headed arrows represent the directional paths from the latent variable to the indicators. We can decompose the variance of each indicator into two parts: the variance that is explained by the latent construct and measurement error variance (Rhemtulla et al., 2020). This figure also illustrates one other implicit assumption of a standard, unidimensional factor model: we assume the correlations among the indicators are completely explained by the variance they share with the

---

[ii]You can also measure latent variables with formative or causal indicators. For a discussion of this approach, see Bollen and Diamantopoulos (2017) or Bollen and Bauldry (2011). However, in this book, we consider only latent variables with reflective indicators, which is by far the more common type of latent variable in the literature.

latent construct. In other words, we assume any covariances among the set of items are due to the latent construct.

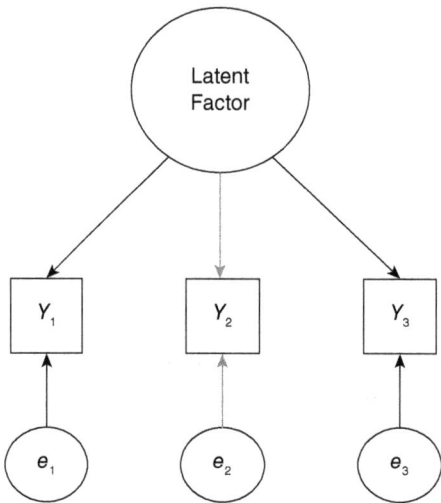

**Figure 4.3** A simple unidimensional factor model

## Factor analysis

To what extent do certain latent constructs explain the pattern of correlations/covariances in the observed variables? The goal of factor analysis is to determine the number of distinct constructs needed to account for the pattern of correlations among a set of measures (Fabrigar & Wegener, 2012). Factor analysis exploits the patterns of correlations among observed variables to make inferences about the existence and structure of latent constructs. Factor analysis provides information about which observed variables are most related to a given factor, as well as how these items relate to the other factors in the solution (Gorsuch, 1997).

Conceptually, standard factor analytic techniques assume that the correlations (covariances) among the observed variables can be explained by the factor structure. In other words, variance in the observed scores can be broken into two pieces: (1) variance that can be explained by the factor and (2) error variance (or uniqueness), which is the variance that is unique to the observed score and is not explained by the latent factor. Factors are the latent constructs of substantive interest that predict shared variance in the observed variables. Factor analysis yields estimates of the strength of the paths (measurement weights) from the latent factors to the indicators, the unique variance in each observed variable (the variance not explained by the factor) and the correlations among the latent variables of interest.

## Types of factor analysis: exploratory and confirmatory factor analyses

The two most common factor analytic techniques are exploratory factor analysis (EFA) and confirmatory factor analysis (CFA). Researchers commonly use EFA to reduce the number of elements from a larger number of observed variables to a smaller number of broader, more generalisable latent constructs (McCoach, Gable et al., 2013). Mathematically, EFA 'seeks the set of equations that maximise the multiple correlations of the factors to the items' (Gorsuch, 1997, p. 533).

One of the major methodological differences between EFA and CFA is the amount of information about the factor structure that is specified a priori. The factor structure represents the linkage between factors and indicators (i.e. which observed variables indicate which factor(s), how many factors are present in the data, etc.). EFA does not require a priori knowledge or specification of the factor structure. In contrast, in standard CFA, the researcher completely specifies the factor structure before undertaking the analysis. Based upon previous literature and experience, researchers clearly articulate the patterns of results they expect to find and then investigate whether and how well the data conform to the hypothesised structure.

CFA permits comparison of several rival models, allows researchers to reject specified models and provides a method to compare several competing models empirically. CFA has many advantages over EFA. These include (a) the ability to yield unique factorial solutions, (b) clearly defining a testable model, (c) assessments of the extent to which a hypothesised model fits the data, (d) information about how individual model parameters affect model fit, (e) the ability to test factorial invariance across groups (Marsh, 1987) and (f) the ability to compare and evaluate competing theoretical models empirically. Standard SEM techniques make extensive use of CFA in the development of the latent variables (i.e. measurement models). For the remainder of this book, we focus exclusively on CFA.

## Measurement models versus structural models

In SEM, *measurement models* and *structural models* are conceptually distinct (Anderson & Gerbing, 1982, 1988). As previously mentioned, latent constructs represent theoretical constructs of interest that cannot be directly measured but that influence scores on the observed variables. To measure such latent constructs, we use multiple observed variables called *indicators*. The *measurement model* specifies the causal relations between the observed variables and the underlying latent variables (Anderson & Gerbing, 1982, 1988). The most common measurement model in SEM is a CFA model. For example, the unidimensional factor model in Figure 4.3 is also an example of a measurement model for a single latent construct.

In the standard conceptualisation of a measurement model, the only directional pathways are from latent variables to observed variables. Therefore, the latent constructs are *exogenous* variables. Generally, in SEM, all exogenous variables correlate with each other. In other words, the latent variables correlate with one another, but they do not predict one another.

The *structural model* is often the model of greatest interest: it specifies the causal or predictive pathways among the conceptual variables of interest. In such cases, the main purpose of the measurement model is to measure the theoretical constructs of interest both more completely and more accurately, using multiple indicators. Multiple indicators enable us to separate the 'true' variance of the latent variable from the measurement error that is inherent in each observed variable. The latent variables are measured without error, so the latent variable model generates unbiased estimates of the structural paths among the conceptual variables of interest. In Figure 4.4, we have reformulated our mediation model (from Figure 4.1) so growth mindset is no longer an observed variable; it is now a latent variable. The model now contains a latent variable for growth mindset as well as the overall structural model in which growth mindset predicts academic persistence, which in turn predicts academic achievement.

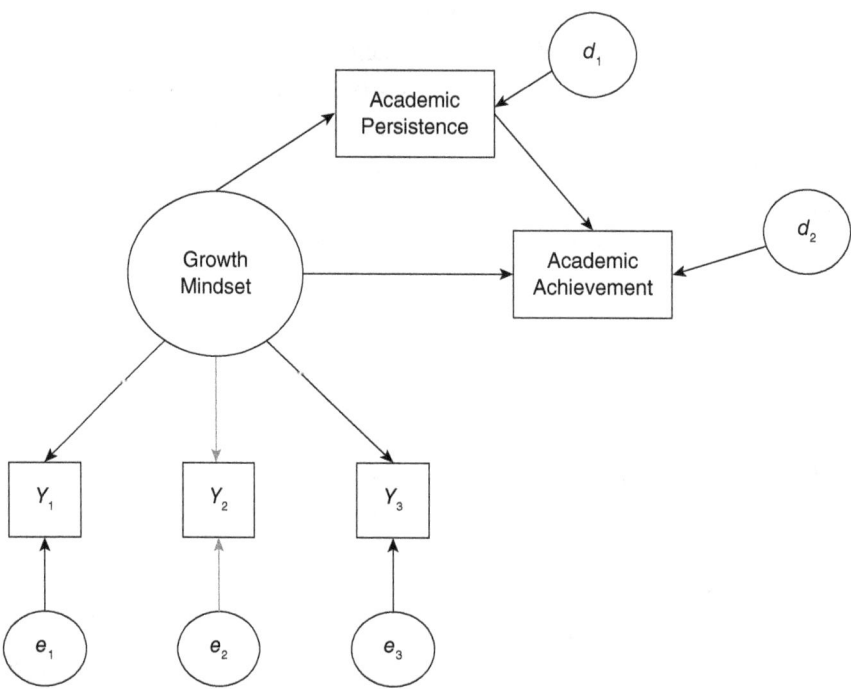

**Figure 4.4** A structural model in which growth mindset (measured with a latent variable) predicts academic persistence, which in turn predicts academic achievement

Misspecified measurement models can lead to errors of inference in the structural part of the model. Therefore, the measurement model must be correctly specified and exhibit adequate fit prior to estimating the structural parameters (Anderson & Gerbing, 1982). We return to this point in Chapter 6, when we describe the model building process in SEM.

## Disturbances and measurement errors

Generally, we refer to residuals for structural endogenous variables as *disturbances* and residuals for endogenous measurement variables as *measurement errors*. The disturbance variance represents the sum of all other causes of the endogenous *structural* variable that are *not* explicitly specified in the structural model. Similarly, error variance in the measurement model represents the sum of all other causes of the indicator variable that are *not* explained by the latent construct (factor). Note the difference between the use of *d*'s in Figure 4.1 and *e*'s in Figure 4.3. In either case, the total variance of any endogenous variable can be partitioned into two pieces: (1) the variance that is explained by its predictor variables and (2) the variance that is unexplained by the predictor variables. As in multiple regression, the proportion of explained variance in an endogenous variable is $R^2$. Therefore, the proportion of unexplained variance in an endogenous variable is $1 - R^2$.

Given that the latent variable has no inherent scale of its own, factor analytic results most commonly report the *standardised* path coefficients for a path from the latent variable to the indicator. These standardised path coefficients are also referred to as **measurement weights/pattern coefficients**, or *factor loadings* in CFA. In Figure 4.3, each of the pattern coefficients estimates the direct effect of the factor on the indicator variable.

The $R^2$ for a unidimensional indicator is simply the square of the standardised factor loading, and $R^2$ represents the proportion of variance in the indicator that is explained by the factor. For multidimensional indicators (where two or more factors predict a given indicator), we still can partition the variance in the indicator into the portion that is explained by the latent constructs and the portion that is unexplained by the latent constructs. The **proportion of variance unexplained by a factor** (or factors) is the measurement error variance for the indicator divided by the total variance of the indicator:

$$1 - R^2 = \frac{Measurement\ Error\ Variance}{Total\ Variance} \tag{4.1}$$

For example, suppose the total variance for an indicator is 100, and in a CFA measurement model, the error variance of that indicator is 20. The proportion of unexplained variance is 20/100 or .20. The $R^2$ for that indicator is 1 – proportion of unexplained variance, so $R^2 = 1 - .20$ or .80.

## Estimation

The goal of estimation in SEM is to find the model parameter estimates that maximise the probability of observing the data. Finding parameter values for an overidentified model is iterative. The computer program repeatedly refines the parameter estimates to minimise the discrepancy between the model-implied variance–covariance matrix and the variance–covariance matrix (Brown, 2015). The ML estimates of the parameters minimise the discrepancy between the variance–covariance matrix and the model-implied variance–covariance matrix. 'ML aims to find the parameter values that make the data most likely' (Brown, 2015, p. 63). The structural equation model has *converged* when a unique set of parameter estimates minimise the difference between the model-implied and sample variance–covariance matrices. (*Note:* We find a unique set of parameter estimates for our specified model. However, that does not mean that our specified model is the only model to fit the data equally well!)

## Model fit and hypothesis testing in SEM

In SEM, we fit our theoretical model to a variance–covariance matrix. The population covariance matrix represents bivariate relationships between the observed variables. The model parameters maximise the likelihood of obtaining the data, given the specified model. The estimated model parameters can then be used to generate the covariance matrix that is implied by the model. Ideally, the parameters in our model should be able to generate a *model-implied covariance matrix* that reproduces the population covariance matrix. The model-implied variance–covariance matrix provides important information about model–data fit. The more closely the parameters reproduce the covariance matrix, the better the 'fit' of the model. The fundamental statistical hypothesis undergirding SEM is $H_0$: $\Sigma = \Sigma(\theta)$. This null hypothesis states that the model-implied variance–covariance matrix is exactly equal to the population variance–covariance matrix. Here, $\Sigma$ is the population covariance matrix, $\theta$ contains the set of parameters or system of equations and $\Sigma(\theta)$ is the model-implied covariance matrix (Paxton et al., 2011). The global fit function ($F$) measures the degree of discrepancy between the model-implied variance–covariance matrix and the actual variance–covariance matrix. The global fit function ($F$) tells us nothing about the predictive utility of the model. Instead, it is a function of the degree to which the model parameters are able to reproduce the covariance matrix.

How do we know if our model fits the data? Hypothesis testing in SEM departs from traditional tests of significance. In most statistical analyses, researchers test the null hypothesis that there is no relationship among a set of variables or that there are no statistically significant differences among a set of variables. Generally speaking, we want to reject the null hypothesis and conclude that there are statistically significant relationships or differences. In SEM, for the global test of model fit, the logic is reversed. We test the null hypothesis that the specified model exactly reproduces the population covariance matrix of observed variables (Bollen & Long, 1993). Assuming the data satisfy distributional assumptions (normality, etc.), the product of the asymptotic distribution of the fit function and the sample size minus 1 ($F * (N - 1)$) is asymptotically distributed as chi-square ($\chi^2$), with degrees of freedom equal to the degrees of freedom of the model. To evaluate exact global model fit, we compare the $\chi^2$ of the specified model to the critical value for $\chi^2$. Under the null hypothesis, the population covariance of observed variables equals the model-implied covariance matrix. Therefore, when the model $\chi^2$ exceeds the critical value of $\chi^2$, we reject the null hypothesis. Rejecting $H_0$ means the specified model does not adequately reproduce the covariance matrix, indicating less than perfect model fit.

However, this approach poses several problems. First, $\chi^2$ is very sensitive to sample size. The larger the sample size, the more likely we are to reject the null hypothesis that the model fits the data. The $\chi^2$ test rejects almost any model with a very large sample size if there is even a miniscule amount of model-data misfit. To correct for this problem, some researchers divide $\chi^2$ by the model degrees of freedom. Ideally, the $\chi^2/df$ ratio should be less than 2. This does not really solve the problem though, as the degrees of freedom are related to model complexity and size, rather than sample size.

As mentioned earlier, ML estimation requires large sample sizes. This creates an obvious tension: we need large sample sizes in SEM, but large sample sizes provide high power to reject the null hypothesis that the model fits the data exactly. Because we hope to fail to reject the null hypothesis, having large sample sizes actually works against us.

Second, the $\chi^2$ test is a test of exact (perfect fit). 'A perfect fit may be an inappropriate standard, and a high $\chi^2$ may indicate what we already know – that $H_0$ holds approximately, not perfectly' (Bollen, 1989, p. 268). Knowing that the model-implied covariance matrix does not exactly fit the population covariance matrix provides no information about the degree to which the model does or does not fit the data. Scientific inquiry generally rewards parsimony and simplicity. Generally, models are simplifications of reality. In some sense, the goal of a model is to capture the essence of a system without completely recreating it. Therefore, it should come as no surprise that model-implied covariance matrices generally fail to exactly reproduce population covariance matrices.

## Alternative measures of model fit

Because $\chi^2$ is notoriously sensitive to sample size and the $\chi^2$ test of model fit tends to be rejected with large sample sizes in SEM, SEM researchers have developed a variety of alternative global model fit measures to evaluate model–data fit (or misfit). The various alternative fit indices attempt to correct the problems that result from judging the fit of a model solely by examining $\chi^2$.

There are two basic types of fit indices: (1) **absolute fit indices** and (2) **incremental fit indices**. Absolute fit indices evaluate the degree to which the specified model reproduces the sample data. Some of the more commonly used absolute fit indices include the root mean square error of approximation (RMSEA) and the standardised root mean square residual (SRMR). One of the most popular fit indices, the RMSEA, is a function of the degrees of freedom in the model, the $\chi^2$ of the model and the sample size. Unlike the $\chi^2$, the value of the RMSEA should not be influenced by the sample size (Raykov & Marcoulides, 2000). In addition, it is possible to compute a CI for the RMSEA. The width of the CI indicates the degree of uncertainty in the estimate in the RMSEA (Kenny et al., 2014). The SRMR represents a standardised summary measure of the model-implied covariance residuals. Covariance residuals are the differences between the observed covariances and the model-implied covariances (Kline, 1998). 'As the average discrepancy between the observed and the predicted covariances increases, so does the value of the SRMR' (Kline, 1998, p. 129). The RMSEA and the SRMR approach 0 as the fit of the model nears perfection. Hu and Bentler (1999) suggest that SRMR values of approximately .08 or below, and values of approximately .06 or below for the RMSEA indicate relatively good model fit.

Incremental fit indices measure the proportionate amount of improvement in fit when the specified model is compared with a nested baseline model (Hu & Bentler, 1999). Some of the most commonly used incremental fit indices include the non-normed fit index (NNFI), also known as the Tucker–Lewis Index (TLI) and the comparative fit index (CFI). Both indices approach 1.00 as the model–data fit improves, and the TLI can actually be greater than 1.00. Generally speaking, TLI and CFI values at or above .95 indicate relatively good fit between the hypothesised model and the data (Hu & Bentler, 1995, 1999) whereas values below .90 generally indicate less than satisfactory model fit. Many factors such as sample size, model complexity and the number of indicators can affect fit indices differentially (Gribbons & Hocevar, 1998; Kenny & McCoach, 2003; Kenny et al., 2014); therefore, it is best to examine more than one measure of fit when evaluating structural equation model. However, the vast array of fit indices can be overwhelming, so most researchers focus on and report only a few. We generally report $\chi^2$, RMSEA, the SRMR, the CFI and the TLI.

## Model comparison

### Chi-Square difference test

The $\chi^2$ difference test compares the model fit of two hierarchically nested models. Two models are hierarchical (or nested) models if one model is a subset of the other. For example, if a path is removed or added between two variables, the two models are hierarchical (or nested) models (Kline, 2015). However, models that simultaneously free one or more parameters while constraining one or more previously freed parameters are not nested. For the $\chi^2$ difference test, we subtract the $\chi^2$ of the more complex model ($\chi_2^2$) from the $\chi^2$ of the simpler model ($\chi_1^2$). We then subtract the degrees of freedom of the more complex model ($df_2$) from the degrees of freedom for the more parsimonious model ($df_1$). We compare this $\chi^2$ difference ($\chi_1^2 - \chi_2^2$) to the critical value of $\chi^2$ with $df_1 - df_2$ degrees of freedom. If this value is greater than the critical value of $\chi^2$ with $df_1 - df_2$ degrees of freedom, we conclude that deleting the paths in question has statistically significantly worsened the fit of the model. If the value of $\chi_1^2 - \chi_2^2$ is less than the critical value of $\chi^2$ with $df_1 - df_2$ degrees of freedom, then we conclude that deleting the paths has not statistically significantly worsened the fit of the model. When deleting paths does not worsen the fit of the model, we choose the more parsimonious model (the one that has fewer paths and more degrees of freedom) as the better model. The $\chi^2$ difference test can only be used to compare hierarchically related models. If observed variables are added or removed from the model (i.e. if the observed variance–covariance matrix changes), the models are not hierarchical models. It is inappropriate to use the $\chi^2$ difference test to compare models that have different numbers of variables or different sample sizes.

Because sample size affects $\chi^2$, sample size also affects the $\chi^2$ difference test. Small differences between the observed and model-implied variance–covariance matrices can produce a large $\chi^2$ when the sample size is very large. Likewise, all else being equal, we are more likely to observe statistically significant $\chi^2$ differences between two hierarchically nested models in a large sample than in a small sample. Therefore, any results should be viewed as a function of the power of the test as well as a test of the competing models.

### Fitting multiple models

Generally, SEM specifies an a priori model, based on previous literature and substantive hypotheses. Unlike traditional statistical techniques, in SEM, it is common to evaluate several models before adopting a final model. Sometimes, after fitting the initial model, a researcher might wish to change certain aspects of the model, a process called **respecification**. There are at least three distinct reasons for estimating multiple structural equation models.

1   Theorists seek the most parsimonious explanation for a given phenomenon. The initial model includes all possible parameters. Subsequent models eliminate unnecessary (non-statistically significant) parameters, a process that is sometimes called *trimming* the model. They test the fit of the new more parsimonious model (with greater degrees of freedom) against the original model using the $\chi^2$ difference test. This practice is far more defensible when eliminating paths that are conceptually expected to be 0 than when model trimming is conducted for purely empirical reasons (i.e. all paths that are non-statistically significant are omitted for purely empirical reasons; Kline, 2015).

2   The researcher compares two or more competing theoretical models. Using SEM, the researcher(s) specify the competing models a priori and then compare the models to determine which model appears to better fit the data.

3   The initial model exhibits poor fit. Subsequent models seek to find a model that provides better fit to the data. Purely empirically motivated model modifications lead down a treacherous path, as we discuss next.

## Model modification and exploratory analyses

### Modification indices

If the model does not exhibit adequate fit, how should the researcher proceed? SEM output may include *modification indices* (sometimes called Lagrange multiplier tests). The modification index for a parameter is the expected drop in $\chi^2$ that would result from freely estimating a particular parameter. Remember, $\chi^2$ drops as we add parameters to our model and lower $\chi^2$ values indicate better fit. If we add a parameter to our model, the $\chi^2$ needs to decrease by at least 3.84 points to be statistically significant at the .05 level. Therefore, some researchers request all modification indices above 4. This provides a list of parameters that could be added to the model that would result in a statistically significant decrease in $\chi^2$. Modification indices are univariate. Therefore, adding two parameters simultaneously would not necessarily result in a change in $\chi^2$ equal to the sum of the two modification indices.

The modification indices suggest which parameters might be added to the model to improve model fit. Parameters with larger modification index values result in larger decreases in $\chi^2$, resulting in greater improvements in model fit. Thus, it can be tempting to use these modification indices to make changes to improve the fit of the model. Proceed very cautiously! Although some suggested model modifications may be conceptually consistent with the research hypotheses, other model modifications may make no conceptual sense. Sometimes the modifications suggested by the SEM program are downright illogical and indefensible. Second, making changes based on modification indices (or model fit more generally) capitalises on chance idiosyncrasies of the sample data and may not be replicable in a new sample. Respecification of structural equation models should be guided by theory, not simply by a desire to improve measures of model fit. A good analyst uses modification indices very cautiously (if at all) and reports and substantively defends each modification.

One of the most common and controversial practices in SEM is model modification. When the model which was specified a priori does not exhibit good fit to the data, the temptation to modify the model to achieve better fit can be irresistible. SEM software programs provide suggested model modifications, based solely on statistical criteria. The researcher is then left to determine what, if any, model modifications are warranted. Although using empirical data to guide decision-making may be helpful for 'simple' modifications, it does not tend to inform 'major changes in structure', and some indications for change may be 'nonsensical' (Bollen, 1989, p. 296). Moreover, when we use the same set of data to both develop a model and evaluate its fit, we undermine the confirmatory nature of our analyses (Breckler, 1990). Further, making modifications based on the desire to improve model capitalises on sampling error; such modifications are unlikely to lead to the true model (Kline, 2015; MacCullum, 1986). If the initial model is incorrect, it is unlikely that specification searches will result in the correct model (Kelloway, 1995). Therefore, we advise against blindly following the brutally empirical suggestions of model modification indices. Models with more parameters may fit the data better simply because of chance fluctuations in the sample data. In essence, we can overfit a model to a set of data, and such models will not replicate well with a new sample. 'A model cannot be supported by a finding of good fit to the data when that model has been modified so as to improve its fit to that same data' (MacCallum, 2001, p. 129). Therefore, replication, not modification, provides the best path to enlightenment in SEM.

## Model fit versus model prediction

A model may exhibit adequate fit, and yet do a poor job of predicting the criterion variable of interest. In fact, a model with no statistically significant parameters can fit well (Kelloway, 1995), whereas a 'poor' fitting model may explain a large amount of the variance in the outcome of interest. In fact, models with highly reliable manifest indicators tend to exhibit worse fit than models with less reliable indicators (Browne et al., 2002). Many researchers who would never neglect to report the $R^2$ for a multiple regression analysis seem to overlook the importance of reporting similar measures of variance explained within SEM. Because SEM places a great deal of emphasis on model fit, some researchers lose sight of the fact that a good fitting model may explain very little variability in the variable(s) of interest. To assess **model prediction** for a given endogenous (dependent) variable, we compute the proportion of variance in the variable that is explained by the model. The ratio of the variance of the disturbance (or error) to the total observed variance represents the proportion of unexplained variance in the endogenous variable. Therefore, $R^2$ is simply 1 minus that ratio (Kline, 2015). Determining the variance explained in non-recursive models is more complex. See Bentler and Raykov (2000) for details on calculating of $R^2$ for non-recursive models.

## When SEM gives you inadmissible results

In addition to examining the parameter estimates, the tests of significance and the fit indices, it is very important to examine several other areas of the output to ensure that the program ran correctly. The variances of the error terms and the disturbances should be positive and statistically significant. As in multiple regression, the stand-ardised path coefficients generally fall between –1.00 and +1.00. Further, the stand-ardised error terms and disturbances should fall in the range of 0.00 to 1.00. Negative error variances and correlations above 1 are called **Heywood cases**, and they indi-cate the presence of an **inadmissible solution**. Heywood cases can be caused by specification errors, outliers that distort the solution, a combination of small sample sizes and only one or two indicators per factor, or extremely high or low population correlations that result in empirical underidentification (Kline, 2015).

Additionally, the SEM program may fail to converge in the allotted number of iterations. Lack of convergence indicates that the algorithm failed to produce an ML solution that minimises the distance between the observed and model-implied covariance matrices. Again, when this happens, the output should not be trusted. Requiring very large or infinite numbers of iterations can be signs of a problem such as an underidentified model, an empirically underidentified model, bad start val-ues, extreme multicollinearity, a tolerance value approaching 0 or other specification error (Kline, 2015). If the program fails to converge, inspect the output for possible errors or clues to the reason for the non-convergence, respecify the model to address the problem, and run the SEM again. It is never advisable to interpret output that contains any Heywood cases, non-convergent or inadmissible solutions.

## Alternative models and equivalent models

In traditional SEM, we specify a particular model a priori, but our hypothesised model is statistically equivalent to a myriad of models. Two models are equivalent if they reproduce the same set of model-implied covariance (and other moment) matrices (Hershberger, 2006; Raykov & Penov, 1999; Tomarken & Waller, 2003). **Equivalent models** have different causal structures but produce identical fit to the data (Hershberger, 2006). Equivalent models produce identical values for the discrepancy between the model-implied matrix and the observed matrix; therefore, they result in identical values for model $\chi^2$ and model fit indices. Perhaps the simplest example of model equivalence is to reverse the causal paths in a path analytic diagram. For example, specifying that $X \to Y \to Z$ is equivalent to specifying that $Z \to Y \to X$. For complex models, there are often dozens (or even hundreds!) of functionally equivalent models that the researcher has not tested. Hershberger (2006), Lee and Hershberger (1990) and Stelzl (1986) demonstrate rules for generating multiple equivalent models. Even when a model fits the data well, any statistically equivalent models would fit the data equally well (Tomarken & Waller, 2003, 2005). Equivalent models can lead to substantially different theoretical or substantive conclusions (Hershberger, 2006; MacCallum et al., 1993; Tomarken & Waller, 2005). Unfortunately, researchers often fail to recognise the existence of equivalent models or consider equivalent models when interpreting the results of their research (MacCallum et al., 1993).

In addition, an untested model may provide even better fit to the data than the researcher's hypothesised model, and there is no fail-safe method to protect against this possibility. Researchers who test an assortment of plausible competing models can bolster their argument for a particular model. However, because the number of rival alternative models may be virtually limitless, testing multiple competing models does not eliminate the possibility that an untested model may provide better fit to the data than does the researcher's model. Therefore, any specified model is a tentative explanation and is subject to future disconfirmation (McCoach et al., 2007).

## A word of caution

SEM is a powerful data analytic technique, but it is not magic. No matter how appealing and elegant SEM may be, it is a data analytic technique, and as such, it is incapable of resolving problems in theory or design (McCoach et al., 2007). The adage 'correlation does not imply causation' applies to SEM as well. Although SEM may appear to imply or specify causal relations among variables, causality is an assumption rather than a consequence of SEM (Brannick, 1995). Wright's original description of the technique still holds true today:

> The method of path coefficients does not furnish general formulae for deducing causal relations from knowledge of correlations and has never been claimed to do so. It does, however, within certain limitations, give a method of working out the logical consequences of a hypothesis as to the causal relations in a system of correlated variables. The results are obtained by a combination of the knowledge of the correlations with whatever knowledge may be possessed, or whatever hypothesis it is desired to test, as to causal relations. (Wright, 1923, p. 254)

Using SEM allows us to ascertain whether a hypothesised causal structure is consistent or inconsistent with the data; however, causal inferences ultimately depend 'on criteria that are separate from that analytic system' (Kazantzis et al., 2001, p. 1080).

SEM can polarise researchers: the technique has been both demonised and canonised (Meehl & Waller, 2002). When applied and interpreted correctly, SEM is an invaluable tool that helps us to make sense of the complexities of our world, and SEM offers several advantages over traditional statistical techniques. However, SEM does not replace the need for good design and sound judgement. Causal inference in SEM requires the same assumptions as any other methods, and the ability to make causal inferences rests firmly on the design of the study. Strong designs allow for stronger causal inferences; weak designs lead to weak causal inferences. Just as 'no amount of sophisticated analyses can strengthen the inference obtainable from a weak design' (Kelloway, 1995, p. 216), no analytic method can replace the need for critical appraisal and common sense.

SEM is both an art and a science. Because structural equation models are so open to modification, SEM allows for a great deal of artistic license on the part of the analyst. SEM allows researchers a great deal of flexibility and control over their analyses, which provides opportunities for both innovation and manipulation. It is this freedom that makes SEM so powerful and so appealing, but also so prone to misuse. With this flexibility comes great responsibility. We must build our models thoughtfully, describe our model building process fastidiously and interpret the results of our SEM analyses cautiously. Producers and consumers of structural equation model should realistically evaluate the strengths and limitations of this technique and should interpret the results of SEM analyses accordingly. In Chapter 6, we provide concrete recommendations for defensible model building processes. However, before doing so, Chapter 5 delves into the important foundational topics of SEM specification and identification.

## Chapter Summary

- Structural equation modelling (SEM) refers to a family of techniques, including (but not limited to) path analysis, confirmatory factor analysis, structural regression models, autoregressive models and latent change models.
- SEM allows researchers to distinguish between observed and latent variables and to explicitly model both types of variables.
- Using latent variables in SEM accounts for potential errors of measurement, allowing researchers to explicitly account for (and model) measurement error (Raykov & Marcoulides, 2000). The ability to separate measurement error or 'error variance' from 'true variance' is one of the reasons that SEM provides such powerful analyses.

- SEM is a regression-based technique. As such, it rests upon four key assumptions and requirements necessary for any regression-based analyses: (1) normality, (2) linearity, (3) independence and (4) adequate variability.
- The basic building block of any structural equation model is the variance–covariance matrix.
- Models with more parameters may fit the data better simply because of chance fluctuations in the sample data. It is possible to overfit a model to a set of data, and such models will not replicate well with a new sample.
- SEM allows researchers a great deal of flexibility and control over their analyses, which provides opportunities for both innovation and manipulation. In SEM, we must build models thoughtfully, describe the model building process fastidiously and interpret the results of analyses cautiously.

## Further Reading

Bollen, K. A. (2002). Latent variables in psychology and the social sciences. *Annual Review of Psychology, 53,* 605–634.
This article discusses the definition and use of latent variables in psychology and social science research.

Kline, R. B. (2015). *Principles and practices of structural equation modeling* (4th ed.). Guilford Press.
This book is an oft-cited introductory guide to structural equation modelling. It provides a non-technical introduction to structural equation modelling as well as several other topics including multilevel structural equation modelling, growth curve analysis and mean and covariance structure analysis.

# 5

# SPECIFICATION AND IDENTIFICATION OF STRUCTURAL EQUATION MODELS

## Chapter Overview

Computing degrees of freedom in SEM .................................................. 96

Model identification: measurement models/CFA ................................... 102

Identification in CFA models .................................................................. 103

Introduction to systems of equations using path diagrams .................... 107

Further Reading ..................................................................................... 115

In Chapter 4, we introduced SEM as a technique for analysing **systems of equations** in which path diagrams pictorially represent systems of equations. This chapter provides an overview of model identification criteria and explicitly links path diagrams to the structural equations that they specify. After explaining the link between path diagrams and structural equations, we demonstrate how to use Wright's rules to derive the **model-implied correlation/covariance** matrix for a given SEM. Appendix 2 provides a more technical introduction to the link between path diagrams and structural equations. Appendices 3 and 4 provide even greater detail on Wright's rules. Appendix 3 demonstrates Wright's standardised tracing rules. Appendix 4 discusses Wrights unstandardised rules and covariance algebra.

SEM involves solving of a set of simultaneous equations in which the known values are a function of the unknown parameters (Kenny & Milan, 2012). In Chapters 4 to 6, our known values consist of the observed variances and covariances because we limit our discussion to models that do not include means or mean structure. However, in Chapter 7, we introduce means and mean structure into our SEMs.

To generate unique estimates for all these parameters, the SEM model must be identified (Kline, 2015). Identification involves demonstrating 'that the unknown parameters are functions only of the identified parameters and that these functions lead to unique solutions' (Bollen, 1989, p. 88). If all parameters in the model are uniquely identified, then the model itself is identified. We provide a brief introduction to identification rules for **recursive structural equation models.** Recursive structural equation models have no feedback loops and no correlated disturbances. Therefore, any variable ($Y$) cannot both be a predictor of and predicted by another variable ($X$). The rules of identification for **non-recursive structural equation models** are far more complicated than the rules for recursive models. Given the introductory nature of this text, we present identification rules for recursive models only. However, readers who are interested in learning more about non-recursive models should read Paxton et al. (2011), which provides a very approachable introduction to non-recursive models. For a fuller discussion of identification issues, see Rigdon (1995), Kline, 2015, Kenny and Milan (2012) or Steiger (2002).

## Computing degrees of freedom in SEM

As mentioned in Chapter 4, the covariance matrix serves as a sufficient statistic for standard SEM: raw data are not necessary. Instead, it is possible to estimate SEM parameters using the covariance matrix. In statistical analyses such as ANOVA and regression, the degrees of freedom are a function of the sample size. In contrast, in SEM, the number of parameters that we can freely estimate is limited by the number of unique elements in the variance–covariance matrix (or the variance–covariance matrix plus means for models that include means). We cannot estimate more parameters than there are unique elements in the variance–covariance matrix, no matter how large our sample size is!

Let's count the number of unique elements for the small variance–covariance matrix shown in Equation (5.1). There are six unique elements: three diagonal elements (the three variances) and three unique off-diagonal elements (the three unique covariances: $cov_{12}$, $cov_{13}$ and $cov_{23}$). (The covariances below the diagonal are identical to the covariances above the diagonal; that is, the covariance between variables 1 and 3 is identical to the covariance between variables 3 and 1, so we count each of these covariances only once.)

$$
\begin{bmatrix}
var_1 & cov_{12} & cov_{13} \\
cov_{21} & var_2 & cov_{23} \\
cov_{31} & cov_{32} & var_3
\end{bmatrix}
\tag{5.1}
$$

The number of unique elements in the variance–covariance matrix is the number of *knowns* (the information that we *know* before we begin our analyses).

If there are many variables in the model, counting the number of unique elements in the variance–covariance matrix is tedious. Luckily, there is an easy formula to compute the number of unique elements in the variance–covariance matrix. The number of unique elements in the variance–covariance matrix (the knowns) equals

$$
\text{Knowns} = \frac{v(v+1)}{2}
\tag{5.2}
$$

where $v$ = the number of observed variables in the variance–covariance matrix. Therefore, if there are 20 observed variables in the variance–covariance matrix, then the number of knowns equals $\frac{20\cdot(20+1)}{2} = \frac{20\cdot21}{2} = 210$. A slight modification to the formula calculates the number of off-diagonal elements of the variance–covariance matrix (i.e. the number of unique **correlations** in a correlation matrix): $v(v-1)/2$, where $v$ is (still) the number of observed variables.

The number of *knowns* places an upper limit on the number of possible *unknowns*, which are the freely estimated parameters in the model. In SEM, we estimate several different types of parameters: exogenous variances and endogenous variances (which can be either disturbances or measurement errors), paths and **covariances/correlations**. For the number of parameters in the model (unknowns), we count the number of freely estimated variances, paths and covariances/correlations. The degrees of freedom in SEM equal the number of knowns (unique elements of the variance–covariance matrix) minus the number of unknowns (freely estimated parameters). A model has positive degrees of freedom if the model contains fewer parameters than there are unique elements in the variance–covariance matrix. Measures of model fit (chi-square, RMSEA, CFI, etc. are only available for models with positive degrees of freedom).

Figure 5.1 depicts a simple path model with four observed variables: (1) parental expectations, (2) growth mindset, (3) academic persistence and (4) academic achievement. Of course, these constructs could be measured using latent variables, and that would be preferable. However, to start simply, we demonstrate tracing rules with a path model. The model estimates paths from parental expectations to growth mindset, academic persistence and academic achievement, from growth mindset to academic persistence and from academic persistence to academic achievement.

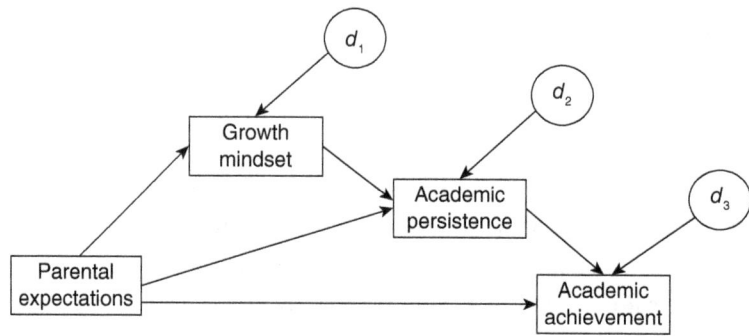

**Figure 5.1**   A simple path model with four observed variables

How many degrees of freedom does this model contain? First, we count the freely estimated parameters. We estimate variances for every exogenous variable and disturbance variances for every endogenous variable in the model. The model in Figure 5.1 estimates one exogenous variance (for parental expectations) and three disturbance variances (for growth mindset, academic persistence and academic achievement). The model also includes five freely estimated paths and zero covariances. Therefore, the number of freely estimated parameters (unknowns) equals 9 (5 paths + 1 exogenous variance + 3 disturbance variances). There are four observed variables, so the number of unique elements in the covariance matrix equals 4 * 5/2, or 10. The degrees of freedom for the model equals the number of knowns (10) minus the number of unknowns (9), or 1 $df$. Why is there 1 $df$? Our model contains no direct effect of growth mindset to academic achievement. There is no path from growth mindset to academic achievement, so that path is constrained to 0. By eliminating that path, in our hypothesised model, the effect of growth mindset on academic achievement is completely mediated by academic persistence. For our model to fit the data, the correlation between growth mindset and academic achievement must be completely explained by their mutual relationships with academic persistence. If this were not true, the hypothesised model (depicted above) would fit more poorly than the model that includes that path. Because this model has only 1 $df$, we know that the $\chi^2$ of the model is completely attributable to the misfit due to the elimination of the path (or direct linkage) from growth mindset to academic persistence.

When the number of knowns equals the number of unknowns, the model is said to be just-identified. A just-identified model contains as many knowns as unknowns, so the parameter estimates can always perfectly reproduce the variance–covariance matrix. Thus, the just-identified model 'fits' the variance–covariance matrix perfectly. Just-identified models always have 0 $df$. Adding a path (direct effect) from growth mindset to academic achievement produces a just-identified (fully saturated) model. For all just-identified models, both the $\chi^2$ and $df$ are 0.

In fact, all multiple regression models are actually just-identified path models, so they have 0 $df$. The knowns are the number of observed variables, which is the sum of the predictors and the outcome variable. The unknowns are the regression coefficients, the exogenous variances for the predictors, the residual variance for the outcome variable and the covariances among all of the exogenous variables (the predictors). Because we allow all exogenous variables to correlate with each other, the number of freely estimated parameters is exactly equal to the number of unique elements of the variance–covariance matrix.

If the specified model requires estimating more parameters than there are unique pieces of information in the variance–covariance matrix, the model has negative degrees of freedom and is underidentified. It is not possible to solve the set of structural equations for underidentified models because there are more unknowns than knowns. Just as it is not possible to find a unique solution to the equation $x + y = 10$ because the number of unknowns is greater than the number of knowns, it is not possible to uniquely identify all of the parameters in a model with negative degrees of freedom. Having non-negative degrees of freedom is a necessary (but not sufficient) condition for model identification: models with positive degrees of freedom can still be underidentified. The problem of underidentification is theoretical rather than statistical (Heise, 1975); it is not data dependent. (*Empirical* underidentification, on the other hand, is data dependent. See Kenny & Milan, 2012, for more information about empirical underidentification.) Therefore, it is important to evaluate whether or not the structural equation models of interest are identified or identifiable during the design phase of the study, prior to data collection.

Luckily, although some of the identification rules for SEM are quite complex, the identification rules for recursive path models are actually quite simple. Recursive path/structural models with non-negative degrees of freedom are always identified. However, non-recursive structural equation models with positive degrees of freedom may not be identified. For more information on the identification of non-recursive models, see Berry (1984), Nagase and Kano (2017) and Paxton et al. (2011). The rules of identification also become more complex for measurement (factor) models, as we shall soon see.

Assuming that the model is not unidentified (inestimable) due to other problems in the specification, a recursive path model with positive degrees of freedom is *overidentified*. An overidentified model uses a smaller number of parameters to estimate all elements of the variance–covariance matrix, resulting in some discrepancy between the available variance–covariance matrix and the parameters to be estimated (Kenny & Milan, 2012).

In the case of overidentification, there is more than one way to estimate one or more of the parameters in the system of equations (Kenny, 2004). An overidentified model is more parsimonious than a just-identified model: it attempts to reproduce all the elements of the variance–covariance matrix with fewer parameters. As such, it is a simplification of reality. However, some level of detail or information is lost in that process. In such a scenario, we favour the solution that produces parameter estimates that maximise the likelihood of observing our data.

SEM model fit is an indication of the degree to which our simplified model reproduces (or fails to reproduce) the variance–covariance matrix (Kenny & Milan, 2012). Measures of model fit (e.g. the $\chi^2$ test) are available for overidentified models; thus, it is possible to evaluate the fit of an overidentified model and test it against other competing models. In fact, it is only possible to examine model fit for models that are overidentified.

As we discussed in Chapter 4, one of the great advantages of using SEM is the ability to incorporate latent variables. The path models that we just described specified structural relationships among variables. However, thus far, our simple path/mediation models have contained observed variables (but not latent variables). Next, we introduce the specification and identification of measurement models, which specify the (causal) relationships among latent and observed variables. Afterwards, we demonstrate the integration of path models and measurement models to estimate hybrid structural equation models with latent variables.

## Model specification of measurement (CFA) models

Path diagrams, visual displays of latent variable models, are the most intuitive way to conceptualise measurement models. Figure 5.2 depicts a measurement model for math and reading ability as a standard CFA model. Recall that in a path diagram, rectangles denote the indicators, or the observed variables, and circles represent latent variables. Each indicator is predicted by both a latent variable and an error. The small circles with $\delta$'s represent the measurement errors, or the residuals. The curved line between the two latent variables indicates a covariance between the two exogenous latent variables.

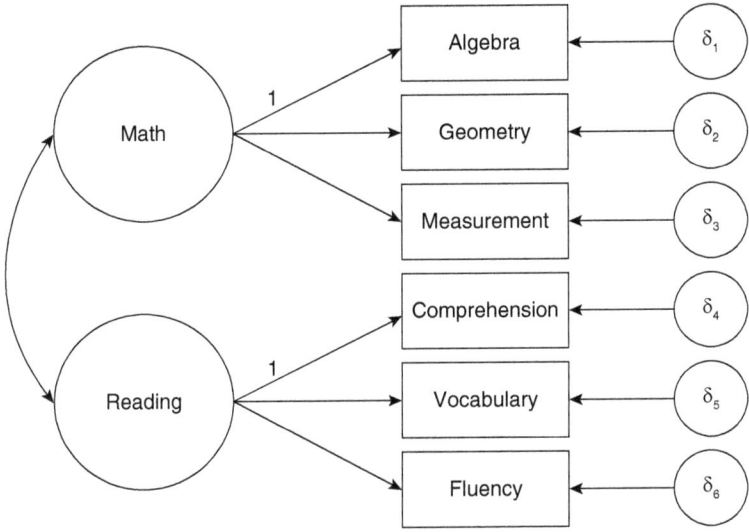

**Figure 5.2** A standard confirmatory factor analysis (measurement) model for math and reading ability

In Figure 5.2, paths connect each of the factors to the three indicators of those factors. Just as in multiple regression and path models, these paths indicate a direct effect from the factor to the indicator (observed variable) *after controlling for* any other variables that also have direct effects on the indicator.

In a standard CFA model, each observed variable is an indicator of only one latent variable, and each observed variable is predicted by both the latent variable and an error. Two sources influence a given indicator – the factor ($F$) and the measurement error term ($\delta$), which encompasses all other unique sources of variance. In other words, a person's response to an item is determined partially as a function of his or her standing on the factor (latent variable) and partially as a function of error, noise or other variables that are not part of the model. The variance in the indicator (the observed variable) consists of two pieces: (1) the variance that can be explained by the latent variable (the factor variance) and (2) the variance that is not explained by the latent variable ($\delta$, the measurement error variance). We distinguish between **measurement error**, residual (error) variance in the measurement model, and disturbance variance, residual variance in the structural model.

A standard CFA model assumes that each factor is **unidimensional**. Conceptually, imagine the attribute being measured by a unidimensional factor falling on a straight line: people who are high on the factor possess more of the attribute and people who are low on the factor possess less of it. Unidimensionality indicates that the statistical dependence among a set of indicators is captured by a single latent construct (Crocker & Algina, 1986; McCoach, Gable et al., 2013).

The standard CFA model assumes that any correlations among the indicators result from their mutual relationship with the latent factor(s). The measurement error terms ($\delta$'s) are independent of each other and the factors. The assumption of **local independence** specifies that after controlling for the factor, the partial correlation between each of the pairs of indicators is 0. All indicators are assumed to be independent of each other, after controlling for the factor(s). Figure 5.2 contains no linkages (i.e. paths or correlations) among the measurement error terms: there are no relationships among the unexplained variances ($\delta$'s) of any of the indicators. In the standard CFA model, each variable is predicted by only one factor. However, it is possible to specify a CFA in which more than one latent variable predicts a given indicator. When two or more latent constructs predict responses on an observed indicator, the indicator is **multidimensional.** For example, a test of word problems is likely to be multidimensional because both mathematics ability and reading ability predict performance on word problem tasks.

Figure 5.2 contains a direct path from math to algebra but no direct path from reading to algebra, which means the direct path from reading to algebra is constrained to be (fixed at) 0. In other words, the model assumes that there is no direct effect of reading on algebra, after controlling for math. This does *not* mean that reading is unrelated to the indicators of mathematics achievement. Rather, the model specifies that the relationship between the reading factor and algebra is indirect: it is a function of the relationship between the reading and math factors and the path from the math factor to the algebra indicator. Because the model contains no direct effect of reading on algebra after controlling for math achievement, the standardised path coefficient from math to algebra is also the model-implied correlation between the math factor and algebra. Squaring the standardised path coefficient from math to algebra computes the proportion of variance in the algebra indicator that is explained by the math factor ($R^2$). The proportion of variance in the indicator that is not explained by the factor is $1 - R^2$, which is also the error variance of the indicator divided by the total variance of the indicator.

## Model identification: measurement models/CFA

Standard CFA estimates parameters for the paths from the latent factors to the observed variables (the factor loadings), the variances of the latent variables, the variances of the measurement errors and the covariances (or correlations) among the latent variables. The factors have no inherent scales because they are latent variables; they are not actually measured or observed. Therefore, to identify the CFA model, we must *scale* the latent variable. Scaling the latent variable provides an anchor to and meaning for the metric of the latent variable. Two common options for scaling the latent variable are the **fixed factor variance strategy** or the **marker variable strategy**.

The *fixed factor variance strategy* constrains the variance of each factor to 1.0. In standard CFA models, the factor's mean is constrained to 0. Therefore, the fixed factor variance strategy results in a standardised solution: each latent variable in the model has a mean of 0 and a variance of 1.

The *marker variable strategy* constrains one unstandardised path coefficient for each factor to 1 and freely estimates the variance of the latent factor. The variable whose unstandardised regression coefficient is fixed to 1 becomes the *marker variable* for the factor, and the factor's freely estimated variance is scaled in the same metric as the marker variable. Figure 5.2 employs the marker variable strategy. Generally, any variable that is reasonably strongly correlated with the factor can be a marker variable. Using an observed variable that is uncorrelated or only weakly correlated with the factor as the marker variable is problematic. Why? Fixing a path coefficient that is very small at 1.0 results in very large unstandardised path coefficients for the other indicators, as their coefficients are computed relative to the coefficient of the poor indicator. Therefore, it is advisable to select one of the strongest indicators as the marker variable. If the variables are measured in different metrics, it is helpful to choose a variable with the most interpretable metric to be the marker variable, given that the latent factor is scaled in the metric of the marker variable.

In standard, single-group CFA, the two scaling methods (fixed factor variance and marker variable) result in statistically equivalent models. The marker variable strategy is the default approach in many statistical software packages (e.g. Mplus, AMOS, lavaan). Furthermore, it is common to use the marker variable approach when conducting multiple-groups CFA to assess the invariance of the model factor structure across different subsets of a sample. So, how does the latent variable scaling method impact the estimated parameters in a CFA model? The standardised parameter estimates are identical, regardless of the scaling technique employed. However, the unstandardised parameter estimates differ across scaling techniques. Using the fixed factor variance strategy, the unstandardised parameter estimates are identical to the standardised parameter estimates. The marker variable approach scales the unstandardised results in the metric of the marker variable for each factor.

## Identification in CFA models

### Freely estimated parameters in CFA

The marker variable strategy constrains one path coefficient (factor loading) per factor to be 1.00 and freely estimates the remaining unstandardised path coefficients. Using the marker variable strategy, we estimate a factor variance for each of our latent variables and an error variance for each of our observed variables. Generally, standard CFA models allow all factors (which are exogenous latent variables) to be intercorrelated;

estimating inter-factor covariances (correlations) for all factors. If we allow any measurement errors to be correlated (covary), then we must count those correlations (covariances) as estimated parameters as well. The number of unknowns (parameters to be estimated) equals the sum of the freely estimated paths, factor variances and covariances, and error variances and covariances. Recall, the number of knowns equals the number of unique elements in the variance–covariance matrix, which can be calculated using the formula: $v(v+1)/2$, where $v$ is the number of observed variables. The degrees of freedom for a given model equals the number of knowns minus the number of unknowns.

How many degrees of freedom does a single-factor model with three indicators have? A single-factor model with three indicators estimates six parameters (unknowns): one factor variance, two paths and three error variances. With three observed variables, the number of unique variances and covariances (the knowns) equals $3 * 4/2 = 6$. The number of knowns (6) equals the number of unknowns (6). Thus, a single-factor model with three indicators is just-identified: it has 0 $df$. A single-factor model with four or more indicators is overidentified. For example, a standard single-factor model with four indicators contains 2 $df$. Why? This model contains eight unknowns (parameters to be estimated): one factor variance, three path/pattern coefficients and four error variances. The number of knowns equals the number of unique elements in the variance–covariance matrix, which is $4 * 5/2$, or 10. There are $10 - 8 = 2$ $df$. See if you can explain why a single-factor model with five indicators has 5 $df$.[1]

How many degrees of freedom does the model in Figure 5.2 contain? Using the fixed factor variance strategy, we estimate six paths, one covariance and six measurement error variances, for a total of 13 parameters. Using the marker variable strategy, we estimate two factor variances, four paths, one covariance and six measurement error variances. Using either strategy, we estimate 13 parameters (unknowns). There are six observed variables in the model, so the variance–covariance matrix contains $6 * 7/2 = 21$ unique elements. Therefore, the model in Figure 5.2 contains 8 $df$ $(21 - 13 = 8)$. Where are these 8 $df$? The model above reproduces 21 variances and covariances (six variances for the six observed variables and all of their covariances) using only 13 freely estimated parameters. Although the source of the degrees of freedom may seem less obvious in the measurement model, the logic is the same: a variance–covariance matrix contains a linkage between every observed variable. Using a fixed factor variance strategy, we estimate all paths and constrain the latent variances to 1. When fixing the factor variances to 1, the number of freely estimated paths and correlations in the measurement model may not exceed the number of unique correlations in the **covariance/ correlation matrix**.

---

[1]We are estimating one factor variance, four paths and five error variances, so the total number of unknowns = 10. The number of knowns = $5 * 6/2 = 15$. $15 - 10 = 5$. Therefore, a single-factor model with five indicators has 5 $df$.

How many observed variables are needed to adequately measure each latent variable? This is a very complex and nuanced issue. Using three or more observed variables is technically sufficient to estimate a single latent variable (factor). A standard one-factor model with three observed variables is just-identified. With four or more observed variables, the single-factor model is overidentified. For models with multiple factors, as few as two observed variables per factor might be technically adequate. However, from a theoretical standpoint, adequately measuring the latent variable of interest may require more indicators than are technically necessary. In general, the more abstract and loosely defined a construct is, the more indicators are necessary to adequately measure the latent variable (Nunnally & Bernstein, 1994).

Models that include multidimensional indicators (observed variables that are predicted by two or more latent variables) can be more difficult to identify/estimate than standard CFA models with only unidimensional indicators (Kenny et al., 1998). This helps explain why more complex CFA models (e.g. multi-trait multi-method matrices; Campbell & Fiske, 1959) are notoriously difficult to estimate (Kenny & Kashy, 1992; Marsh & Grayson, 1995; Marsh & Hocevar, 1983). For CFA models that include correlated errors, in addition to ensuring that the number of knowns is equal to or greater than the number of unknowns, 'each latent variable needs two indicators that do not have correlated errors and every pair of latent variables needs at least one indicator that does not share correlated errors' (Kenny & Milan, 2012, p. 153).

In summary, standard CFA models (with unidimensional items and no correlated errors) are identified if the number of knowns is equal to or greater than the number of unknowns. However, the identification rules for models with correlated errors and models with multidimensional indicators are more complicated (Brown, 2015). For more details about the identification of such CFA models, see Kenny et al. (1998), who provide a thorough treatment of identification issues in CFA models.

## Degrees of freedom for hybrid SEM

How many degrees of freedom does the hybrid SEM model in Figure 5.3 contain? There are seven observed variables: three indicators of academic persistence, three indicators of growth mindset and one indicator of academic achievement. Therefore, the number of knowns equals 7 * 8/2, or 28. How many parameters are freely estimated? Using the **fixed factor variance strategy**, we estimate six pattern coefficients (paths from latent variables to observed indicators of their respective factors), six measurement error variances, two disturbance variances and three structural paths (the paths among the conceptual variables of interest: growth mindset, academic persistence and academic achievement). Therefore, this model has 17 free parameters (unknowns). With 28 knowns and 17 freely estimated parameters, there are 11 *df* (28 – 17). All 11 *df* come from the measurement portion of the model. There are three structural variables: (1) growth mindset, (2) academic persistence and (3) academic achievement, and there are linkages among these three structural variables. Therefore, the structural portion of the model is just-identified. This means that any model misfit is due to misspecifying the measurement portion of the model. We return to this issue in Chapter 6, when we describe the model building process.

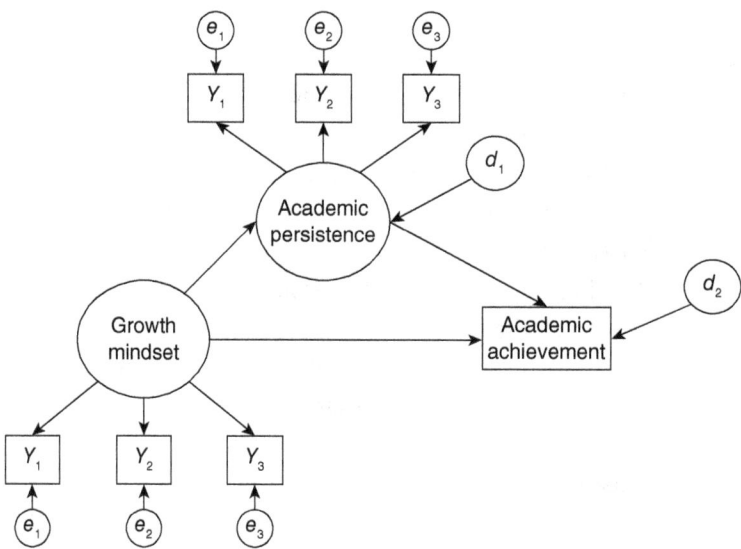

**Figure 5.3** A hybrid structural equation model: A path model that includes latent variables

## Equations for a measurement (CFA) model

Using the path diagram in Figure 5.2, we can represent the measurement model as a system of equations, as shown in Exhibit 5.1. The system of equations captures all the direct pathways among the variables but does not include (non-directional) correlations among variables. To start, let's write an equation for each of the endogenous variables in our model. The equation for each endogenous variable is analogous to a regression equation. The endogenous variable always appears on the left-hand side of the equation. Any (observed or latent) variable with a single-headed arrow leading to the endogenous variable appears on the right-hand side of the equation. Because standard structural equation models are linear models, the terms are additive.

**Exhibit 5.1** Systems of equations corresponding to Figure 5.2

Algebra = $\lambda_1$Math + $\delta_1$

Geometry = $\lambda_2$Math + $\delta_2$

Measurement = $\lambda_3$Math + $\delta_3$

Comprehension = $\lambda_4$Reading + $\delta_4$

Vocabulary = $\lambda_5$Reading + $\delta_5$

Fluency = $\lambda_6$Reading + $\delta_6$

## Introduction to systems of equations using path diagrams

### Getting started with Wright's rules

**Wright's tracing rules**, developed in the 1910s and 1920s by biologist Sewall Wright (Heise, 1975), provide the basic principles of path analysis. Wright's **standardised tracing rules** provide the most intuitive method to generate the model-implied correlations from the standardised parameters for recursive path/ structural equation models.

It is also possible to generate the model-implied variance–covariance matrix from the unstandardised parameter estimates using the **unstandardised tracing rules**. However, the standardised tracing rules are both more straightforward and more common than the unstandardised tracing rules, and the standardised path coefficients and correlations are generally easier to interpret. In Appendix 3, we provide the technical details that undergird our discussion of Wright's standardised tracing rules In Appendix 4, we discuss Wright's rules for generating the model-implied covariances using the unstandardised path coefficients for CFA and path models and we demonstrate the equivalence of using either unstandardised tracing rules or **covariance algebra** for generating model-implied covariances. These technical details provide a deeper understanding of the mathematical underpinnings of SEM.

### Standardised tracing rules

Wright's tracing rules for standardised variables (Wright, 1918, 1934) are a set of rules for tracing the model which implies distinct correlations between two variables based on the structural relations between variables in a path diagram. The model-implied correlation matrix is essentially the standardised version of the model-implied covariance matrix. Using Wright's standardised tracing rules (Loehlin, 2004; Wright, 1918, 1934), we can determine the model-implied correlation between any two variables in a proper (recursive) path diagram using three simple rules:

1   No loops (you cannot pass through the same variable twice in one trace)
2   No going forward then backward within a given trace (but you can go backward then forward)
3   A maximum of one curved arrow per path

Rule 2 states that traces can go backward and then forward, but not forward and then backward, which may seem confusing and capricious at first glance. Why can we go backward and then forward but not forward and then backward? Conceptually, rule 2 accounts for linkages due to common causes, but not linkages due to common effects. We illustrate this idea with two simple three-variable systems of equations, depicted in Figures 5.4 and 5.5.

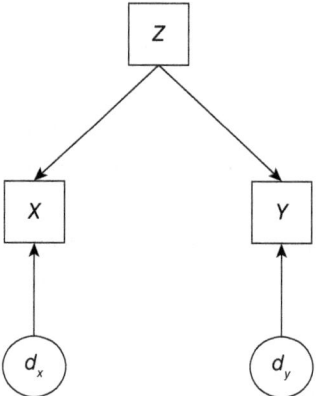

**Figure 5.4**   Tracing rule: You can go backward and then forward. This figure illustrates a linkage due to common causes (upstream variables)

*Note.* If *X* and *Y* are both predicted by *Z*, then they must be related to each other.

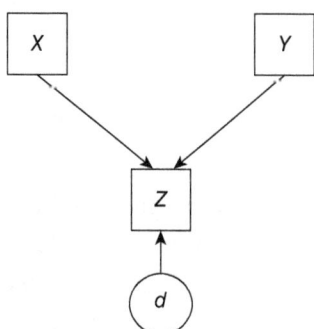

**Figure 5.5**   Tracing rule: No going forward and then backward. There is NO linkage due to common effects (downstream variables)

*Note.* *X* and *Y* can both predict *Z* and still be uncorrelated with each other. (They each predict different portions of the variance in *Z*.)

In Figure 5.4, $Z$ predicts both $X$ and $Y$. Because both $X$ and $Y$ share variance in common with $Z$, they must share variance in common with each other. Tracing backward from $Y$ to $Z$ and then forward from $Z$ to $X$ accounts for the variance that $Y$ and $X$ share, given that they share a common predictor ($Z$). In Figure 5.5, both $X$ and $Y$ predict $Z$. Two exogenous variables can predict the same endogenous variable without being related to each other. Similarly, two variables can be correlated with a third variable without being correlated with each other (Wright, 1934). In this case, $X$ and $Y$ are uncorrelated with each other (there is no curved arrow connecting $X$ and $Y$), so each variable explains different portions of the variance in $Z$. The prohibition on tracing forward and then backward prevents counting linkages due to common effects when determining model-implied correlations.

Using these three rules, we can determine the model-implied (or expected) correlations among all the variables in the model. To do so, we sum all the **compound paths**, or traces, between two variables (i.e. direct and indirect as described in Chapter 4; Loehlin, 2004; Neale & Cardon, 1992). A compound path (trace) is a pathway connecting the two variables following the three rules above and is the product of all constituent paths (Loehlin, 2004). However, there may be many compound paths that connect the same set of two variables. To compute the model-implied correlation between two variables, first, take the product of all elements within each compound path. Then sum all the compound paths. In other words, the model-implied correlation involves multiplying each of the coefficients in a trace and summing over all possible traces (each trace is referred to as a compound path and we sum over the compound paths). Using these rules, we can compute the model-implied correlation between any two variables in a path diagram.

## Example of the standardised tracing rule

Figure 5.6 is a path diagram with standardised path coefficients. Using the tracing rules, we can compute the model-implied correlations among all pairs of variables in the model (Table 5.1). The model-implied correlation between parental expectations and academic achievement is the sum of the compound paths (traces) connecting the two variables. What are all the potential traces from parental expectations to academic achievement? Using the tracing rule, there are three *traces* from parental expectations to academic achievement. The first is the direct effect of parental expectations on academic achievement: that path = .1. The second is the indirect effect of parental expectations on academic achievement through academic persistence: .3 * .5 = .15. The third is the indirect pathway through growth mindset and academic persistence: .2 * .4 * .5 = .04. The sum of these three compound paths (.1 + .15 + .04 = .29) is the model-implied correlation between parental expectations and academic achievement. In this case, it is also the total effect of parental expectations on academic achievement. The total effect and the model-implied correlation are identical when all traces involve only paths (i.e. none of the traces include correlations).

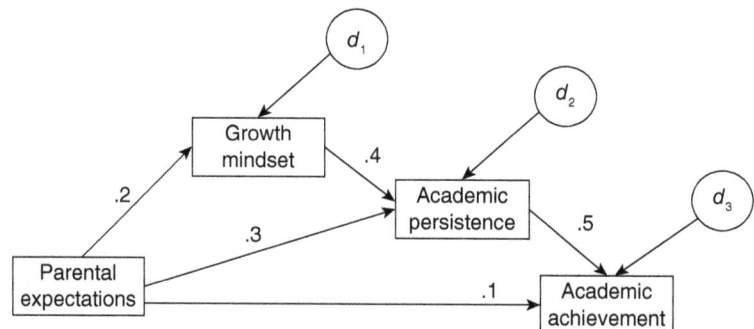

**Figure 5.6** Using the tracing rules to compute model-implied correlations

**Table 5.1** Model-implied correlations among the four observed variables in Figure 5.6

| Variable | Parental Expectations | Growth Mindset | Academic Persistence | Academic Achievement |
|---|---|---|---|---|
| Parental Expectations | 1.0 | | | |
| Growth Mindset | .2 | 1.0 | | |
| Academic Persistence | .38 | .46 | 1.0 | |
| Academic Achievement | .29 | .25 | .54 | 1.0 |

What is the model-implied correlation of academic persistence and academic achievement? Again, we compute the model-implied correlation as the sum of the traces connecting the two variables. Using Wright's rules, there are three distinct traces that link academic persistence to academic achievement. The first is the most obvious: the direct effect of academic persistence on academic achievement has a standardised coefficient of .5. Rule 2 states that traces can go backward through arrowheads and then forward (but not forward and then backward). The second trace goes backward from academic persistence to growth mindset ($b = .4$), then backward from growth mindset to parental expectations ($b = .2$) and then forward from parental expectations to academic achievement. The product of these three paths is .4 * .2 * .1 = .008. The third trace goes backward from academic persistence to parental expectations ($b = .3$), and then forward from parental expectations to academic achievement (.1). The product of these two paths is .3 * .1 = .03. The model-implied correlation between academic persistence and academic achievement is the sum of these three traces (compound paths): .5 + .03 + .008 = .538 (which rounds to .54).

The model-implied correlation between growth mindset and academic persistence is .46. Why? There are two distinct traces (compound paths) linking growth mindset and persistence: a direct pathway ($b$ = .40) and a trace that goes backward through parental expectations ($b$ = .20) then forward from parental expectations to persistence (.30). That compound path is .06. The sum of the two compound paths results in a model-implied correlation of .46 (.40 + .06).

Finally, even though there is no direct effect of growth mindset on academic achievement, the model-implied correlation between growth mindset and academic achievement is not 0. Why? Following Wright's rules, there are actually three compound paths (traces) that link growth mindset and academic achievement. The first is the indirect effect of growth mindset on academic achievement via academic persistence, which is .4 * .5 = .20. The second compound path traces backward from growth mindset to parental expectations ($b$ = .2) and then forward from parental expectations to academic achievement ($b$ = .1). This compound path = .2 * .1 = .02. The third compound path traces backward from growth mindset to parental expectations ($b$ = .2), then forward from parental expectations to academic persistence (.3), and then forward from academic persistence to academic achievement (.5), which equals .2 * .3 * .5 = .03. Summing these three compound paths (traces) (.20 + .02 + .03) yields the model-implied correlation, which is .25. In other words, even though the model constrains the direct effect of academic persistence on academic achievement to 0, the model-implied correlation between academic persistence on academic achievement is .25. Using the tracing rules above, confirm that the model-implied correlation between parental expectations and academic persistence is .38.

## Standardised tracing rules for measurement models

We can apply the tracing rules to compute model-implied correlations in a standard CFA model. Each indicator is predicted by only one factor and there are no correlations among measurement errors, which greatly simplifies the tracing rules. Because there is only one compound path connecting any two variables, the model-implied correlation between the two variables of interest is simply the product of the paths and correlations connecting the two variables. If the factor model adequately explains the data, the correlation between any two indicators of the same factor should equal the product of the paths connecting them. The correlation between two indicators on two different factors should equal the product of the paths connecting each indicator to its respective factor multiplied by the correlation between the two factors.

Figure 5.7 contains standardised path coefficients and correlations for our CFA model. Using the standardised tracing rules, we can estimate the model-implied correlation between two observed variables in our model using the path (pattern) coefficients and correlation coefficients. For two observed variables within the same factor, the model-implied correlation is the product of the paths from the factor to each of the observed variables because we can go backward from one variable to the factor and then forward from the factor to the other variable. For example, the correlation between the algebra and geometry scores is the product of the standardised path coefficients for the two paths leading from the Math factor to these respective scores: .80 * .70 = .56. Likewise, the model-implied correlation between algebra and measurement is .8 * .6 = .48 and the model-implied correlation between geometry and measurement is .7 * .6 = .42.

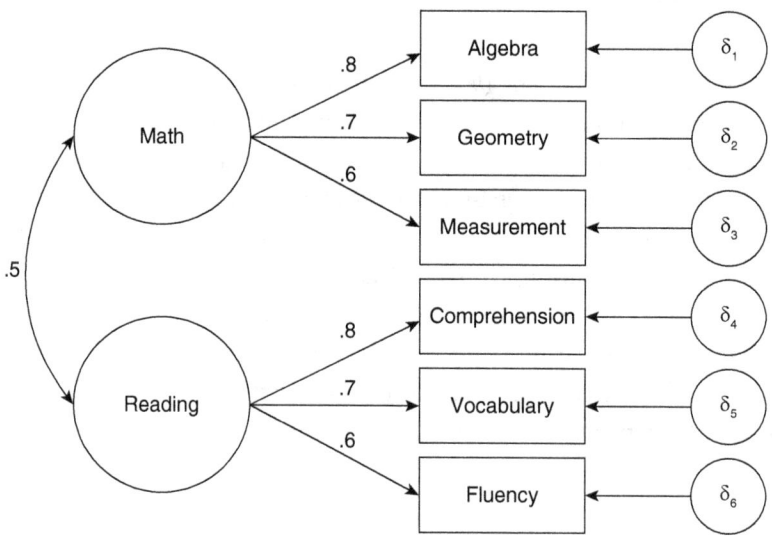

**Figure 5.7** A two-factor CFA model with standardised path coefficients and correlations

To estimate the model-implied correlation between two observed variables from two different factors, we trace backward from observed variable 1 to its factor, then we trace through the correlation between the two factors (the curved arrow), and then we trace forward from factor 2 to observed variable 2. For example, the standardised path from the math factor to algebra scores is .80, the correlation between the math and reading factors is .50, and the standardised path coefficient from the reading factor to comprehension scores is .80. Therefore, the model-implied correlation between algebra scores and comprehension scores is .80 * .50 * .80, or .32. The model-implied correlation between geometry and fluency is .7 * .5 * .6, which equals .21. Be sure that you can compute all the model-implied correlations among the six observed variables in Table 5.2 using Figure 5.7.

**Table 5.2**  Model-implied correlations among the six observed variables in Figure 5.7

| Variable | Algebra | Geometry | Measurement | Comprehension | Vocabulary | Fluency |
|---|---|---|---|---|---|---|
| Algebra | 1.0 | | | | | |
| Geometry | .56 | 1.0 | | | | |
| Measurement | .48 | .42 | 1.0 | | | |
| Comprehension | .32 | .28 | .24 | 1.0 | | |
| Vocabulary | .28 | .245 | .21 | .56 | 1.0 | |
| Fluency | .24 | .21 | .18 | .48 | .42 | 1.0 |

Using the tracing rule, we can also compute the model-implied correlations between the factors and the observed variables. In factor analysis, these are generally referred to as the *structure coefficients*. The model-implied correlation between the math factor and comprehension scores is the product of the correlation between the math factor and the reading factor (.50) and the standardised path from the reading factor to comprehension scores (.80) = .80 * .50 = .40. So even though the direct path from the math factor comprehension scores is 0, the model-implied correlation between the math factor and reading comprehension scores is .40 (and the model-implied correlation between algebra and reading comprehension scores is .32).

In a standard CFA model, because the measurement error terms ($\delta$'s) are independent of each other and of the factors, they do not contribute to the estimation of the model-implied correlations among the observed variables in the standard CFA model. Forcing the measurement errors to be uncorrelated with each other and with the factors specifies that the error variances in the observed indicators (the residual observed score variance that is not explained by their respective factors) are independent (uncorrelated with each other). Allowing the errors of two variables to correlate with each other indicates that factor structure does not adequately capture the correlation between the two variables: the two variables are either more or less correlated with each other than would be predicted by the CFA model. Usually, the correlation between the error variances is positive, indicating that the observed correlation of the two observed variables is higher than would be explained by the factor structure: the two indicators share something in common that is not explained by the factor(s). Generally, correlated errors are indicative of unmodelled multidimensionality. Correlated errors may signal the presence of method effects (e.g. mode of data collection, wording effects) or substantive similarities among variables that are not fully captured by the specified factor structure. Therefore, adding correlated errors to a model should be substantively motivated and conceptually defensible. Although we recommend correlating errors sparingly, sometimes it is appropriate or even necessary to correlate errors. For example, when conducting longitudinal CFAs, it is common practice to allow the measurement error terms of the same indicator to correlate across time. Why? If the exact same measure is administered at multiple time points, it seems quite plausible that the unique variance in that measure would correlate across time, even after controlling for the latent variable at each time point. Correlating the errors allows the unexplained variance in an indicator at one time point to be related to the unexplained variance in the same indicator, measured at a different time point.

What happens when we add correlated errors to a CFA/measurement model? Imagine that we add a correlation between the measurement errors of the comprehension and vocabulary scores (Figure 5.8). Both indicators load on the same latent variable (reading) and they are both unidimensional (only one factor predicts each of the indicators). Therefore, the two indicators (a) load on the same factor, (b) have no cross-loadings with any other factors and (c) do not have correlated errors with any other indicators. In this restrictive scenario, adding a correlation between the two error terms perfectly reproduces the correlation between the two indicators (observed variables). Why? The correlation between the residuals can take on any positive or negative value (between −1 and 1) that best reproduces the correlation between the two indicators. Regardless of the parameter estimates of the two direct paths (factor loadings), it is possible to specify an error correlation that perfectly reproduces the correlation between the observed variables. Under such conditions, adding a correlation between two measurement errors of two indicators of the same factor eliminates the correlation between those two variables as a source of

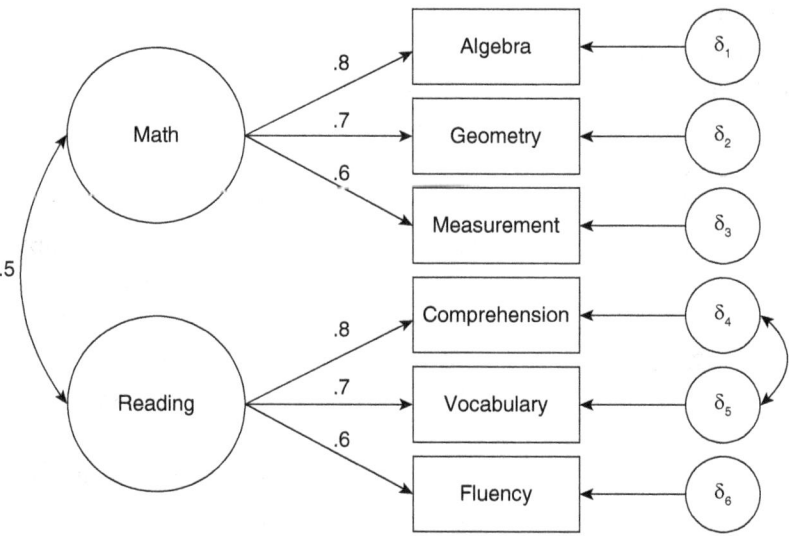

**Figure 5.8** A two-factor CFA model with standardised parameters and a correlated error

information that helps us to understand the factor structure of that latent construct. (However, if the indicators load on more than one factor, including cross-loadings, or if either of the indicators shares a correlated error with another indicator as well, then correlating their measurement errors does not exactly reproduce the correlation between the indicators.)

In Chapter 6, we outline a sequence of model building steps. Then, using all that we have learned, we fit and interpret a latent variable (hybrid) structural equation model.

## Chapter Summary

- Identification involves demonstrating 'that the unknown parameters are functions only of the identified parameters and that these functions lead to unique solutions' (Bollen, 1989, p. 88).
- The number of knowns places an upper limit on the number of freely estimated parameters in the model (the unknowns). These unknowns are the parameters that we wish to estimate.
- A just-identified model contains as many knowns as unknowns, so the parameter estimates can always perfectly reproduce the variance–covariance matrix.
- An overidentified model uses a smaller number of parameters to estimate all elements of the variance–covariance matrix, resulting in some discrepancy between the available variance–covariance matrix and the parameters to be estimated.
- If the specified model requires estimating more parameters than there are unique pieces of information in the variance–covariance matrix, the model has negative degrees of freedom and is underidentified. It is not possible to solve the set of structural equations for underidentified models because there are more unknowns than knowns.
- In standard, single-group CFA, two scaling strategies result in statistically equivalent models: the fixed factor variable and marker variable strategies.

## Further Reading

Kenny, D. A. (2004). *Correlation and causality* (Rev. ed.). Wiley-Interscience. http://davidakenny.net/doc/cc_v1.pdf
Kenny's classic book is out of print, but the pdf is available at the address above. This book explains identification rules and covariance algebra in great detail.

Kenny, D. A., & Milan, S. (2012). Identification: A nontechnical discussion of a technical issue. In R. H. Hoyle (Ed.), *Handbook of structural equation modeling* (pp. 145–163). Guilford Press.

As the title states, this book chapter by Kenny and Milan provides a non-technical discussion of the technical issue of identification. The chapter provides an overview and rules of thumb for identification. Furthermore, the chapter discusses identification for path analysis models without feedback (recursive), with feedback (non-recursive) and with omitted variables as well as latent variable models and latent growth models.

# 6

# BUILDING STRUCTURAL EQUATION MODELS

## Chapter Overview

The two-step approach to fitting latent variable structural models ........ 118

Suggested model fitting sequence for hybrid structural
equation model ..................................................................................... 120

Example: mediation example using a structural equation model
with latent variables ............................................................................. 126

Further Reading ..................................................................................... 142

## The two-step approach to fitting latent variable structural models

In this chapter, we provide a two-step strategy for fitting latent variable structural models. Before analysing the full structural model, we first examine the adequacy of the measurement model (Anderson & Gerbing, 1988; Kline, 2015). Next, we recommend a sequence of structural models. Finally, we provide a detailed example in which we apply our strategy to fit a *hybrid* latent variable structural model and we interpret the results. The term *hybrid* signifies that the model contains both a measurement and a structural model (Kline, 2015). In Chapter 5, Figure 5.3 contained a hybrid model. A structural model without latent variables is usually called a path model, and a latent variable model without structural paths is usually referred to as a measurement model or a factor model. A model that combines both paths and latent variables is therefore a 'hybrid' of a structural/path model and a measurement/factor model.

*Measurement model.* As mentioned in Chapter 4, conceptually, latent variable structural equation models consist of both a **measurement model** and a **structural model**. The measurement model depicts the relationships between the observed variables and the underlying latent variables (Anderson & Gerbing, 1982, 1988; Kline, 2015). The most common type of measurement model is a CFA model; however, we use the term *measurement model* to refer to a model that includes *all* observed variables contained in the conceptual model, whether or not they are indicators of latent variables. In other words, the measurement model includes all structural variables of interest, regardless of whether those structural variables are latent or observed. The structural model depicts the predictive paths and consists of the structural paths between and among the latent variables as any observed variables that are not indicators of an underlying variable.

The measurement model imposes a factor structure to explain the relationships between the observed indicators and the latent and structural variables. The measurement model includes correlations among all exogenous structural (conceptual) variables. Because the measurement model allows correlations among all structural variables, the structural portion of the measurement model is *just-identified*. Therefore, any model misfit is due to the measurement portion of the model (the relationships between the structural variables and the indicators).

## Why fit the measurement model first?

We recommend engaging in a two-step modelling process. Before analysing the full structural model, first examine the adequacy of the measurement model (Anderson & Gerbing, 1988; Kline, 2015). Why is it important to fit the measurement model first?

First, if the fit of the measurement model is unsatisfactory, the fit of the full model will also be unsatisfactory (Gerbing & Anderson, 1993): the full hybrid model cannot fit better than the measurement model does. Therefore, it is important to address any problems in the measurement model prior to estimating the hybrid structural model.

In addition, separately estimating the measurement model prior to estimating the structural model provides some degree of protection against interpretational confounding. Interpretational confounding occurs when the empirical meaning of a latent variable is different from its a priori conceptual meaning (Burt, 1976). Unfortunately, the empirically derived meaning of a latent variable may change considerably, depending on the specification of the structural model (Anderson & Gerbing, 1988). When this occurs, 'inferences based on the unobserved variable then become ambiguous and need not be consistent across separate models' (Burt, 1976, p. 4). If alternate structural models lead to appreciably different estimates of the pattern coefficients within the measurement model, this is indicative of interpretational confounding (Anderson & Gerbing, 1988). Estimating the measurement model separately prior to the estimation of the full structural model minimises the potential for interpretational confounding because the relationships among all conceptual variables of interest are freely estimated; such models place no constraints on the structural parameters (Anderson & Gerbing, 1988). Therefore, if the standard measurement model is adequately specified, the pattern coefficients from the measurement model should remain fairly consistent across the final measurement model and subsequent full structural models (Anderson & Gerbing, 1988). Finally, we recommend fitting the pure measurement model prior to examining the structural model because using this process eliminates the temptation to examine the structural model paths and their significance while respecifying the measurement model (D. A. Kenny, personal communication, July 7, 2020).

## Structural model

For standard, conventional recursive structural models, regardless of whether they are path models or latent variable structural models, we also advocate utilising a multistep, sequential process to fit the structural model.[i] The system works equally well with path models (that contain no latent variables) or with structural models that contain one or more latent variables. However, it is not appropriate for most non-recursive structural models (which cannot be identified using just-identified models) and some types of longitudinal models. We have found the recommended sequence that we outline in this chapter to be invaluable for fitting standard, 'garden variety' structural models, with or without latent variables.

---

[i]The steps that we outline resemble those outlined in Kline (2015) and Anderson and Gerbing (1988); however, there are aspects of the system that we describe that do not explicitly appear in those sources in the exact way that we outline in this chapter. They are, however, the steps that David A. Kenny outlined in his SEM class, when the first author was a student of his, and they are the steps that she continues to teach in her SEM classes. They are also outlined on David Kenny's website http://davidakenny.net/cm/pathanal.htm.

In our experience, researchers often fit their **conceptual structural model** first. After discovering that their conceptual model does not fit, they engage in a process of ad hoc model modifications to try to improve the fit of the model. Such an approach is akin to *p*-hacking, and it is unlikely to produce replicable (or reproducible) results. Instead, we suggest using a sequential approach to fitting and comparing a set of hierarchically nested models. The first model is a just-identified model: it includes the **conceptually omitted (deleted) paths** that are specified to be 0 in the conceptual structural model. Conceptually omitted paths (which are theoretically specified to be 0) that are in fact 0 provide stronger support for the conceptual model than finding that specified paths are non-zero (Kenny, 2019). The system that we outline below is easy to implement and document, and the logic of the sequence is defensible.

## Suggested model fitting sequence for hybrid structural equation model

### Phase 1: fit the measurement model

We recommend following the sequence of steps for fitting the measurement model as part of a larger structural equation model.

1  (Optional but recommended for very large measurement models). For models with a very large number of indicators, it may be advantageous to specify a CFA measurement model separately for each of the instruments or for logical subsets of the factors/items from the overall model. Evaluate the fit of the measurement model, and compare the model fit and parameter estimates to those that have been previously reported in the research literature or in the validation reports for the instruments. At this point, if there are problematic variables, decide how to handle them. For example, if one of the items is completely unrelated to the other items on that factor, eliminating that item from further analyses may be warranted. If two items are highly correlated with each other and their measurement errors are correlated, even after accounting for the factor structure, it is possible to allow the measurement errors of the two variables to correlate. Just remember that correlated measurement errors imply the existence of unmodelled multidimensionality. In some cases, with very large numbers of items, you may want to consider parcelling the items into variables that include multiple items. Parcelling can be advantageous in certain situations (imagine you have given a set of 300 items to 250 people – you cannot possibly analyse your data at the item level), but it is not without controversy. One major drawback to parcelling is that it can conceal problems within your measurement model. An incredible amount of research exists on parcelling. For those who are interested in learning more about parcelling, we recommend Bandalos (2002), Cole et al. (2016), Little et al. (2002), Little et al. (2013), Marsh et al. (2013), Rhemtulla (2016) and Sass and Smith (2006), among others.

2  Specify the full-hypothesised measurement model that includes all latent variables and all observed variables. Include single-indicator latent variables and conceptual observed variables as well. Because the measurement model is just-identified structurally, the $\chi^2$ for any of the structural models that build on the measurement

model can never be lower than the $\chi^2$ for the measurement model itself. So if measurement model exhibits poor fit, then the full model must also exhibit poor fit: the fit of the measurement model places an upper limit on the fit of the hybrid structural model.

3   If the fit of the measurement model is poor, examine the residual variance–covariance matrix and the model modification indices (MIs) to try to determine why the fit is poor. As discussed in Chapter 4, the MI for a parameter indicates the approximate expected decrease in chi-square for a model that includes (freely estimates) the parameter. The covariances with the largest standardised (or normalised) residuals are the covariances that are most poorly reproduced by the model. After diagnosing the likely source(s) of the misfit, determine whether it is possible to modify the measurement model without compromising the theoretical framework and conceptualisation of the model.

4   In addition to evaluating the adequacy of the model from a model fit perspective, examine the estimated parameters to ensure that the measurement model is admissible (does not contain negative variances or correlations above 1) and that the model exhibits discriminant validity (i.e. separately specified factors are actually separate factors). Factor correlations approaching 1.0 suggest that separately specified factors are so highly related that they may indeed be measuring the same construct. Generally, factor correlations below .80 are not indicative of discriminant validity issues, whereas factor correlations above .90 are usually considered problematic. Also, examine the magnitudes of the factor loadings (are the indicators well explained by their factors?).

5   It is essential to document all modifications. This description should include the list of modifications, the order in which the modifications were made and a description of the theoretical and empirical justification for each of the model modifications. We recommend creating a table that documents the sequence of modifications, the rationale for engaging in each of the modifications and the change in model fit that resulted from each of the modifications.

6   The final measurement model from phase 1 becomes the starting point for phase 2, specifying the structural model. Of course, you may never move to phase 2. This could happen for at least two different reasons: (1) your measurement model is unsalvagable or (2) your final model is a measurement model – there is no structural model of interest.

## Phase 2: specify and fit the structural model

We recommend a three-step process for evaluating standard recursive structural models. Before describing our approach, we must qualify our recommendations with a few caveats. If the structural model has a very specific form, as is often the case with longitudinal models, do not fit a just-identified model. Also, the just-identified structural model is likely to be unidentified for most fully non-recursive models, so we do not recommend fitting just-identified non-recursive models. However, for standard, basic, recursive path and latent variable structural models, we recommend the following approach.

The **three steps of structural model specification** are:

*Step 1: Fit a just-identified structural model.* Fit the conceptual model plus any *conceptually omitted* paths. The *conceptual model* for the **hybrid structural equation model** depicts the researcher's hypothesis about the interrelations of both the latent and the observed variables. Using the conceptual model as a base, fit the **just-identified structural equation model**, which includes all conceptually omitted paths. This is a just-identified structural model, so its $\chi^2$ and degrees of freedom are equal to that of the final measurement model (that allows all structural variables to correlate with each other).

Figure 6.1 illustrates a situation in which the conceptual structural model contains two conceptually omitted paths. Figure 6.2 includes the two conceptually omitted paths as dashed arrows. The chi-square difference test between the models in Figures 6.1

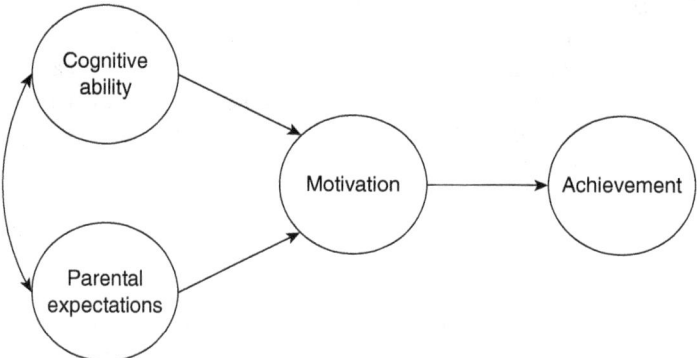

**Figure 6.1**   A conceptual model in which motivation completely mediates the pathways from cognitive ability and parental expectations to achievement. This model does not allow for direct effects from cognitive ability and parental expectations to achievement

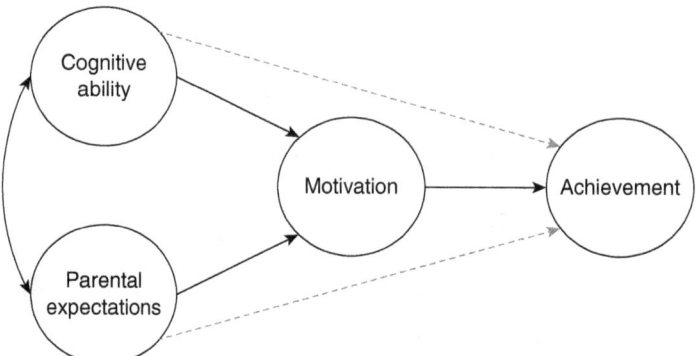

**Figure 6.2**   The conceptual model in which motivation completely mediates the pathways from cognitive ability and parental expectations to achievement. The dashed lines represent conceptually omitted paths. The just-identified structural model includes all paths in Figure 6.2

and 6.2 is a joint test of whether including the two paths improves the fit of the model (or deleting the two paths worsens the fit of the model), and therefore has 2 *df*. However, it is possible that the addition of one of the two paths improves the fit of the model and adding the second path does not. Therefore, we can use the just-identified model to evaluate each of the paths separately.

Why is it important to include conceptually omitted paths? The researcher believes that there is no direct effect from cognitive ability to academic achievement, after controlling for motivation. By not including the direct effect from cognitive ability to achievement, the standardised path coefficient for the effect of motivation on academic achievement is equivalent to the simple bivariate correlation between motivation and achievement. It does not control for cognitive ability. Without including the path from cognitive ability to achievement, the path from motivation to achievement could be spurious: it is possible that the entire effect of motivation on achievement could be explained by the relationship of both variables to cognitive ability. Estimating the conceptual model provides no information about whether motivation actually explains additional variance in academic achievement, over and above (or after accounting for) cognitive ability. Therefore, it is important to include paths whenever it is possible that there could be a causal effect (or a predictive pathway) (Heise, 1975).

How can we determine whether we can safely eliminate one of the conceptually omitted paths? Traditionally, researchers have relied upon tests of statistical significance. However, there are drawbacks to relying solely on tests of statistical significance. Remember that tests of statistical significance are influenced by sample size, and structural equation models generally require large sample sizes. Also, when evaluating more than one conceptually omitted path, it is appropriate to correct for the number of statistical significance tests (Mulaik, 2009). Finally, and perhaps most importantly, researchers should consider which is the graver error in their own research: including a path that is not necessary or deleting a path that is. Researchers who are more concerned about eliminating paths that might be necessary should consider setting a more liberal alpha (i.e. .10) for the tests of conceptually omitted paths. On the other hand, researchers who are testing many conceptually omitted paths have a large sample size and place great value on parsimony may wish to employ a more conservative alpha of .01 when evaluating conceptually omitted paths. In either case, we also consider the magnitude of the estimated path and the proportion of variance that the conceptually omitted path explains in the outcome variable. We recommend pairing the statistical test with an effect size measure such as the proportion of variance in the outcome variable ($R^2$) that is explained by the conceptually omitted path. Again, the interpretation of this $R^2$ measure is context dependent. In some research scenarios, a path that explains less than 1% of unique variance in the outcome variable would be regarded as trivial. In other research contexts, explaining .5% of unique variance in the outcome variable is of great practical import. Although the decision criteria may differ depending on the research context, it is essential to decide upon a defensible system a priori (prior to running your structural equation model analyses) and to fully describe this process in the methodology section of the research paper.

Given that the measurement model and the just-identified structural model are statistically equivalent, why do we fit both models? Fitting the just-identified structural model allows us to determine whether either of the conceptually omitted paths is necessary. We could compare the fit of the conceptual structural model to the fit of the measurement model using the $\chi^2$ difference test. If the fit of the conceptual model were not (statistically significantly) worse than the fit of the measurement model, then we would favour the conceptual structural model (because it is more parsimonious and fits equally well). However, if the fit of the conceptual model is worse than the fit of the measurement model, and if the conceptual model contains two or more conceptually omitted paths, we find ourselves in a bit of a quagmire. The $\chi^2$ difference test comparing the just-identified (or measurement) model to the conceptual model tells us that the just-identified model fits better than the conceptual model, but it does not tell us *why* the just-identified model fits better. In such a scenario, it is possible that eliminating only one of the multiple conceptually omitted paths causes the decrement in model fit. By estimating the just-identified structural model, we can separately evaluate each of the conceptually omitted paths. If the path coefficient for a conceptually omitted path is very small and not statistically significantly different from 0, then eliminating the path from the model generally increases $\chi^2$ by a negligible or very small amount. If the path coefficient for conceptually omitted path is more substantial and statistically significantly different from 0, then the model that excludes the conceptually omitted path has a higher $\chi^2$ than the model that includes the path. Therefore, fitting the just-identified structural model allows us to evaluate each of the conceptually omitted paths separately in addition to evaluating them jointly (using the $\chi^2$ difference test).

*Step 2*: Delete any of the conceptually omitted paths from step 1 that prove unnecessary and rerun the model. This model includes all conceptual paths plus all conceptually omitted paths that proved to be necessary from step 1. After evaluating whether the conceptually omitted paths can in fact be omitted from the conceptual model, we fit the conceptual model (with any necessary conceptually omitted paths included). If none of the theoretically deleted paths in the just-identified structural model are necessary, then we estimate the conceptual model. However, if including any of the conceptually omitted paths substantially improves the fit of our model and/or the paths are non-zero, we retain those paths in the model, and we estimate a variation of our conceptual model that includes the addition of the necessary conceptually omitted paths.

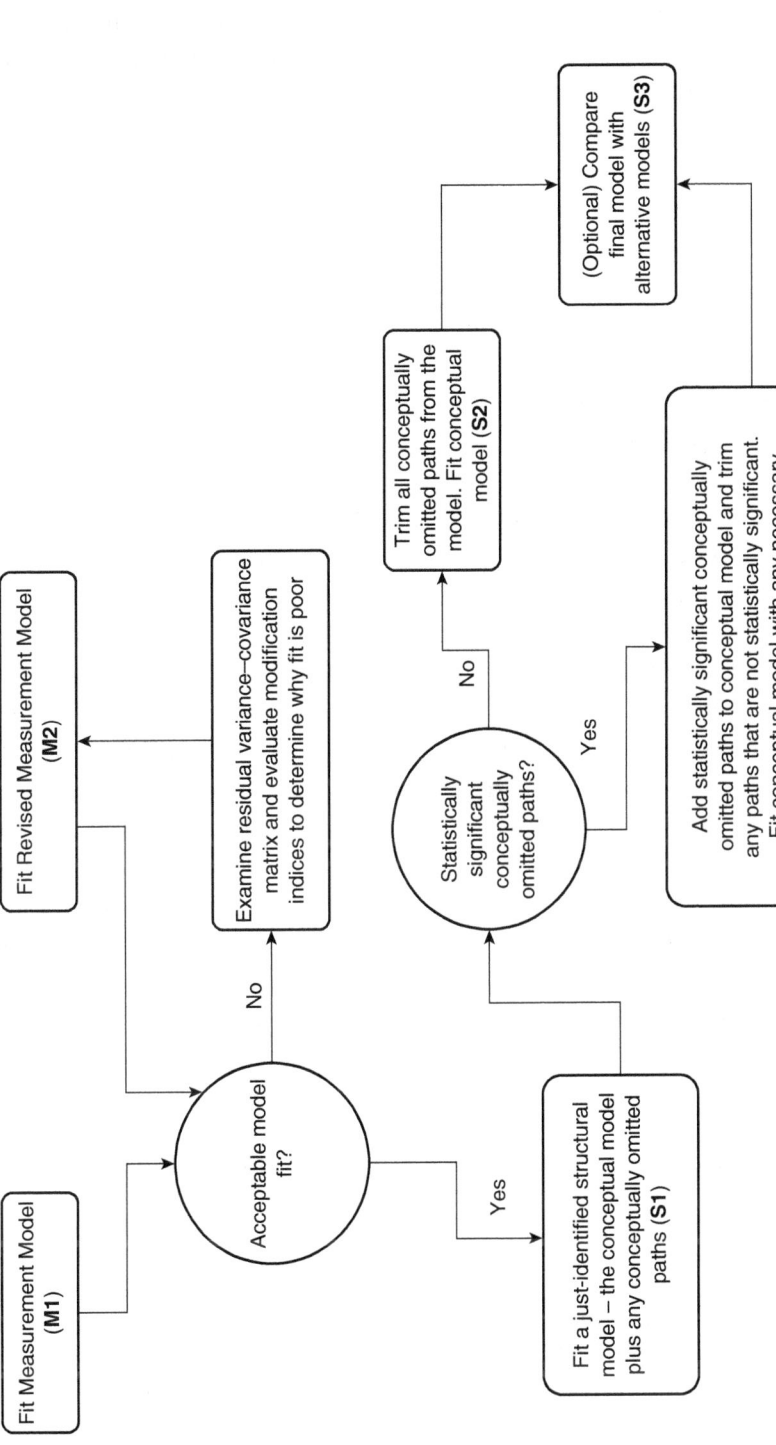

**Figure 6.3** Flowchart for model building steps with hybrid structural equation modelling

The following are the text contents of the flowchart:

Fit Measurement Model (**M1**)

Fit Revised Measurement Model (**M2**)

Examine residual variance–covariance matrix and evaluate modification indices to determine why fit is poor

Acceptable model fit?

No

Yes

Fit a just-identified structural model – the conceptual model plus any conceptually omitted paths (**S1**)

Statistically significant conceptually omitted paths?

No

Yes

Trim all conceptually omitted paths from the model. Fit conceptual model (**S2**)

Add statistically significant conceptually omitted paths to conceptual model and trim any paths that are not statistically significant. Fit conceptual model with any necessary conceptually omitted paths (**S2**)

(Optional) Compare final model with alternative models (**S3**)

*Step 3 (Optional): Fit competing conceptual models (if there are any).* Finally, we can compare our conceptual model to any other competing models or alternative models (that we have specified a priori). Alternatively, sometimes researchers choose to trim non-statistically significant, non-necessary conceptual paths. However, trimming is generally unnecessary, and it is inadvisable if you want to be able to compare your model to other researchers' estimates of the same model.

A flowchart in Figure 6.3 outlines the basic steps in the hybrid structural equation model building sequence.

## Example: mediation example using a structural equation model with latent variables

We illustrate the analysis and interpretation of a structural equation model using data from a study that examined the impact of a differentiated mathematics curricula on the achievement of grade 3 students (McCoach et al., 2014). The current analysis includes only the treatment students from the larger study. In this mediational model, prior ability and prior mathematics achievement predict performance on the classroom achievement; classroom achievement, prior mathematics achievement and cognitive ability predict subsequent mathematics achievement (see Figure 6.4). The most substantively interesting path in the model is the path from classroom mathematics achievement to standardised mathematics achievement at post-test. A direct effect from classroom math achievement to standardised math achievement would indicate that classroom performance on the math intervention units predicts subsequent standardised mathematics achievement (scores on the ITBS-PT [Iowa Test of Basic Skills post-test]), even after controlling for prior standardised mathematics achievement (ITBS-PR [Iowa Test of Basic Skills pretest]) and prior ability. The mediation example below is a latent variable (hybrid) structural equation model. Using the model building steps outlined earlier, we explore the following three research questions:

- *Research Questions 1 (RQ1):* How well do treatment students' ability and prior mathematics achievement predict their ITBS post-test mathematics scores? To answer this question, we examine both the direct and total effects of ability and ITBS on post-test mathematics achievement.
- *Research Question 2 (RQ2):* How well does classroom math achievement at post-test predict ITBS post-test mathematics scores, after controlling for prior standardised mathematics achievement (ITBS-PR) and ability? To answer this question, we examine the direct effect of the post-classroom assessment on post-test mathematics achievement.
- *Research Question 3 (RQ3):* Does classroom mathematics achievement mediate the effect of the prior standardised mathematics achievement (ITBS-PR) on post-standardised mathematics achievement (ITBS-PT)? To answer this question, we examine the indirect effect of pre-ITBS on post-ITBS.

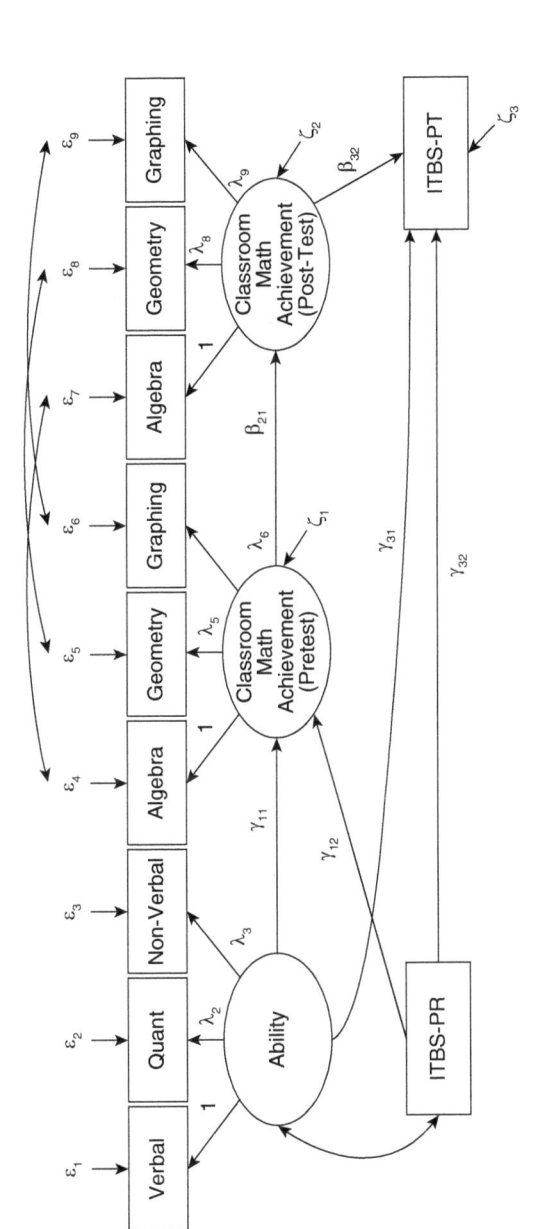

**Figure 6.4** The conceptual hybrid structural equation model for our example. Note that there are three conceptually omitted paths in the structural portion of the model: (1) the path from ITBS-PR to classroom post-test mathematics achievement, (2) the path from classroom pretest mathematics achievement to ITBS-PT and (3) the path from ability to classroom post-test mathematics achievement

Note. ITBS-PR = Iowa Test of Basic Skills pretest; ITBS-PT = Iowa Test of Basic Skills post-test.

## Data

The National Research Center on the Gifted and Talented provided the data for this example. The data include standardised mathematics achievement test data, unit test data and group cognitive ability scores for 897 treatment students who were part of a multisite cluster randomised trial study. Table 6.1 contains the list of variables and descriptive statistics for the current analysis. Table 6.2 contains the matrix of correlations among the variables, presented in standard lower matrix form. The ITBS, a norm-referenced standardised assessment of mathematics achievement, serves as the main outcome variable of interest. (For more information about the ITBS and the larger student data set, see McCoach et al., 2014). We conducted all analyses using Mplus version 8 (Muthén & Muthén, 1998–2017).

**Table 6.1** Descriptive statistics for the observed variables in Figure 6.4

| Variable | N | Mean | Standard Deviation |
|---|---|---|---|
| VERB | 897 | 108.99 | 14.23 |
| QUAN | 897 | 108.76 | 13.78 |
| NONV | 897 | 107.22 | 14.15 |
| ITBS-PR | 897 | 186.83 | 21.25 |
| ITBS-PT | 897 | 209.38 | 22.23 |
| ALG-PR | 897 | 8.20 | 5.46 |
| ALG-PT | 897 | 15.76 | 5.42 |
| GEO-PR | 897 | 13.95 | 5.37 |
| GEO-PT | 897 | 23.10 | 4.44 |
| GRA-PR | 897 | 5.55 | 2.71 |
| GRA-PT | 897 | 10.13 | 2.97 |

Note. VERB = Verbal scores on Ability test; QUAN = Quantitative scores on Ability test; NONV = non-verbal scores on Ability test; ITBS-PR (exogenous), ITBS-PT (endogenous) = Iowa Test of Basic Skills pretest and post-test scores, respectively; ALG-PR, ALG-PT = Algebra scores on a pre- and post-classroom assessment, respectively; GEO-PR, GEO-PT = Geometry scores on a pre- and post-classroom assessment, respectively; and GRA-PR, GRA-PT = Graphing scores on a pre- and post-classroom assessment, respectively.

## Measurement model

Our model includes the following latent variables: Ability, Classroom Math Achievement (at pretest) and Classroom Math Achievement (at post-test). Ability is a latent variable with three manifest indicators: (1) verbal CogAT scores (Verbal), (2) quantitative CogAT scores (Quant) and (3) non-verbal CogAT scores (Non-Verbal). Each of the Classroom Math Achievement latent variables also contains three

**Table 6.2** Correlation matrix for the observed variables in Figure 6.4

| Variable | | | | | | | | | | |
|---|---|---|---|---|---|---|---|---|---|---|
| VERB | 1.00 | | | | | | | | | |
| QUAN | .71 | 1.00 | | | | | | | | |
| NONV | .51 | .67 | 1.00 | | | | | | | |
| ITBS-PR | .56 | .65 | .53 | 1.00 | | | | | | |
| ITBS-PT | .58 | .66 | .57 | .66 | 1.00 | | | | | |
| ALG-PR | .44 | .51 | .48 | .50 | .52 | 1.00 | | | | |
| ALG-PT | .44 | .57 | .51 | .56 | .61 | .62 | 1.00 | | | |
| GEO-PR | .34 | .44 | .43 | .45 | .39 | .42 | .37 | 1.00 | | |
| GEO-PT | .49 | .60 | .54 | .53 | .60 | .50 | .62 | .50 | 1.00 | |
| GRA-PR | .38 | .47 | .39 | .44 | .40 | .44 | .46 | .48 | .45 | 1.00 |
| GRA-PT | .41 | .53 | .45 | .46 | .51 | .46 | .60 | .37 | .56 | .55 | 1.00 |

*Note.* The 1 on the diagonal is because the same variable that is represented on the row is represented in the corresponding column. That is, the correlation of a variable with itself is 1. VERB = Verbal scores on Ability test; QUAN = Quantitative scores on Ability test; NONV = non-verbal scores on Ability test; ITBS-PR (exogenous), ITBS-PT (endogenous) = Iowa Test of Basic Skills pretest and post-test scores, respectively; ALG-PR, ALG-PT = Algebra scores on a pre- and post-classroom assessment, respectively; GEO-PR, GEO-PT = Geometry scores on a pre- and post-classroom assessment, respectively; and GRA-PR, GRA-PT = Graphing scores on a pre- and post-classroom assessment, respectively.

indicators (three classroom tests at pretest and post-test). As part of the study, students completed three instructional units: (1) Algebra, (2) Geometry and Measurement and (3) Graphing and Data Analysis. Before each of the three units, students completed a pretest; after each of the three units, students completed a post-test, which was identical to the pretest. The three criterion-referenced mathematics unit pretests (Algebra, Geometry and Graphing) served as the indicators for the Pre-Mathematics Achievement latent variable; the same three unit tests administered at post-test served as the indicators for the Post-Mathematics Achievement latent variable. The measurement model includes three error covariances between each unit assessment at pretest and the identical assessment at post-test. The correlated errors indicate that the same unit tests, administered at pretest and at post-test are likely to be more correlated with each other than would be predicted by the standard CFA model. These correlations were specified a priori and are included in the initial measurement model. See Little (2013) or Grimm et al. (2016) for more details about longitudinal CFA.

Students' scores on the ITBS standardised mathematics achievement test were not available at the item level. Instead, our data includes students' composite standardised mathematics achievement scores, administered in the fall (at pretest) and in the spring (at post-test). These ITBS scores are observed, not latent; however, they are conceptual variables of interest. All the other observed variables in this model are indicators of latent variables, and the latent constructs represent conceptual variables of interest. The measurement and structural models include all these variables.

## Conceptual model

The conceptual model, displayed in Figure 6.4 depicts the hypothesised structural paths between variables.

## Model building steps

As outlined earlier, the model building process for a hybrid structural equation model involves (a) fitting and testing the measurement model (M1); (b) if necessary, revising the measurement model (M2); (c) estimating the just-identified structural model (which is the conceptual model with conceptually omitted paths added) (S1); (d) estimating the conceptual model that includes any necessary conceptually omitted paths (S2); and (e) (optional S3) comparing the model from step 4 (S2) to any alternative models. Although conceptually we distinguish between the measurement and structural portions of the model, the entire hybrid structural equation model is estimated simultaneously.

## Measurement model

Figure 6.5 depicts the measurement model. (*Note:* The parameters in the model are omitted for simplicity.)

The measurement model maps the observed measures onto the theoretical constructs. In addition to the factors and indicators that are included in a traditional CFA model, the measurement model includes all manifest structural variables of interest. In the measurement model, the only unidirectional paths are from the latent variables to the indicators (observed variables). There are five structural variables in this model: ability, classroom math achievement (pretest), classroom math achievement (post-test), ITBS-PR math score and ITBS-PT math score. Three of the five structural variables are latent: ability, classroom math achievement (pretest) and classroom math achievement (post-test). Two of the five structural variables are manifest (observed) variables: ITBS-PR math and ITBS-PT math. In the measurement model, all structural variables are exogenous; we allow all structural variables to correlate with each other.

In this example, the measurement model contains three latent variables, nine endogenous observed variables and two exogenous observed variables. There are two

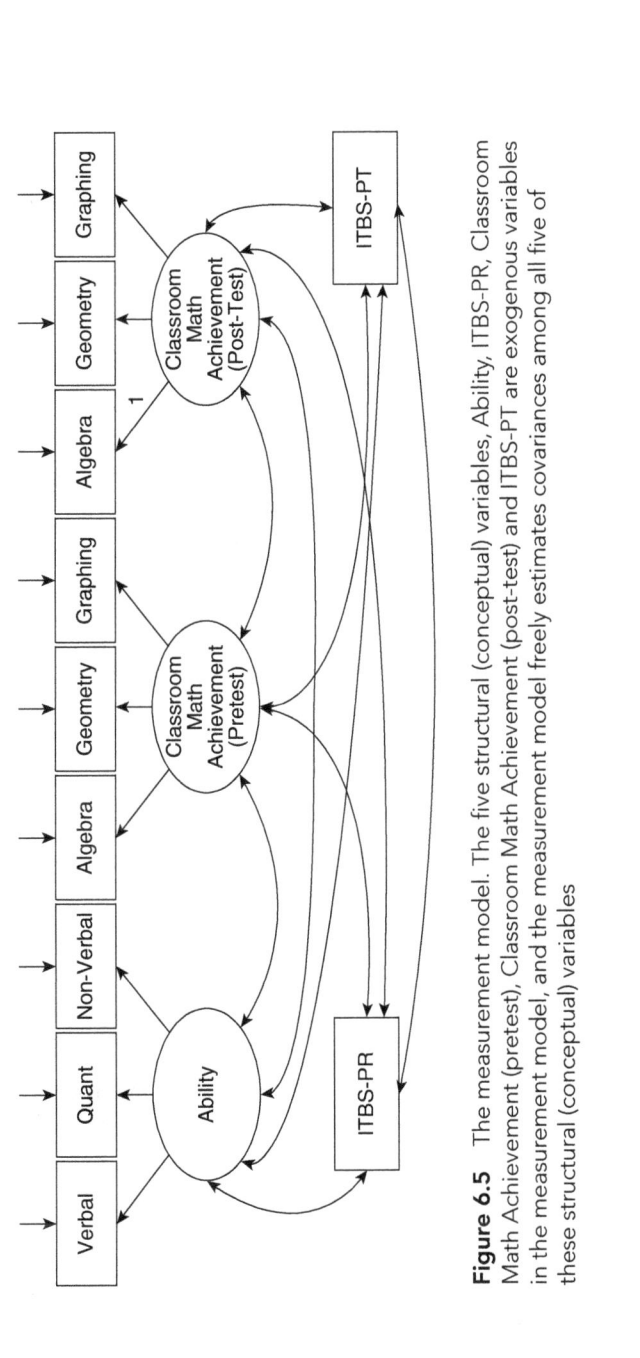

**Figure 6.5** The measurement model. The five structural (conceptual) variables, Ability, ITBS-PR, Classroom Math Achievement (pretest), Classroom Math Achievement (post-test) and ITBS-PT are exogenous variables in the measurement model, and the measurement model freely estimates covariances among all five of these structural (conceptual) variables

different types of observed variables in the model: indicators ($p = 9$) and exogenous observed structural variables ($s = 2$). There are no endogenous observed structural variables in the measurement model. Because the model contains 11 observed variables ($v$), there are $(v(v + 1))/2 = (11(11 + 1))/2 = 66$ unique elements in the variance–covariance matrix. (The number of knowns is 66.)

How many free parameters does the measurement model contain? Our measurement model contains 33 free parameters (i.e. unknowns): nine measurement error variances (one for each of the endogenous indicators); three error covariances; six pattern coefficients or factor loadings (three loadings are marker variables: their paths are fixed to one); 10 covariances or correlations between the three latent structural variables and the two observed structural variables; and five variances (i.e. three for the three latent exogenous variables and two for the two exogenous observed variables). Therefore, the degrees of freedom in the model = knowns – unknowns = 66 – 33 = 33 *df*. The model is recursive and has positive degrees of freedom, so it is (over)identified. Notice that all 33 *df* come from the measurement portion of the model: the structural portion of the model is just-identified.

Table 6.3 includes fit statistics for the models discussed below. Table 6.4 contains statistics and unstandardised and standardised results for the measurement model.

**Table 6.3** Model fit information for the measurement and structural hybrid models

| Model | $\chi^2$ (*df*) | CFI/TLI | SRMR | RMSEA 90% CI |
|---|---|---|---|---|
| M1: Measurement | 122.58 (33) | 0.98/0.97 | 0.02 | 0.06 [0.05, 0.07] |
| S1: Just-identified structural | 122.58 (33) | 0.98/0.97 | 0.02 | 0.06 [0.05, 0.07] |
| S2: Conceptual + Necessary omitted | 125.71 (35) | 0.98/0.97 | 0.02 | 0.05 [0.04, 0.06] |
| S3: Original conceptual | 149.48 (36) | 0.98/0.97 | 0.03 | 0.06 [0.05, 0.07] |

*Note.* $\chi^2$ is the chi-square test of model fit. Degrees of freedom (*df*) in parenthesis. All $\chi^2$ in the table are statistically significant ($p < .001$). CFI = comparative fit index; TLI = Tucker–Lewis index; SRMR = standardised root mean square residual; RMSEA = root mean square error of approximation; CI = confidence interval.

**Table 6.4** Measurement model results (step M1)

| | Unstandardised | | | Standardised[a] | | |
|---|---|---|---|---|---|---|
| | Est. | SE | p | Est. | SE | p |
| Pattern coefficients | | | | | | |
| CLASS-PR | | | | | | |
| Algebra | 1.00 | 0.00 | — | 0.71 | 0.02 | <.01 |
| Geometry | 0.87 | 0.06 | <.01 | 0.63 | 0.03 | <.01 |
| Graphing | 0.46 | 0.03 | <.01 | 0.66 | 0.02 | <.01 |

| | Unstandardised | | | Standardised[a] | | |
|---|---|---|---|---|---|---|
| | Est. | SE | p | Est. | SE | p |
| CLASS-PT | | | | | | |
| Algebra | 1.00 | 0.00 | — | 0.80 | 0.02 | <.01 |
| Geometry | 0.82 | 0.03 | <.01 | 0.79 | 0.02 | <.01 |
| Graphing | 0.49 | 0.02 | <.01 | 0.72 | 0.02 | <.01 |
| Ability | | | | | | |
| Verbal | 1.00 | 0.00 | — | 0.76 | 0.02 | <.01 |
| Quantitative | 1.16 | 0.04 | <.01 | 0.91 | 0.01 | <.01 |
| Non-verbal | 0.96 | 0.04 | <.01 | 0.73 | 0.02 | <.01 |
| Variances | | | | | | |
| CLASS-PR | 451.13 | 21.30 | <.01 | 1.00 | 0.00 | — |
| CLASS-PT | 493.74 | 23.31 | <.01 | 1.00 | 0.00 | — |
| Ability | 15.00 | 1.36 | <.01 | 1.00 | 0.00 | — |
| ITBS-PR | 18.54 | 1.37 | <.01 | 1.00 | 0.00 | — |
| ITBS-PT | 117.24 | 9.00 | <.01 | 1.00 | 0.00 | — |
| Covariances | | | | | | |
| Ability ↔ CLASS-PR | 33.39 | 2.52 | <.01 | 0.80 | 0.02 | <.01 |
| Ability ↔CLASS-PT | 37.89 | 2.59 | <.01 | 0.81 | 0.02 | <.01 |
| CLASS-PR ↔ CLASS-PT | 14.25 | 1.11 | <.01 | 0.86 | 0.02 | <.01 |
| Ability ↔ ITBS-PR | 165.44 | 10.91 | <.01 | 0.72 | 0.02 | <.01 |
| Ability ↔ITBS-PT | 177.39 | 11.57 | <.01 | 0.74 | 0.02 | <.01 |
| CLASS-PR ↔ ITBS-PR | 57.52 | 4.18 | <.01 | 0.70 | 0.02 | <.01 |
| CLASS-PR ↔ITBS-PT | 57.48 | 4.39 | <.01 | 0.67 | 0.03 | <.01 |
| CLASS-PT ↔ITBS-PR | 61.70 | 4.21 | <.01 | 0.68 | 0.02 | <.01 |
| CLASS-PT ↔ITBS-PT | 71.65 | 4.57 | <.01 | 0.75 | 0.02 | <.01 |
| ITBS-PR ↔ITBS-PT | 311.91 | 18.89 | <.01 | 0.66 | 0.02 | <.01 |
| Error variances | | | | | | |
| Verbal | 84.98 | 4.69 | <.01 | 0.42 | 0.03 | <.01 |
| Quantitative | 31.66 | 3.35 | <.01 | 0.17 | 0.02 | <.01 |
| Non-verbal | 92.71 | 5.04 | <.01 | 0.46 | 0.03 | <.01 |
| Algebra PR | 14.75 | 0.92 | <.01 | 0.50 | 0.03 | <.01 |
| Algebra PT | 10.54 | 0.70 | <.01 | 0.36 | 0.03 | <.01 |
| Geometry PR | 17.38 | 0.99 | <.01 | 0.61 | 0.03 | <.01 |
| Geometry PT | 7.41 | 0.47 | <.01 | 0.37 | 0.03 | <.01 |
| Graphing PR | 4.18 | 0.24 | <.01 | 0.57 | 0.03 | <.01 |
| Graphing PT | 4.29 | 0.24 | <.01 | 0.49 | 0.03 | <.01 |

*(Continued)*

**Table 6.4** (Continued)

|  | Unstandardised | | | Standardised[a] | | |
|---|---|---|---|---|---|---|
|  | **Est.** | *SE* | *p* | **Est.** | *SE* | *p* |
| Error covariances | | | | | | |
| Algebra PR ↔Algebra PT | 3.72 | 0.58 | <.01 | 0.30 | 0.04 | <.01 |
| Geometry PR ↔ Geometry PT | 2.10 | 0.48 | <.01 | 0.19 | 0.04 | <.01 |
| Graphing PR ↔ Graphing PT | 1.21 | 0.18 | <.01 | 0.29 | 0.04 | <.01 |

*Note.* Est. = estimate; *SE* = standard error; CLASS-PR = classroom math achievement (pretest); CLASS-PT = classroom math achievement (post-test); ITBS = Iowa Test of Basic Skills. The double-headed arrow represents a correlation.

[a]STDYX standardisation was used.

The chi-square test of model fit is statistically significant ($\chi^2$ = 122.58 with 33 *df*): the model-implied variance–covariance matrix does not adequately reproduce the variance–covariance model, indicating some degree of model–data misfit. Given that chi-square is notoriously sensitive to sample size (Bentler, 1990; Bentler & Bonett, 1980), it is helpful to examine additional fit indices to determine the degree of model misfit, especially in large samples. The incremental fit indices (CFI and TLI) for the measurement model are above .97. The SRMR is 0.02 and the RMSEA was 0.06 with a 90% confidence interval of [0.05, 0.07]. The fit statistics for the measurement model exceed the traditional benchmarks for adequate model fit: CFI/TLI > .95 (Hu & Bentler, 1999), SRMR < 0.08 (Hu & Bentler, 1999) and RMSEA < 0.08 (MacCallum et al., 1996).

Table 6.4 contains the results of the measurement model. Generally, we report the standardised pattern (path) coefficients in a measurement model. For unidimensional items, the square of the pattern coefficient represents the proportion of variance in the indicator explained by the factor. For example, in Table 6.4, ability explains 83% of the variance in the quantitative CogAT ($.91^2$), 58% of the variance in the verbal CogAT score ($.76^2$) and 54% of the variance in the non-verbal CogAT score ($.73^2$). The class pre-unit test latent variable explains 39.7% of the variance in geometry scores prior to instruction ($.63^2$), but the class unit post-test latent variable explains 62.4% of the variance in Geometry scores after instruction ($.79^2$).

The correlations among the structural variables in the model (in the covariances section under standardised results) are quite high, ranging from .66 to .86. The highest correlation is between the class math achievement latent variable (CLASS) at pretest and the class math achievement variable (CLASS) at post-test. Given that the two latent variables are measured by the same set of three unit tests at two different time points (prior to instruction and after instruction), this makes sense. The correlation between the ability latent variable and the mathematics pre-unit and post-unit class achievement variables are above .80, suggesting that students who have higher ability tend to have higher class mathematics achievement, both prior to and after receiving instruction.

The error variance for each of the unit tests represents the variance in the test that is not accounted for by the CLASS factor. All the error variances are positive (as they should be, given that they are variances) and statistically significantly different from 0, which is good. This means that the CLASS factor does not completely explain the variance in any of the unit tests. A factor that completely predicts one of the observed variables suggests complete redundancy between the factor and the observed variable. Such a finding would cast doubt on the appropriateness of the latent variable model for capturing the latent construct of interest. The correlations among the measurement errors for the matched pre and post-unit tests were relatively small: they ranged from .19 (for Geometry) to .30 (for Algebra). These correlated errors indicate that the covariance between the unit test at pretest and the same unit test at post-test cannot be completely explained by the factor structure. The measurement model did not require any respecifications.

After evaluating the measurement model, we estimated the series of hybrid structural models, following the steps outlined above: (1) the just-identified model, (2) the conceptual model with any necessary conceptually omitted paths identified in step 1 included and (3) the original conceptual model. Tables 6.5, 6.6 and 6.7 illustrate the model building process. Figure 6.4 illustrates the hypothesised structural relationships or paths associated with each of the three research questions. (Remember, a hybrid structural equation model simultaneously estimates the measurement and structural models.)

## Just-identified structural model

Figure 6.6 depicts the just-identified structural model. The just-identified structural model contains one linkage between each of the structural variables, even if some of these linkages are not part of the conceptual model. The conceptually omitted paths are ones hypothesised to be unnecessary. In Figure 6.6, we denote the conceptually omitted (deleted) paths with a *d*. In step 1, our goal is to determine which (if any) of the conceptually omitted (deleted) paths we should include when estimating the conceptual model.

Table 6.3 contains the fit statistics for the just-identified structural model in the second row (model labelled, S1). The fit of the just-identified structural model is identical to the fit of the final, full-measurement model because both models contain a linkage between each pair of structural variables. However, in the measurement model, all of these linkages are bivariate correlations, whereas in the structural model, some of these linkages are paths.

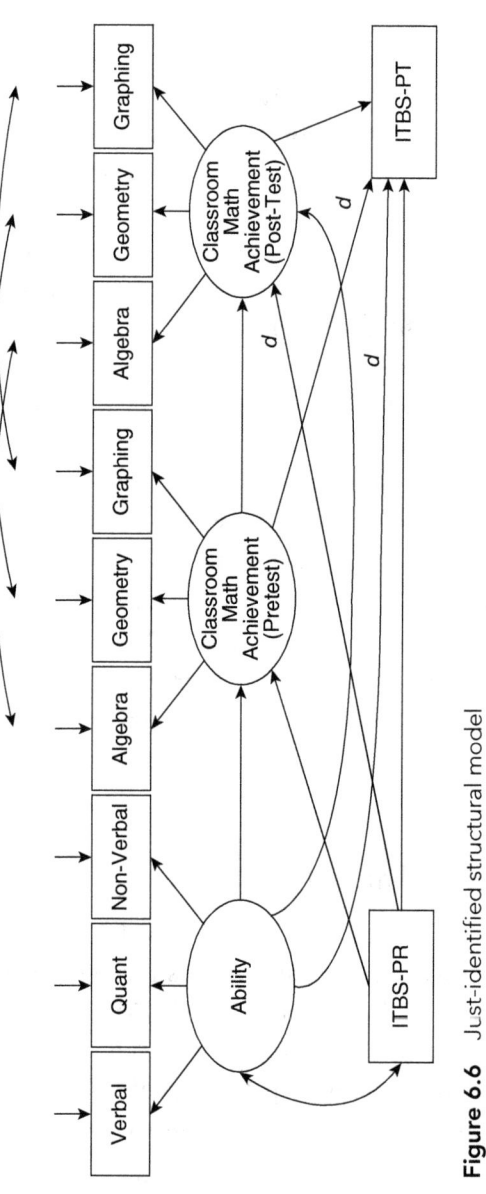

**Figure 6.6** Just-identified structural model

The left-hand columns of Table 6.5 report the results of the just-identified structural model. Table 6.6 documents our decisions about whether to delete the three paths that were not included in the original conceptual model. As seen in Tables 6.5 and 6.6, the direct path from ability to post-test classroom math achievement is statistically significant ($p < .001$) and has a substantial path coefficient (.343). Even after controlling for pretest class math achievement, students with higher ability tend to have higher post-test math achievement. In addition, the direct effect of pretest CLASS scores on post-test CLASS scores is substantially lower in the model that included ability.

**Table 6.5** Standardised results for the just-identified structural model and conceptual + necessary conceptually omitted paths model

| | Just-Identified Structural Model | | | Conceptual + Necessary Conceptually Omitted Paths Model | | |
|---|---|---|---|---|---|---|
| | Est. | SE | p | Est. | Lower 2.5 | Upper 2.5 |
| Structural paths | | | | | | |
| Ability → CLASS-PR | 0.61 | 0.05 | <.01 | 0.60 | 0.51 | 0.70 |
| **Ability → CLASS-PT** | **0.34** | **0.06** | **<.01** | **0.36** | **0.23** | **0.48** |
| CLASS-PR →CLASS-PT | 0.55 | 0.07 | <.01 | 0.57 | 0.45 | 0.69 |
| Ability → ITBS-PT | 0.30 | 0.06 | <.01 | 0.28 | 0.16 | 0.40 |
| ITBS-PR → CLASS-PR | 0.26 | 0.05 | <.01 | 0.28 | 0.18 | 0.37 |
| CLASS-PR → ITBS-PT | −0.11 | 0.08 | .17 | | | |
| ITBS-PR → CLASS-PT | 0.04 | 0.04 | .30 | | | |
| CLASS-PT → ITBS-PT | 0.45 | 0.07 | <.01 | 0.38 | 0.27 | 0.49 |
| ITBS-PR → ITBS-PT | 0.22 | 0.04 | <.01 | 0.21 | 0.13 | 0.27 |
| Exogenous covariances | | | | | | |
| Ability ↔ ITBS-PR | 0.72 | 0.02 | <.01 | 0.72 | 0.69 | 0.76 |
| Error covariances | | | | | | |
| Algebra PR ↔ Algebra PT | 0.30 | 0.04 | <.01 | 0.19 | 0.22 | 0.37 |
| Geometry PR ↔ Geometry PT | 0.19 | 0.04 | <.01 | 0.19 | 0.11 | 0.27 |
| Graphing PR ↔ Graphing PT | 0.29 | 0.04 | <.01 | 0.29 | 0.22 | 0.35 |

*Note.* Est. = estimate; SE = standard error; CLASS-PR = classroom math achievement (pretest); CLASS-PT = classroom math achievement (post-test); ITBS = Iowa Test of Basic Skills. The double-headed arrow represents a correlation.

**Table 6.6** Table of conceptually omitted paths

| Path | Statistically Significant | Discard |
|---|---|---|
| Ability → CLASS-PT | Yes | No |
| ITBS-PR → CLASS-PT | No | Yes |
| CLASS-PT → ITBS-PT | No | Yes |

*Note.* The single-headed arrow is used to represent the path from one variable to another. The correct vernacular, for example, Ability CLASS-PT would be that classroom math achievement post-test is regressed on ability or that ability predicts classroom math achievement post-test. The nomenclature used throughout the current example is simply the effect of variable 1 on variable 2.

## Conceptual + necessary conceptually omitted paths model

Next, we estimated the conceptual + necessary conceptually omitted paths model (labelled S3 in Table 6.3), shown in Figure 6.7. The conceptual + necessary conceptually omitted paths model evaluates the research questions of interest, including the conceptually omitted paths that proved necessary, even if they were not part of the original conceptual model. In this model, the additional conceptually omitted path ($\gamma_5$) represents the direct effect of ability on classroom math achievement post-test. The right-hand columns of Table 6.5 contain the standardised results from fitting this model, and Table 6.7 contains the total, direct and indirect effects of interests. To evaluate the indirect effects, we empirically derived the 95% confidence intervals using bootstrapping. (If the 95% bootstrap CI does not contain 0, the estimate of the indirect effect is statistically significant.)

The conceptual + necessary conceptually omitted paths model fit as well as the just-identified model. Eliminating the two paths increased the chi-square from 122.58 with 33 *df* to 125.71 with 35 *df*. (Remember, the simpler model always has fewer estimated parameters, resulting in more degrees of freedom.) The chi-square difference test was not statistically significant ($\Delta\chi^2 = 3.13(2)$, $p = .21$), so the fit of the conceptual model with the additional path from ability to post-unit mathematics achievement is no worse than (just as good as) the fit of the just-identified model. In such a situation, we prefer the more parsimonious model because it fits as well as the more complex model using fewer parameters. Therefore, we favour the conceptual model with the additional path from ability to post-unit mathematics achievement over the just-identified model.

Next, we compared this modified conceptual model to our original conceptual model. These two models differ by 1 *df*: the modified conceptual model includes the path from ability to post-math class achievement whereas the original conceptual model does not. The $\chi^2$ difference test between these two models was statistically significant ($\Delta\chi^2 (1) = 26.90$, $p < .001$). The model $\chi^2$ decreased by 26.90 points after adding the path from ability to post-test mathematics achievement, indicating better fit of the model that includes the path from ability to post-test class mathematics achievement. Therefore, our preferred and final model is the 'revised' conceptual model that includes the additional path from ability to post-test mathematics achievement. Conceptually, this path indicates that ability helps to explain post-test classroom mathematics achievement, even after controlling for pretest mathematics achievement. Higher ability students have better classroom mathematics achievement, and this effect cannot be explained by mathematics pre-unit achievement or prior standardised mathematics achievement.

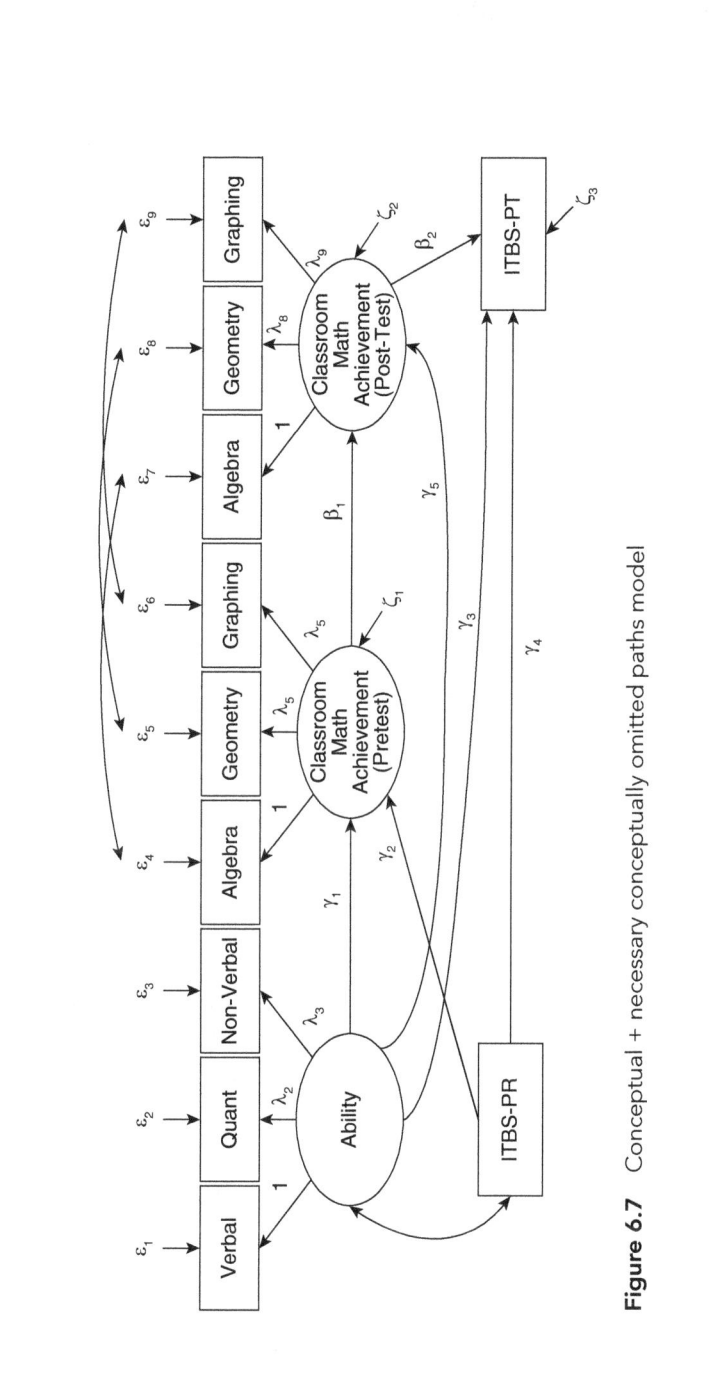

**Figure 6.7** Conceptual + necessary conceptually omitted paths model

## Direct, indirect and total effects

The total effect of pretest standardised mathematics achievement (ITBS-PR) on post-test standardised mathematics achievement (ITBS-PT) was .27, indicating that after controlling for ability, every standard deviation unit increase on standardised mathematics pretest scores (ITBS-PR) predicts a 0.27 *SD* increase in ITBS-PT scores. The direct effect of ITBS-PR on ITBS-PR was 0.21. The indirect effect of ITBS-PR on ITBS-PR (.06) is the product of the direct effect of pre-achievement on the class pretest (ITBS-PR → CLASS-PR) (.28), the direct effect of the class pretest on the class post-test (CLASS-PR → CLASS-PT) (.57) and the direct effect of the class post-test on the ITBS post-test (CLASS-PT → ITBS-PT) (.38). (Table 6.5 contains these standardised path coefficients.)

**Table 6.7**  Standardised total, total indirect, specific indirect and direct effects for Model S2 (the conceptual + significant deleted paths model) with 95% bootstrap CIs

|  | Est. | Lower 2.5 | Upper 2.5 |
|---|---|---|---|
| ITBS-PR to ITBS-PT |  |  |  |
| Total | 0.27 | 0.19 | 0.34 |
| Total indirect | 0.06 | 0.03 | 0.10 |
| Specific indirect |  |  |  |
| ITBS-PR → CLASS-PR → CLASS-PT → ITBS-PT | 0.06 | 0.03 | 0.10 |
| Direct |  |  |  |
| ITBS-PR → ITBS-PT | 0.21 | 0.13 | 0.27 |
| Ability to ITBS-PT |  |  |  |
| Total | 0.55 | 0.48 | 0.62 |
| Total indirect | 0.27 | 0.19 | 0.36 |
| Specific indirect |  |  |  |
| Ability → CLASS-PR → CLASS-PT → ITBS-PT | 0.14 | 0.09 | 0.19 |
| Ability → CLASS-PT → ITBS-PT | 0.13 | 0.08 | 0.22 |
| Direct |  |  |  |
| Ability → ITBS-PT | 0.28 | 0.16 | 0.40 |
| CLASS-PT to ITBS-PT |  |  |  |
| Total | 0.38 | 0.27 | 0.49 |
| Total indirect | 0.00 | 0.00 | 0.00 |
| Direct |  |  |  |
| CLASS-PT → ITBS-PT | 0.38 | 0.27 | 0.49 |

*Note.* CIs = confidence intervals; Est. = estimate.

***$p < .01$. **$p < .05$. *$p < .10$.

Using the parameter estimates from the final model, we can interpret the direct, indirect and total effects. A specific indirect effect is an indirect effect that passes through a particular sequence of variables. For instance, in our example, there is an indirect effect of ability on math class post-test scores on ITBS post-test scores. There can be multiple sets of indirect effects of one variable on another. The total of the indirect effects through all pathways is the total indirect effect. Summing the total indirect effect and the direct effect produces the total effect. Table 6.7 contains the direct, specific indirect, total indirect and total effects for our final model.

Returning to our original research questions, how well does classroom math achievement at post-test (CLASS-PT) predict ITBS post-test mathematics scores (ITBS-PT) after controlling for ITBS pretest mathematics scores and ability? The direct effect of CLASS-PT on ITBS-PT (.38) indicates that even after controlling for prior ability and prior achievement CLASS-PT positively predicts post-test stand-ardised mathematics achievement, suggesting that skills students developed in the mathematics unit may actually enhance their performance on subsequent mathematics achievement tests. At the very least, we can say that students who do better on the post-math assessments tend to do better on the subsequent mathematics standardised achievement test, even after controlling for prior ability and achievement.

How well does ability predict students' ITBS post-test mathematics scores, after controlling for prior mathematics achievement? Ability positively predicts standard-ised post-test mathematics achievement (ITBS-PT): the standardised total effect of ability on post math achievement is .55. After controlling for prior achievement, ability has both a direct effect (.28) and specific indirect pathways through the class math pretest and post-test assessments (0.13 and 0.14, respectively). Classroom math achievement at pretest and post-test partially mediate the effect of ability on subse-quent standardised mathematics achievement. Every unit increase in ability results in an expected 0.55 increase in ITBS-PT scores, after controlling for prior standardised mathematics achievement (ITBS-PR). About half of that effect is direct, and about half of the effect is mediated by pretest, post-test and ITBS-PR.

Finally, does classroom mathematics achievement mediate the effect of the pre-ITBS mathematics achievement on post-ITBS mathematics achievement? The stand-ardised indirect effect of pre-ITBS on post-ITBS through the classroom assessments is small (.06), but it is statistically significantly different from 0. After controlling for ability, students with higher pre-ITBS mathematics achievement do better on the class pretest (.28). The direct effect of class pretest on class post-test is .55. The direct effect of class post-test on ITBS post-test math scores is .38. The indirect effect of pre-ITBS on post-ITBS through the class math tests is .28 * .55 * .38 = .06. So, class unit math tests do partially mediate the effect of the pre-ITBS math on post-ITBS math. However, the direct effect of pre-ITBS on post-ITBS is .21. After controlling for ability, students who do better on the pre-ITBS tend to do better on the class math assessments, and they tend to do better on the post-ITBS. Having higher pre-ITBS math achievement does predict better class achievement, which does in turn predict higher post-ITBS math achievement.

This concludes our overview of SEM. Although Chapters 4 to 6 introduced many fundamental concepts, our book, is, by design, fairly introductory. For more thorough introductions to the theory and practice of SEM, we recommend Bollen (1989), Kline (2015) and Hoyle (2012).

The final section of this book introduces growth curve modelling as a method to model systematic change over time. Growth curve models can be fit using either SEM or MLM. Therefore, we present growth curve models in both frameworks. Comparing the application of SEM and MLM to growth curve models reinforces many of the concepts that we have introduced thus far and also helps to illustrate the fundamental linkages between MLM and SEM. Therefore, the final two chapters on growth modelling represent in some sense, the grand finale of our short tour of modern modelling techniques.

### Chapter Summary

- This chapter describes a two-step strategy for fitting latent variable structural models. First, assess the adequacy of the measurement model. Afterwards, fit a sequence of structural models, beginning with the just-identified model (see Figure 6.3).
- A detailed example illustrates how to fit and interpret a structural equation model. The example is a mediational model where prior ability and prior mathematics achievement predict performance on the classroom achievement; classroom achievement, prior mathematics achievement and cognitive ability predict subsequent mathematics achievement (see Figure 6.4).

## Further Reading

Bollen, K. A. (1989). *Structural equations with latent variables*. Wiley.
This is a classic book on structural equation modelling. Even though the book is a few decades old, the conceptual foundations of structural equation modelling presented in the book remain relevant. The book provides a comprehensive and technical, as well as comprehensible, introduction to structural equation modelling.

MacCallum, R. C., Wegener, D. T., Uchino, B. N., & Fabrigar, L. R. (1993). The problem of equivalent models in applications of covariance structure analysis. *Psychological Bulletin, 114*, 185–199.

This article is a nice extension to this chapter as it covers an important methodological issue with structural equation modelling in practice. Mainly, the authors discuss the problem of equivalent models in structural equation modelling and provide recommendations about how to confront the issue in practice.

# 7

# LONGITUDINAL GROWTH CURVE MODELS IN MLM AND SEM

## Chapter Overview

Conceptual introduction to growth modelling ......................................... 148

Multilevel model specification for individual growth models .................. 154

Estimating linear growth from an SEM perspective................................. 159

Incorporating time-varying covariates ..................................................... 164

Piecewise linear growth models................................................................ 169

Polynomial growth models........................................................................ 174

Fitting higher order polynomial models ................................................... 180

Building growth models ............................................................................ 183

Further Reading ........................................................................................ 186

Baltes and Nesselroade (1979) defined longitudinal research as 'repeated, time-ordered observation of an individual or individuals with the goal of identifying processes and causes of intra-individual change and of interindividual patterns of intra-individual change' (p. 7). *Longitudinal modelling* is a very broad, general term for a wide variety of statistical models to analyse observations collected on the same units (i.e. individuals) over time. These models include autoregressive or Markov chain models, latent transition models, individual **growth curve models**, latent change score models, dynamic models, time-series models, survival models and growth mixture models, just to name a few (McArdle & Nesselroade, 2014; Muthén & Muthén, 1998–2017; Nesselroade & Baltes, 1979; Singer & Willett, 2003; Willett, 1997). Each of these longitudinal models focuses on different aspects of the change process and makes very different assumptions about the underlying mechanisms that influence the stability of observations across time.

However, selecting an appropriate model is not simply a statistical decision: it requires attention to both theoretical and practical research contexts. Hence, Jack McArdle (2009) aptly asked the essential question: 'what is your model for change?' (p. 579). Determining the correct model for the analysis of longitudinal data requires a substantive theory about whether and how the data should change over time, as well as some understanding of how observations across time should relate to each other. In addition, the purpose and nature of the research questions help to determine the correct longitudinal model.

A concrete example may help to illustrate the importance of McArdle's insightful question. Imagine Dr Moody wants to investigate how people's moods change over time. How do moods fluctuate from day to day? What predicts fluctuations in mood? How well is current mood predicted by prior mood (Does yesterday's mood predict today's mood?)? Dr Moody collects longitudinal mood data on adults every day for 3 months. Dr Moody expects to see day-to-day changes in mood but no systematic increase or decrease in mood across time. Instead, she hypothesises that a person's mood today is predicted by their mood yesterday. Further, she believes that a person's mood 2 days ago (mood at $t - 2$) would be completely mediated by yesterday's mood (mood at $t - 1$). In other words, there are no lingering effects of a person's mood the day before yesterday (when $t - 2$) when predicting today's mood (mood at time $t$), over and above (or after controlling for) yesterday's mood (when $t - 1$). Dr Moody needs a model that allows for correlations between adjacent time points, and a model that allows for longitudinal mediation, but she does not require a model that allows for systematic growth over time. Autoregressive models are quite common in the SEM literature (Bast & Reitsma, 1997; Biesanz, 2012; Kenny & Campbell, 1989; Marsh, 1993). In the most common autoregressive model, mood at time $t$ is modelled as a function of mood at one (or more) earlier time points and a random disturbance term (Biesanz, 2012; Bollen & Curran, 2006). Dr Moody's research questions more closely align with autoregressive models than with growth models. Although growth models and autoregressive models are two common families of longitudinal models, they are certainly not the only models that exist for modelling longitudinal data.

Chapters 7 and 8 focus on one specific type of longitudinal model: the individual growth curve model. Why? First, growth models appeal to developmental, educational, and behavioural researchers because they allow for the estimation of systematic growth or decline over time. Growth can be positive, where outcomes increase across time, or negative, where outcomes decrease across time. Not all negative growth is undesirable. An intervention meant to decrease participants' depression or anxiety would be successful if scores decreased across time.

Second, the popularity of these models has skyrocketed over the past two decades. We conducted a brief search of the keyword 'growth curve model*' in the PsycINFO database over a 40-year time span. From 1980 to 1989, the term appeared in only six entries; between 2010 and 2019, it appeared in 2707 entries (see Figure 7.1). Third, individual growth models can be specified using either MLM or SEM. Therefore, growth modelling provides an ideal opportunity to explicitly forge the underlying connections between SEM and MLM.

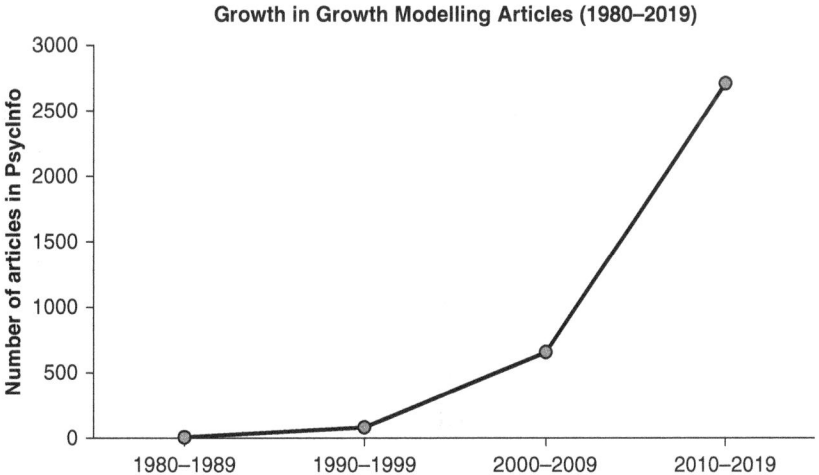

**Figure 7.1** Growth in the number of articles in PsycINFO that mention growth modelling (1980–2019)

This chapter provides a brief introduction to individual growth curve modelling in both the MLM and SEM frameworks and discusses the benefits and drawbacks of each framework for growth curve modelling. In reality, we cannot introduce the entire family of potential individual growth models within one introductory chapter.

Instead, we provide an overview of three of the simplest and most commonly used individual growth models: (1) the **linear growth model**, (2) the **polynomial growth model** and (3) the **piecewise growth model**. Chapter 8 provides an applied example, fitting these growth curve models using both SEM and MLM. For more details on growth modelling and lucid introductions to a variety of other growth models (including truly **non-linear growth models**), we highly recommend Grimm et al. (2016), who present growth models using both SEM and MLM. Fuller and more technical treatments of this topic from a multilevel perspective appear in Raudenbush and Bryk (2002), Hoffman (2015) and Singer and Willett (2003). Fuller and more technical treatments of this topic from an SEM perspective appear in Bollen and Curran (2006), Little (2013) and McArdle and Nesselroade (2014).

## Conceptual introduction to growth modelling

Many of the most interesting research questions in education, psychology and the social sciences involve the measurement of change. Learning rates, trends across time and growth are all examples of change measures. The individual growth curve model is well suited to capture systematic growth in an outcome variable of interest across time (e.g. academic achievement, antisocial behaviour or health and well-being). Growth models are especially important in educational research because 'the very notion of learning implies *growth* and *change*' (Willett, 1988, p. 346) and in developmental psychology, a field devoted to the study of how and why people change over time. However, growth models have broad applicability in sociology, economics, public health and the broader social science arena.

Research questions about growth or decline implicitly involve the measurement of systematic change over time. For example, we may wonder how students' reading skills develop between kindergarten and fifth grade. Is this growth steady or does the rate of growth change over time? In general, to understand how people change across time, we need to consider several key questions: What is the expected shape of this growth trajectory? Is this rate of change steady, or does it vary across time? Do people grow in the same way or is there variability between people in their rates of change? What factors predict inter-individual differences in rates of change? Using growth curve models, we can estimate each individual's **initial status** and the growth rate. In other words, growth curve models allow participants to start at different levels and change at different rates.

## The importance of theory and data

Clearly, strong substantive theory, knowledge of the content area and thorough reviews of prior literature are essential for researchers to begin developing hypotheses to address the questions posited above – they are critical elements of the modelling process. Before fitting growth curve models (and ideally, before collecting the data), we must have a sense of the shape of the expected growth *trajectory*. A *trajectory* is a curve that a body describes in space (e.g. the trajectory of a missile) or a path, progression or line of development (*Merriam-Webster*); a growth trajectory captures the shape of change across time.

Imagine that we were interested in studying how children's expressive vocabulary grows over the first 3 years of their lives. What shape would the expected trajectory be? When babies are born, they cannot speak. Over the first year of life, expressive vocabulary grows slowly. Most children can speak a few words or less by the time they are 1. By age 2, the average child has a 200- to 300-word expressive vocabulary and by age 3, the average child knows approximately 1000 words! Over the first 3 years, expressive vocabulary does not grow at a constant rate; vocabulary growth accelerates. Further, there is a great deal of variability between children in terms of how quickly they acquire vocabulary. Some children sound like miniature adults by age 3. Other children can produce very few words by age 3. Our model of vocabulary growth needs to reflect all this information. A reasonable model might hypothesise that at birth, the average vocabulary is 0, and the variance in that initial vocabulary size is also 0. (After all, no babies, no matter how brilliant, speak at birth!) The rate of change in babies' vocabulary growth is not steady: expressive vocabulary grows slowly at first, but then the rate of vocabulary growth increases across time. Finally, not all children grow at the same rate. Some grow more slowly, and some grow more quickly. This means there is considerable between-child variability in vocabulary growth. This basic theory provides valuable information with which to design our study and conduct our analyses.

However, even strong theories can be wrong. No modelling technique, no matter how novel or sophisticated, can substitute for a solid understanding of the data. Therefore, prior to running any statistical models, we recommend visually inspecting both the means on the outcome variable across time as well as a sample of individual growth trajectories (as depicted in Figure 7.2). Although examining the average growth trajectory can provide a quick sense of the average growth across time, 'the shape of the average change trajectory may not mimic the shape of the individual trajectories from which it derives' (Singer & Willett, 2003, p. 35).

Graphs of growth plot time on the *x*-axis and the outcome variable on the *y*-axis. Figure 7.2 plots a sample of 20 students' scores on a reading test administered at ages 6, 8 and 10. These individual trajectories reveal a great deal about students' growth. Generally, students seem to exhibit positive growth. However, students' growth rates vary a great deal. For example, student 1 grows very quickly; students 2 and 3 grow very slowly, students 9 and 10 have nearly flat growth trajectories – they do not appear to grow at all. Student 9's growth trajectory appears to be slightly negative and

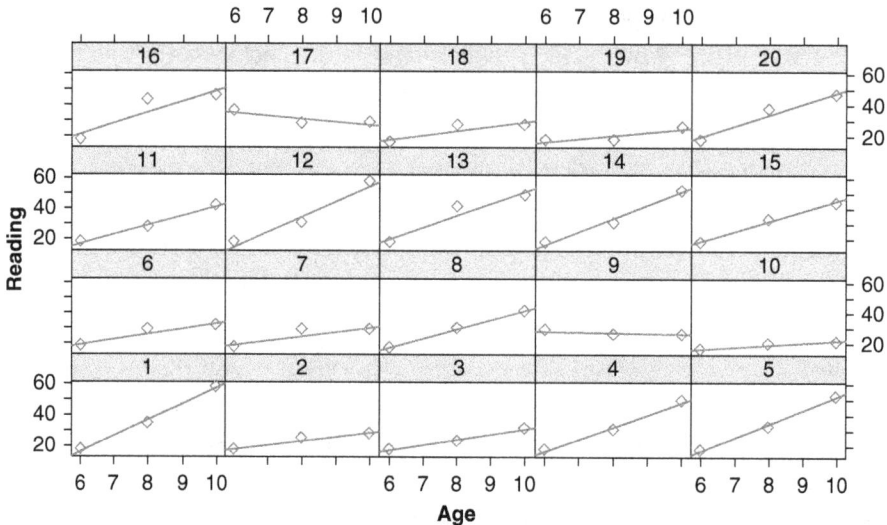

**Figure 7.2** Individual growth plots for a sample of 20 students' scores on a reading test administered at ages 6, 8 and 10

student 10's growth trajectory appears to be slightly positive, but their trajectories are virtually flat, meaning they exhibit no (0) growth. Student 17 exhibits slightly negative growth: this student's scores decrease slightly across the 4 years! Individual growth plots depict within-person change. Comparing the growth plots across panes provides information about similarities and differences in people's growth. Accurately modelling the within-person growth trajectory and understanding between-person differences in the within-person growth trajectory are key issues in growth modelling. Therefore, these individual growth plots offer invaluable guidance on building the growth model, and they provide a preview of the likely results.

## Requirements for growth models

There are several requirements for modelling individual growth curve models in either SEM or MLM. Ideally, we should ensure that we can address these requirements during the research design phase, prior to beginning data collection.

First, to study change using individual growth modelling requires collecting data from the same individuals across multiple time points. Cross-sectional data, which compares the scores of different people, do not meet this requirement.

Estimating even the simplest linear growth trajectories requires observations from at least three time points. Scores on any observed variable represent a combination of true score and error. Using the difference between scores at two time points to estimate change confounds 'true' change and measurement error. The change score (difference) could either overestimate or underestimate the degree of 'true' change. To extricate the confounded nature of the measurement error and true change requires at least three data points (Singer & Willett, 2003).

Curvilinear growth trajectories (as shown in Figure 7.3) require observations across four or more time points. Generally speaking, more complex trajectories require more repeated observations. Therefore, increasing the number of observations per person affords more flexibility and provides opportunities to fit more complex growth functions. When designing longitudinal studies, researchers should carefully consider both the number and the spacing of data collection points necessary to accurately capture the change process of interest. When data points are spaced too far apart or when there are too few data points, it may not be possible to adequately model the functional form of the true underlying change process. On the other hand, repeated observations that occur in rapid succession may be redundant, may waste precious resources and may not capture the full range or cycle of the phenomenon of interest. Hence, planning the data collection schedule for a longitudinal study requires careful consideration of the hypothesised model for change. Determining the optimal collection schedule is not a trivial matter, and these decisions can make or break the entire longitudinal study. We cannot overemphasise the importance of considering these issues carefully at the outset of the study.

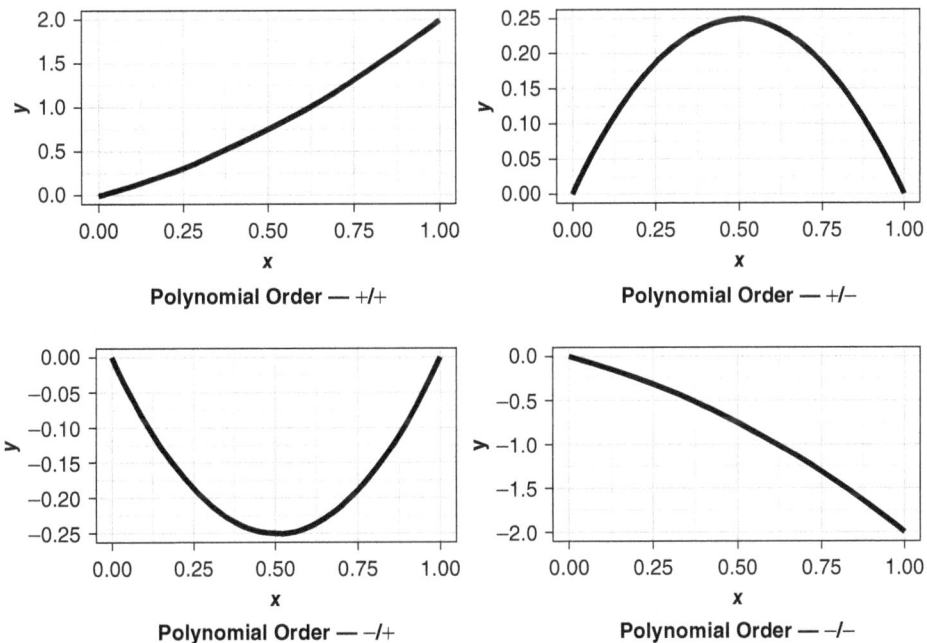

**Figure 7.3** Four different quadratic growth trajectories

Second, correctly modelling the functional form of growth across time requires knowledge about how much time has elapsed between repeated measurements. Accurately capturing the distance between measurement occasions is necessary to plot scores from the outcome variable on the $y$-axis at the correct locations on the time continuum ($x$-axis). Multilevel growth models include at least one 'time' variable. A time variable is a level-1 predictor that 'clocks' how much time has elapsed between time points. Some common metrics for time variables include the number of weeks/months/years between measurement occasions or the student's age in months (or years) for each administration of the assessment (McCoach, Rambo et al., 2013).

Third, the outcome variable must measure the same underlying construct on repeated administration, and scores on the outcome variable must be equatable or comparable across time (Singer & Willett, 2003). Finally, the scale of the assessment must remain consistent across occasions so that scores can be directly compared across time. Two ways to meet these requirements are (1) to use the same assessment at each administration or (2) to use a vertically scaled assessment (Singer & Willett, 2003). **Vertical scaling** equates assessment scores across age or grade levels, rendering the scores directly comparable across time, even if examinees of different ages, grades or ability levels complete different assessments. Conceptually, the vertical scaling procedure places all scores on the same equal-interval 'ruler' so that growth is measured in the same metric. In the absence of vertical scaling, we cannot use scores on two different, unlinked scales to measure growth or change (McCoach & Rambo, 2018).

However, simply using a scale that provides a common metric across measurement occasions may not be sufficient for capturing growth over time. The assessments must produce psychometrically sound data. The scores should demonstrate evidence of reliability, both at a given time point and across time. Likewise, the validity of the assessment must also remain consistent across multiple administrations of the assessment. For example, a multiplication and division assessment might be an adequate measure of mathematics achievement for a third grader, but it would not be a valid indicator of mathematics achievement for a high-school student. Assessments must be sensitive to changes in the outcome variable and appropriate to measure the outcome across the entire time span of data collection (Singer & Willett, 2003).

## Time-structured versus time-unstructured data

Researchers gather information from participants according to some data collection schedule. If all units are measured at the same time points, the data are considered time structured. With **time-structured data**, all participants follow the same data collection schedule, so the intervals between data collection points are the same for the entire sample (Kline, 2015).

In contrast, it is also possible for each participant to have his or her own unique data collection schedule. Data are time-unstructured if time intervals can vary both within and across people (Singer & Willett, 2003). If data are time-unstructured, the data collection schedule can be completely different for every person in the sample (Skrondal & Rabe-Hesketh, 2008). Forcing **time-unstructured data** to follow a time-structured pattern can decrease precision and increase bias in the parameter estimates. Therefore, understanding the time structure of the data is crucial, and it may even influence the decision about whether to build growth models using MLM or SEM.

## Growth modelling using multilevel versus SEM frameworks

Both SEM and MLM enable the flexible estimation of growth curve models (Kline, 2015). Therefore, for most individual growth models, the choice between SEM and MLM is one of preference rather than of necessity (Grimm et al., 2016). However, each framework does possess certain advantages. For example, MLM handles time-unstructured data more easily, can model nested data and more easily handles fully non-linear growth models (e.g. models with non-linear or exponential, slope parameters). In fact, some growth models that are fully non-linear cannot currently be estimated in standard SEM packages. Finally, REML estimation is available in MLM, whereas only FIML is available in SEM. When the number of level-2 units is small, REML estimates of the level-2 variance–covariance components are less biased than the FIML estimates because the REML estimates adjust for the uncertainty about fixed effects, whereas the FIML estimates do not (Raudenbush & Bryk, 2002). Therefore, MLM is a better option for fitting growth models in small samples.

On the other hand, SEM provides several advantages over standard MLM approaches. First, it is easier to specify a wide variety of error covariance structures within the SEM framework. Second, SEM provides the ability to build a measurement model for the dependent variable, allowing for the incorporation of latent variables, measured by multiple indicators at each time point. Third, structural equation models are multivariate, allowing simultaneous or sequential estimation of growth models for multiple outcomes. The intercept and slope are latent variables in SEM and can easily serve as latent predictors or outcome variables in larger SEMs. Fourth, SEM allows for more flexibility in the incorporation of time-varying covariates (TVCs) into the growth model. Finally, because the information about time is contained in the factor loadings, when there are at least four time points, it is possible to fix two factor loadings and free the rest of the loadings, allowing for a very flexible expression of the growth trajectory. Thus, SEM provides a great deal of flexibility in terms of its actual modelling capabilities.

The MLM and SEM frameworks also differ in terms of model specification and data format. In Figure 7.2, each student's reading score was measured across three time points. MLMs conceive of repeated observations as univariate: measures of the same variable across time are nested within individuals. In MLM, time enters the model as an explicit independent variable. Therefore, multilevel approaches to growth curve analyses require long, univariate, person–period data files. Because each repeated observation represents a unique row in the data set, each person (or unit) occupies multiple rows within the person–period data file (Singer & Willett, 2003). In contrast, in SEM repeated observations are multivariate. Therefore, SEM requires data in wide format, such that each row denotes a unique person, and measures of different outcomes appear as separate variables. Therefore, it is important to be able to restructure longitudinal data from long to wide format, and vice versa.

The remainder of the chapter is organised as follows: First, we demonstrate how to specify a simple linear growth model in both MLM and SEM. Next, we explain how to model non-linearity using two common methods: (1) piecewise models and (2) polynomial models. The key to correctly modelling piecewise and polynomial models is correctly incorporating (coding) time into the analyses. The coding of time follows the exact same logic in MLM and SEM. Finally, we briefly discuss how to incorporate other TVCs within MLM and SEM.

## Multilevel model specification for individual growth models

### The two-level multilevel model for linear growth

From the multilevel perspective, a simple growth curve model has two levels: the **within-individual level** (level 1) and the **between-individual level** (level 2). Observations across time, the level-1 units, are nested within people, the level-2 units. As such, the level-1 model captures the shape of an individual's growth trajectory over time and includes variables that vary across occasions, within individuals. Multilevel growth models contain (at least) one time variable to clock the passage of time. Because time is an independent variable, each unit can have its own unique values on the time variable, which facilitates easy modelling of time-unstructured data. The unconditional linear growth model (Equation 7.1) contains the time variable, but no other covariates (at level 1 or level 2):

Level 1:

$$y_{ti} = \pi_{0i} + \pi_{1i}(time_{ti}) + e_{ti}$$

Level 2:

$$\pi_{0i} = \beta_{00} + r_{0i}$$

$$\pi_{1i} = \beta_{10} + r_{1i}$$

Combined:

$$y_{ti} = \beta_{00} + r_{0i} + (\beta_{10} + r_{1i})(time_{ti}) + e_{ti}$$

$$y_{ti} = \beta_{00} + \beta_{10}(time_{ti}) + r_{1i}(time_{ti}) + r_{0i} + e_{ti}$$

(7.1)

The dependent variable ($y_{ti}$), the outcome for person $i$ at time $t$, and $y_{ti}$, is a function of $\pi_{0i}$ (i.e. the intercept for person $i$), $\pi_{0i}$ * $time_{ti}$ (i.e. the time slope for person $i$), and some degree of time-specific individual error, $e_{ti}$. In the level-1 equation, the intercept ($\pi_{0i}$) represents the predicted value on the outcome variable ($y_{ti}$) for person $i$ when time = 0. The time slope, $\pi_{1i}$ represents the linear rate of change in the outcome ($y_{ti}$) over time.

The level-2 equations in the unconditional linear growth model convey the overall expected initial level and average growth trajectory across people. Specifically, the randomly varying intercept ($\pi_{0i}$) for individual ($i$) is predicted by an overall intercept ($\beta_{00}$) and $r_{0i}$. The level-2 residual for the intercept, $r_{0i}$, represents the difference between the model-predicted intercept ($\beta_{00}$) and person $i$'s intercept ($\pi_{0i}$). Likewise, the randomly varying linear **growth slope** ($\pi_{1i}$) for each individual ($i$) is predicted by the intercept for the slope ($\beta_{10}$) and $r_{1i}$. The level-2 residual for the linear growth slope, $r_{1i}$, represents the difference between the overall model-predicted linear growth slope and person $i$'s growth slope.

Imagine a simple scenario in which time is coded 0, 1, 2 (i.e. time is centred at the initial time point) and the outcome variable is reading achievement. The intercept ($\pi_{0i}$) represents the predicted reading achievement of person $i$ at the start of the study (when time = 0), and $\beta_{00}$ is the expected value of the intercept. Therefore, $\beta_{00}$ represents the average (expected or model-predicted) reading achievement at the start of the study, and $r_{0i}$ denotes the difference between average (or expected) reading achievement and each person's individual reading achievement. If $r_{0i}$ is positive, then individual $i$'s intercept is higher (more positive) than the overall intercept. $\beta_{10}$ is the expected (or average) growth slope: as time increases by 1 unit, expected reading achievement increases by $\beta_{10}$ units. The residual $r_{1i}$ represents the discrepancy between the model-predicted growth rate and each individual's growth slope ($\pi_{1i}$ represents the linear growth rate for person $i$). Again, if $r_{1i}$ is positive, then person $i$'s reading growth slope is more positive than the overall expected reading growth slope, and if $r_{1i}$ is negative, then person $i$'s rate of reading growth is more negative than the overall (expected) reading growth rate.

*Centring Time.* Because the intercept represents the predicted value on the outcome variable when time = 0, we must code or *centre* time so that the intercept is an interpretable parameter of substantive interest. It is important to centre time around some meaningful value that occurs during the study, but there are many possible ways to do so. One common approach is to centre time at the beginning of the study period (or at the first data collection point). In that case, the intercept represents the expected value on the outcome at the first time point (at the beginning of the study). We could also centre time at the final time point in the study, or we could centre time in the middle of the data collection period.

Alternatively, each participant's age at each time point can serve as the time variable. Figure 7.2 plots students' reading growth from ages 6 (when most children are first learning to read) to 10. Centring age (the time variable) at 6 years produces a centred age/time variable that is 0 when each participant is 6 years old. The intercept represents the model-predicted outcome score *at age 6* for each child. This approach has the added advantage of *controlling for age* in addition to centring time. How does using age (instead of time point 1) control for age? Imagine collecting the first wave of data on children's reading scores in spring of kindergarten. Some children in the sample are 5½ years old, some are 6 years old, and some are 6½ years old. (Because time is a variable in the MLM, we can use students' *exact* ages.) Table 7.1 illustrates this point for four sample students. To centre age at age 6, subtract 6 from each student's age. We use the new, transformed variable, *AGE-6*, as the time variable in our growth models. For a 6.5-year-old, *AGE-6* at the first data collection point is .5, whereas for a 5.5-year-old, *AGE-6* equals –.5. The intercept is the expected reading score when time/age = 0, so the intercept is the model-predicted reading score if the student were to take the reading test at exactly age 6. For the student who was 5.5 at the first assessment (Amelia), the intercept represents her predicted reading score 0.5 years *after* she took the assessment. Therefore, assuming Amelia's expected growth slope is positive, Amelia's intercept (her predicted score at age 6) is higher than the score that she received at the first assessment point (i.e. we would expect Amelia to do better at age 6 than she did at age 5.5). For the student who was 6.5 years at the first assessment (Dedrick), the intercept represents his predicted reading score 0.5 years *before* he took the assessment. Assuming Dedrick's expected growth slope is positive, his intercept (predicted score at age 6) is lower than his score at the first assessment point (i.e. we would expect Dedrick's reading score to be lower at age 6 than at age 6.5).

**Table 7.1** Demonstration of how centring using age controls for age

| Student | Age at Wave 1 | AGE-6 | $\pi_{0i}$ (Predicted Reading Score When AGE-6 = 0) |
|---|---|---|---|
| Amelia | 5.5 | −0.5 | Predicted reading score 0.5 years *after* the student took the assessment |
| Boris | 5.75 | −0.25 | Predicted reading score 0.25 years *after* the student took the assessment |
| Ceci | 6.0 | 0 | Predicted reading score when the student took the assessment |
| Dedrick | 6.5 | 0.5 | Predicted reading score 0.5 years *before* the student took the assessment |

In this simple linear growth model, the time slope or growth parameter represents the constant (linear) rate of change in the outcome over time, measured in the units of the time variable. The growth slope $\beta_{10}$ is the expected change in $Y$ per-unit change in time. Time can be measured in any interpretable metric – milliseconds, hours, days, months, weeks and so on. For example, in Table 7.1, the time variable is measured in years: A 1-unit change in time represents a 1-year change, so $\beta_{10}$ represents the expected yearly change in the growth slope.

## Variance components

Both the slope and intercept terms contain the subscript, $i$, indicating that each individual ($i$) has a separate slope and intercept. The error term, $e_{ti}$, captures the deviation of a particular observation from the model-predicted growth trajectory. Thus, $e_{ti}$ represents the within-person measurement error associated with individual $i$'s data at time point $t$. The variance of $e_{ti}$ ($\sigma^2$) (Raudenbush & Bryk, 2002) is the pooled within-person error variability around individuals' trajectories (Raudenbush & Bryk, 2002). In MLM, the pooled within-person error variance ($\sigma^2$) is generally assumed to be constant across repeated observations. In other words, the amount of prediction error should not vary as a function of time. When fitting growth models in SEM, constraining all the error variances to be equal produces an equivalent model. However, in SEM, it is fairly common to allow the error variances to vary across time.

The variance of $r_{0i}$ ($\tau_{00}$) represents between-person variability in the intercept. Centring the intercept around the initial time point, $\tau_{00}$ represents the between-person variability in initial status. In that case, all other things being equal, the greater the between-person variance in the intercept (i.e. the larger $\tau_{00}$), the more people differ from each other in terms of where they start. Our choice of centring affects not just the expected value of the intercept but also the between-person variance in the intercept ($\tau_{00}$). When time is centred at age 6, $\tau_{00}$ represents the between-person variability in reading scores at age 6. Centring reading scores at age 5 would decrease $\tau_{00}$, whereas centring reading scores at age 7 would increase $\tau_{00}$. Why? There is much less between-person variance in reading scores at age 5 (when most students cannot yet read) than there is at age 7 (when some students are fluent readers and some students are struggling readers).

Likewise, the variance of $r_{1i}$ ($\tau_{11}$) represents the between-person variability in the time slope (Raudenbush & Bryk, 2002). All other things being equal, greater between-person variance in the slope (i.e. larger $\tau_{11}$) indicates greater variability in people's growth rates. In linear growth models, because the growth rate is constant across time, the centring of time has no effect on the expected growth slope or $\tau_{11}$, the between-person variance in the growth slope. If the growth rate varies across time (as is the case in higher order polynomial models, as we shall see later in this chapter), then the choice of centring does impact both the growth slope and the between-person variance in the growth slope ($\tau_{11}$).

Inclusion of $r_{0i}$ and $r_{1i}$ in these level-2 equations allows for between-person variability in peoples' intercepts and growth slopes. Generally, we allow the randomly varying intercepts and slopes to co-vary (i.e. $\tau_{01}$ is freely estimated). The standardised $\tau_{01}$ estimate from the **unconditional growth model** (which does not yet include any level-1 or level-2 predictors other than time) provides the correlation between level (i.e. the level of the intercept when time $t = 0$) and rate of change (the growth slope). If time is centred at initial status, then, $\tau_{01}$ quantifies the relationship between where people start and how fast they grow. Again, the choice of centring affects the estimate of $\tau_{01}$. Conceptually, this makes sense. For instance, the correlation between reading scores at age 6 and reading growth ($\tau_{01}$ when the time variable is centred at age 6) is not equal to the correlation between reading scores at age 10 and reading growth ($\tau_{01}$ when the time variable is centred at age 10).

Regardless of the number of time points, a standard unconditional growth model estimates six parameters: two fixed effects (the expected intercept and the expected slope) and four variance components (the between-person variance in the intercept, the between-person variance in the slope, the covariance between the slope and the intercept and the pooled within-person residual). Earlier in the chapter, we discussed vocabulary growth over the first 3 years of life. If we collected data on vocabulary development from birth to age 3 and centred our growth model at age 0 (birth), our intercept, $\beta_{00}$, expected expressive vocabulary at age 0 would be 0. Moreover, $\tau_{00}$, the between-person variance in the intercept would also be 0 because every infant would have the same vocabulary score of 0 at time 0. The correlation between the slope and the intercept ($\tau_{01}$) would be 0. (If a variable does not vary, then its correlation with any other variable would be 0.) As such, centring time at age = 0 wastes three of our precious growth parameters (the intercept's mean and its variance and its covariance with the slope). Generally, we should centre the time variable so that there is at least some between-person variance in the intercept.

## Estimating linear growth from an SEM perspective

Now let's fit the same growth curve model within the SEM framework. SEM uses latent variables to model individual growth curves. In fact, the latent growth model is a measurement (CFA) model: the intercept and the slope are latent variables. There is one notable difference between a standard CFA model and a latent growth model. In a standard CFA model, to freely estimate the variance of the factor, we constrain one path from each latent variable to be 1.0 for identification purposes (a technique that we referred to as the marker variable strategy). In a standard latent growth model, we constrain all paths (factor loadings). These paths contain the information about time, and they allow us to generate expected (model-predicted) means at each observed time point. Figure 7.4 presents a graphical depiction of a linear growth model with observed data collected at three time points (i.e. T1, T2 and T3).

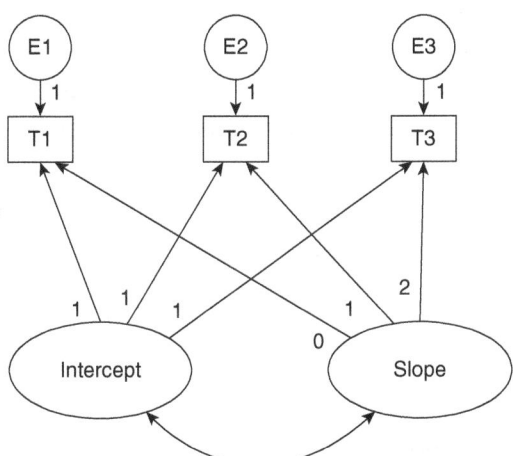

**Figure 7.4**   A linear growth model in the structural equation modelling framework

   In SEM, we can model the mean structure as well as the structure of the variance–covariance matrix. However, all our prior examples focused exclusively on modelling the variance–covariance structure. Growth curve models explicitly model the change in level across time. Therefore, modelling growth with SEM requires estimating a model that includes means (and intercepts), as well as variances and covariances.

## Means and intercepts

Before explaining the specification and identification of growth models within an SEM framework, we pause to explain the concept of mean structure in SEM. Prior to this chapter, although we borrowed heavily from multiple regression when we presented structural equation models, we never discussed intercepts or means. In fact, all the models discussed in Chapters 4 to 6 never examined means (or intercepts).

As mentioned in the introduction to MLM, in standard regression analyses, the intercept ($b_o$) is often a' 'throw away' parameter. A standard regression model has the form $Y = b_o + b_1(X) + e_i$. The model-predicted value of $Y$ is $\hat{Y}$, and $\hat{Y} = b_o + b_1(X)$. In these equations, $b_o$, the intercept, is the expected (predicted) value of $Y$ when $X = 0$. In multiple regression models, the intercept is the expected (predicted) value of $Y$ when all the predictor variables ($X$s) are held constant at 0. A standard regression model does not include a parameter for the mean of $Y$. Instead, it contains a parameter for the intercept, the expected value of $Y$ when all predictor variables are 0.

Similarly, the variance of $Y$ is not a parameter in our regression model. Instead, a regression model includes a parameter for the residual variance of $Y$, the variance in $Y$ that cannot be explained by the predictor(s). SEM parameters include variances of exogenous variables and residual variances of endogenous variables: disturbance variances for endogenous structural variables or measurement error variances for endogenous measurement variables. Analogously, structural equation models with mean structure include parameters for the means of exogenous variables and the intercepts of endogenous variables. The models that include only observed variables (e.g. path models with no latent variables) estimate a mean for every exogenous variable and an intercept for every endogenous variable. Therefore, the mean structure is fully saturated or just identified. In such a scenario, a model that includes the mean structure has the same number of degrees of freedom as a model that does not include the mean structure.

In latent variable models, the specification of the mean structure is somewhat more complicated. Just as it is not possible to estimate more parameters than unique elements of the variance–covariance matrix, it is not possible to estimate more means and intercepts than observed means. In single-group models, the maximum number of mean and intercept parameters equals $V$ (the number of observed variables). There are two ways to specify a model with means and intercepts in standard measurement (CFA) models. Option 1 freely estimates the intercepts for all observed variables and constrains the means of all latent variables to 0. This resembles the fixed factor variance strategy for identifying latent variables. Using the fixed factor variance strategy, we estimate all the paths (factor loadings), but we constrain the factor variance to equal 1. Fixing the factor variance to 1 and the factor mean to 0 scales the factor in a z-score metric, as z-scores are normally distributed with a mean of 0 and a standard deviation (and variance) of 1.

Option 2 uses a *marker intercept* approach. Constraining a path from the latent variable to one of its observed indicators to 1 enables us to freely estimate the variance of the latent variable. The indicator with the fixed path is the *marker variable*. Similarly, if we constrain the intercept for the marker variable to 0, we can estimate the latent mean. Regardless of whether we use the marker variable approach or the fixed factor mean approach, the mean structure of standard latent variable models is fully saturated (just-identified) because the number of estimated mean and intercept parameters equals the number of observed variable means.

---

### Box 7.1

#### How to Specify Means Within a Standard Latent Variable Model

*Option 1:* Freely estimate intercepts for all the observed variables. Constrain the means for all latent variables to equal 0.

*Option 2:* Choose a marker variable for each latent variable. Constrain the intercept for the marker variable to equal 0. Freely estimate the factor mean for each latent variable.

---

## SEM specification of growth models

The SEM approach is analogous to the multilevel approach. In traditional SEM, time is introduced through the factor loadings for the latent variable representing the linear growth slope (Stoel & Garre, 2011). However, because the information about the passage of time is contained in the path coefficients (not a separate variable), everyone must have the same basic data collection schedule. For this reason, traditional SEM analyses require time-structured data, whereas multilevel models can accommodate time-unstructured data. There are multiple ways to fit growth models with time-unstructured data in SEM. However, given space constraints, we cannot discuss them in this book. We highly recommend Chapter 4 of Grimm et al. (2016) and Muthén and Muthén (1998–2017) for more information about methods of incorporating time-unstructured data into latent growth curve models within the SEM framework.

## Degrees of freedom for SEM growth models

To compute the degrees of freedom, we again count the number of unique pieces of information (knowns) and compare that to the number of estimated parameters (unknowns). When modelling only the variance–covariance structure, the knowns equals the number of unique elements of the variance–covariance matrix, easily computed using $v(v + 1) / 2$, where $v$ = the number of observed variables. In models that include both the mean structure and the variance–covariance structure, the number of unique pieces of information equals the number of unique elements of the variance–covariance matrix plus the number of observed means. Because each observed variable has one observed mean, the number of knowns for a model that includes mean structure equals $v(v + 1) / 2 + v$, where $v$ is the number of observed variables. We can simplify this expression to $v(v + 3) / 2$.

For models that include mean structure, we must consider identification separately for the mean and covariance structures. If both the mean structure and the covariance structure are identified, then the overall model is identified. If either the mean structure or the covariance structure is underidentified, then the model is not identified. Both the mean portion of the model and the variance–covariance portion of the model must have adequate degrees of freedom and be identified. Figure 7.4 contains three observed variables. There are six unique elements in the variance–covariance matrix (3 variances and 3 covariances) and three observed means. Therefore, the total number of knowns = 9. To identify a model that includes a mean structure, the degrees of freedom for the mean structure must be non-negative: the number of observed means must be equal to or greater than the number of freely estimated parameters in the mean structure. Therefore, with three observed variables, we can estimate no more than three means and intercepts.

How many freely estimated parameters are there? Let's start with the mean structure: we estimate a mean for the **latent intercept factor** and a mean for the **latent slope factor**. We cannot estimate intercepts for each of the observed variables across time ($v$) as well as means for the two latent variables (2). Doing so would require $v + 2$ means, and we have only $v$ means. Instead, the standard latent growth model constrains the intercepts for the outcome variables across all time points to 0. With three time points, the model estimates two mean parameters using three observed means. The mean structure is overidentified: it has 1 $df$. Thus, the mean structure can be a source of model–data misfit. Conceptually, the intercept and the growth slope parameters should be able to reproduce the expected scores on the outcome across time. Model misfit from the mean structure indicates that the latent means are unable to adequately reproduce the expected outcome scores.

How many variance–covariance parameters does the model contain? The model parameters include a variance for each latent variable (an intercept variance and a slope variance), a covariance between the slope and the intercept, and one or more error variances for the outcome variables. The variances of the slope and intercept factor and the covariance between the factors are equivalent to the elements of the tau matrix in MLM. The standard multilevel model estimates a pooled residual variance ($\sigma^2$). To fit an equivalent model in SEM, we constrain the error variances of the outcome variable to be equal across time points. Such a model estimates four variance–covariance parameters: two latent variances (for the slope and the intercept), a latent covariance between the slope and the intercept and a time-specific error variance (constrained to be equal across time points).

To recap, the parameters of the unconditional growth model include the means, variances and covariances of the latent variables and the time-specific measurement error(s). The SEM equivalent of the MLM unconditional linear growth model estimates six parameters: two latent means, two latent variances, the latent covariance and one error variance (assuming all time-specific measurement errors are constrained to be equal).

## Interpreting the parameters in SEM growth models

The latent variable means are the expected values for the intercept and slope parameters. The mean of the intercept factor depicts the expected (average) level (when time = 0) and the mean of the slope factor describes the expected (average) rate of change across time. The variances of the latent variables capture the between-person variability in the growth slope and level (when time = 0). For instance, a slope variance of 0 indicates that everyone is growing at the same rate, a common assumption in repeated-measures ANOVA.

In Figure 7.4, the factor loadings for the intercept factor are fixed to 1 across all time points. The factor loadings for the slope variable contain information about the passage of time. Figure 7.4 centres time at the first observation. The path from the slope factor to T1 (the observed outcome variable measured at time 1) equals 0. Consequently, the mean of the intercept factor represents the expected value of the outcome variable at time 1. Why? Because the factor loading for the slope is 0 at time 1, the expected time 1 score equals the mean of the latent intercept. The remainder of the path coefficients from the slope factors to the outcome variables (T2 and T3) indicate the amount of time that has elapsed since time 1 (when time = 0). If time is measured in years, fixing the T2 path to 1 indicates that 1 year elapsed between the collection of T1 and T2 data. Likewise, the T2 path is fixed at 2 indicating that 2 years have elapsed between T1 and T3. In Figure 7.4, the three data points are spaced a year apart. Given that time is measured in years, the mean of the slope represents the yearly growth rate.

The time-specific residual variances represent time-specific measurement error in the outcome variable (Bollen & Curran, 2006; Muthén & Muthén, 1998–2017). Conventional multilevel models pool this measurement error across time points and people; constraining all of the error variances to be equal fits an equivalent growth model in SEM. However, it is perfectly acceptable (and not uncommon) to allow the error variances to vary across time.

## Model-predicted scores

The model-predicted scores at each time point are a function of initial status (the mean intercept) and the rate of change across time (the mean slope). Table 7.2 demonstrates how to use the parameter estimates for the intercept and slope means to generate the expected (model-implied) means at each time point.

**Table 7.2** Demonstrates how the means and loadings for the intercept and slope are used to compute the expected mean at each time point

| Wave (w) | $M_I$ (Intercept Mean) | $L_{Iw}$ (Intercept Loading) | $M_s$ (Slope Mean) | $L_{sw}$ (Slope Loading) | $M_I * L_{Iw} + M_s * L_{sw}$ | Expected Mean at Wave w |
|---|---|---|---|---|---|---|
| 1 | 50 | 1 | 10 | 0 | (50 * 1) + (10 * 0) | 50 |
| 2 | 50 | 1 | 10 | 1 | (50 * 1) + (10 * 1) | 60 |
| 3 | 50 | 1 | 10 | 2 | (50 * 1) + (10 * 2) | 70 |
| 4 | 50 | 1 | 10 | 3 | (50 * 1) + (10 * 3) | 80 |
| 5 | 50 | 1 | 10 | 4 | (50 * 1) + (10 * 4) | 90 |

*Note. $M_I$ represents the mean of the intercept; $L_{Iw}$ represents the factor loading from the intercept factor to the observed score at wave w; $M_s$ represents the mean of the slope; and $L_{sw}$ represents the factor loading of the slope factor to the outcome at wave w. We use the formula $M_I * L_{Iw} + M_s * L_{sw}$ to compute the expected mean.*

## Incorporating time-varying covariates

### Incorporating time-varying covariates in MLM

Time is not the only variable that can vary across time. Growth models can also include other time-varying predictors. **Time-varying covariates** (TVCs) are level-1 variables that can vary across time as well as across people. Thus, TVC can have different values for the same person at different time points throughout the study period. Although the time variable itself is technically a TVC, the term *TVC* usually refers to a variable other than time that varies both within and across people. Examples of TVC include mood/affect, caloric intake, body mass index, employment status and so on. The slope parameter for the TVC indicates the expected change in the outcome variable for each 1-unit increase in the TVC. TVCs are level-1 variables; they contain within-person variance.

Level 1:

$$y_{ti} = \pi_{0i} + \pi_{1i}(time_{ti}) + \pi_{2i}(TVC_{ti}) + e_{ti}$$

Level 2:

$$\pi_{0i} = \beta_{00} + r_{0i}$$
$$\pi_{1i} = \beta_{10} + r_{1i}$$
$$\pi_{2i} = \beta_{20} + r_{2i}$$

(7.2)

In Equation (7.2), $TVC_{ti}$ is a time-varying covariate and $\pi_{2i}$ is the coefficient for the 'effect' of the TVC on $y_{ti}$, after controlling for any other variables in the model. The subscript $i$ indicates that the $\pi_{2i}$ parameter can differ across people. In addition, the random effect for $\pi_{2i}$ ($r_{2i}$) indicates that the TVC slope randomly varies across people. In other words, for some people the effect of the TVC on the dependent variable could be quite strong and positive, whereas for others it could be weak, or even negative.

In the unconditional growth model, the intercept is the expected value of the outcome variable when time = 0. More generally, the intercept is the expected value of the outcome variable holding all other variables in the equation constant at 0. The interpretation of the intercept and growth slope change with a TVC. With the addition of a TVC, the intercept is the expected value of the outcome variable when both time and the TVC are 0. Similarly, the growth slope is now conditional on the TVC; it is the expected growth, holding the TVC constant at 0.

The interpretation of the growth parameters also depends on the centring or coding of the TVC. It is common to dummy code a dichotomous TVC. With continuous variables, it is essential to carefully consider the best approach to centring the TVC. Chapter 2 of this book described two approaches to centring level-1 variables: (1) grand mean centring and (2) group mean centring. Grand mean centring involves subtracting the overall mean of the covariate from each person's score. Group mean centring involves subtracting the cluster (level-2) mean of the TVC from each person's score. However, in growth models, observations across time are clustered within people, and the cluster mean is the individual's person-specific average on the TVC across all time points. Therefore, a group mean–centred TVC represents an individual's deviation from their own personal average.

For example, imagine a study that models change in anxiety across time and includes stress as a TVC. Time is centred at the initial data collection period. To group mean centre stress, first we compute each person's mean stress: this is the average stress across time for person $i$. Then, we subtract person $i$'s mean stress level from each of their time-specific stress scores. Group mean–centred stress for person $i$ at time $t$ represents the deviation in stress from the person's own average stress level. Therefore, the coefficient for the stress slope indicates the expected change in anxiety if stress increases by 1 unit. Because stress is group mean centred, $\pi_{0i}$ represents predicted anxiety at the start of the study when person $i$'s stress is at their own personal average across time, and $\beta_{00}$ is expected anxiety. The growth slope, $\pi_{1i}$, is the expected change in anxiety per unit change in time, assuming individual $i$'s stress level is at their own personal average stress level (i.e. it is the growth in anxiety, holding stress constant at individual $i$'s own mean), and $\beta_{10}$ is the expected change in anxiety across time (holding stress constant within person, but not between people). In contrast, when grand mean centring stress at level 1, the intercept ($\beta_{00}$) is the predicted initial value when stress is at the overall average, both across time and across people. The growth slope ($\beta_{10}$) is the expected growth rate for a person of average stress.

The TVC slope ($\beta_{20}$) is the predicted change in anxiety as stress increases by 1 unit. In both the group mean– and grand mean–centred equations, the TVC slope represents the expected change in anxiety for a unit change in stress. However, using group mean centring, the change in stress is the deviation from a person's average level, whereas in grand mean centring, the change in stress is the deviation from the sample average.

Because group mean centring eliminates the between-cluster variance in the predictor variable, if the average level of the TVC differs across people, it is important to include the mean value of the TVC as a person-level predictor at level 2. For instance, if stress is group mean centred, it is important to include mean stress level as a grand mean–centred time-invariant predictor at level 2. As seen in Equation (7.3), mean stress level predicts people's initial anxiety ($\beta_{01}$ predicts between-person differences in the randomly varying intercept, $\pi_{0i}$), change in anxiety across time ($\beta_{11}$ predicts between-person differences in the randomly varying growth slope, $\pi_{1i}$) and occasion-specific changes in anxiety ($\beta_{21}$ is the effect of mean stress on the individually varying slope of the within-person changes in stress on anxiety, $\pi_{2i}$).

Level 1:

$$y_{ti} = \pi_{0i} + \pi_{1i}(time_{ti}) + \pi_{2i}(Stress_{ti}) + e_{ti}$$

Level 2:

$$\pi_{0i} = \beta_{00} + \beta_{10}(Mean\,stress) + r_{0i}$$
$$\pi_{1i} = \beta_{10} + \beta_{11}(Mean\,stress) + r_{1i}$$
$$\pi_{2i} = \beta_{20} + \beta_{21}(Mean\,stress) + r_{2i}$$

(7.3)

## Variance components

The slope of the TVC randomly varies and co-varies with both the time slope and the intercept, so the tau matrix has six parameters: three variances (the intercept variance, growth slope variance and TVC slope variance) and three covariances (the covariance between the growth slope and the intercept, the covariance between the intercept and the TVC slope and the covariance between the growth slope and the TVC slope). (See Equation 7.4.) Standardizing these covariances produces three correlations between initial status and growth ($\tau_{10}$), initial status and the relationship between TVC and the outcome ($\tau_{20}$) and growth and the relationship between TVC and the outcome ($\tau_{21}$).

$$\tau = \begin{bmatrix} \tau_{00} & & \\ \tau_{10} & \tau_{11} & \\ \tau_{20} & \tau_{21} & \tau_{22} \end{bmatrix} \tag{7.4}$$

The $TVC_{ti}$ is a level-1 variable, and the value of this covariate changes across time; however, for a given individual ($i$) the effect of the TVC on the dependent variable ($\pi_{2i}$) is *constant* across time. In other words, although the effect of the TVC varies across people, for each person, the effect of the TVC is constant across time. For example, in a study of vocabulary growth in children, the weekly hours of parent–child talk could be a TVC. At every time point, the researcher measures both the dependent variable (expressive vocabulary) and the independent variable (the number of hours that the parent talks to the child each week). Although weekly parent–child talk hours can change at each data collection point and the effect of parent–child talk on vocabulary can differ across individuals, the estimated within-person relationship between parent–child conversation and vocabulary development remains constant across the entire study.

We can ease the assumption that the effect of the TVC remains constant across time by allowing time to moderate the effect of the TVC on the outcome variable. To do so, we add an interaction between time and the TVC by creating a new observed variable that equals the product of the time and TVC variables (Singer & Willett, 2003), as in Equation (7.5). Generally, centring continuous TVCs makes parameter estimates for the interaction and the TVC more easily interpretable. We can follow the same logic and guidelines we use when creating interaction terms in multiple regression models. Generally continuous variables are centred around their means, and the interaction term is computed as the product of the two mean-centred variables.

Level 1:

$$y_{ti} = \pi_{0i} + \pi_{1i}(time_{ti}) + \pi_{2i}(TVC_{ti}) + \pi_{3i}(time_{ti} * TVC_{ti}) + e_{ti}$$

Level 2:

$$\pi_{0i} = \beta_{00} + r_{0i}$$
$$\pi_{1i} = \beta_{10} + r_{1i}$$
$$\pi_{2i} = \beta_{20} + r_{2i} \tag{7.5}$$
$$\pi_{3i} = \beta_{30} + r_{3i}$$

The level-2 parameter for the interaction term ($\beta_{30}$) captures the expected differential effect of the TVC across time. If $\beta_{30}$ is positive, then the effect of the TVC becomes increasingly positive across time; in contrast, if $\beta_{30}$ is negative, the effect of the TVC becomes more negative as time passes (assuming time is centred at initial status). If the interaction term randomly varies across people (includes $r_{3i}$), the slope of the time-varying effect of the TVC differs across people. In other words, including $r_{3i}$ allows between-person variance in the degree to which the effect of the TVC on the outcome varies as a function of time.

It is important to carefully consider the necessity of including random effects for such interaction terms. Although introducing a randomly varying interaction between time and a given TVC provides great flexibility, it also increases the number of estimated parameters in the model. For example, excluding the randomly varying interaction yields a model with one random intercept, two random slopes and three covariance parameters (see Equation 7.2). But the addition of the randomly varying error term ($r_{3i}$) for the interaction effect adds four additional covariance parameters to the model: one variance parameter ($\tau_{33}$) and three covariance parameters ($\tau_{03}$, $\tau_{13}$ and $\tau_{23}$). Therefore, the model depicted by Equation (7.5) has a total of 15 parameters: four fixed effects ($\beta$s) and 10 variance–covariance parameters (i.e. 4 variances and 6 covariances) in the tau matrix, and $\sigma^2$.

## Incorporating time-varying covariates into SEM latent growth models

In SEM, the easiest way to incorporate TVC into latent growth models is to simply add the TVCs as additional exogenous predictors of the outcome variables at the appropriate times/waves. Because SEM growth models are multivariate, each of the outcome variables and each of the TVCs are separate variables. This results in separate paths from each TVC (at time $t$) to each outcome variable (at time $t$). In such a model, the effect of the TVC naturally varies across time; no interaction terms are required. To force the effect of the TVC on the outcome variable to the same across time, constrain all the path coefficients from the TVC to the outcome variables to be equal. Comparing the model fit of these two models (one that constrains the effect of the TVC to be constant across time and one that allows the effect of the TVC to vary across time) helps determine whether such a constraint is reasonable. In addition to allowing the effect of the TVC to vary across time, SEM easily allows for cross-lagged TVC paths. Additionally, unlike MLM, it is not necessary to observe TVC at each time point. Thus, SEM allows for a great deal of flexibility in terms of incorporating TVC into latent growth models.

However, adding TVCs as exogenous time-specific predictor variables is not the same as adding a TVC in MLM: there is no analogue of the random effect. Therefore, TVC slopes do not vary across people. Remember, MLM utilises random effects to model between-person variance in growth parameters, whereas SEM utilises latent variables (factors) to model between-person variance in growth parameters. Therefore, in SEM, to allow between-person variance in the effect of the TVC requires including a separate factor for the TVC into the latent growth model. For more details about this approach, see Curran et al. (2012) and Curran et al. (2014). For an excellent discussion on the disaggregation of within-person and between-person effects in longitudinal models of change, see Curran and Bauer (2011).

Although the inclusion of TVCs creates complexities in the growth model building process, incorporating TVCs can be critical to understanding the process of change. Thoughtfully coded time-varying variables may more accurately reflect the process of change, as well as correlates of that change. Incorporating TVCs can also be an effective strategy for modelling discontinuities or non-linearity in growth trajectories (McCoach & Kaniskan, 2010), as we discuss next.

## Piecewise linear growth models

Often, growth trajectories are not modelled well by a single linear slope or rate of change, even after adjusting for TVCs. Rather, growth may occur in phases, each of which could have a different growth rate. Thus, the overall growth trajectory consists of multiple growth segments corresponding to fundamentally different patterns of change (Collins, 2006). Piecewise models split the growth trajectories into separate linear components (Raudenbush & Bryk, 2002). Each piece (linear component) can have a different slope parameter, and these growth slopes may randomly vary across people. Piecewise linear growth models (also called spline growth models) allow researchers to determine whether growth rates are the same or different across time periods and to investigate differences in substantive predictors of growth across different growth phases. For example, imagine that a researcher collects multiple observations before and after implementing a new intervention designed to increase students' mathematics achievement. If the intervention is effective, then mathematics achievement should grow more quickly during the intervention period than it did during the period prior to intervention. Conceptually, this suggests one growth rate (slope) prior to the intervention and another growth rate (slope) during the intervention.

## How many individually varying slopes can my data support?

MLM accommodates individual variation in the growth parameters through random effects. The variances of these random effects capture the between-person variability in the growth parameters. SEM accommodates individual variability using latent variables (factors). The variances of the factors capture the between-person variability in the parameters. If a linear growth model requires three time points, how many time points are required to fit a piecewise model with a randomly varying intercept and two randomly varying slopes?

As a general rule, MLM can accommodate one fewer random effect than there are time points ($r_{max} = t - 1$). For example, a linear growth model that allows the intercept and slope to randomly vary across people requires two random effects (one for the intercept and one for the growth slope). This explains why fitting a linear model requires at least three time points. However, a piecewise model that contains two randomly varying slopes (and a randomly varying intercept) requires at least four observations (time points) per person, and a piecewise model that contains three randomly varying slopes (and a randomly varying intercept) requires at least five observations per person. Why?

A model with a randomly varying slope and a randomly varying intercept contains four variance components: $\sigma^2, \tau_{00}, \tau_{01}, \tau_{11}$. (There are also two fixed effects for the slope and the intercept, but this exposition focuses on the identification of the variance components). Under the assumption of homogeneity, those four variance components reproduce or re-create the elements of the covariance matrix of composite residuals ($\Sigma_r$), which has the same dimension as the variance–covariance matrix of the outcome variable across time. When there are three time points, the observed variance–covariance matrix has six unique elements (three variances and three covariances). Therefore, a model estimating the four variance component parameters is overidentified: the standard linear growth model estimates two fewer variance–covariance components (unknowns) than there are observed variances and covariances (knowns). Now, imagine trying to fit a model with *three* random effects and only three time points. A model with three random effects contains seven variance components: $\sigma^2, \tau_{00}, \tau_{01}, \tau_{02}, \tau_{11}, \tau_{12}, \tau_{22}$ (three variances for the three level-2 residuals, three covariances for the three level-2 residuals and the level-1 residual variance). However, the composite variance–covariance matrix contains only six unique pieces of information. Such a model is underidentified: it has –1 *df* because the number of parameters to estimate (unknowns = 7) exceeds the number of unique elements in the variance–covariance matrix (knowns = 6). However, with four time points, the composite variance–covariance matrix contains 10 unique elements (four variances and six covariances, see Exhibit 7.1). The model with three random effects contains only seven unknown parameters. Therefore, with four time points, the model is overidentified; it has 3 *df*. However, it is not possible to estimate a model that contains four random effects with four time points. A model with four random effects contains 10 elements in the variance–covariance (tau) matrix (four variances and six covariances) plus the level-1 residual variance parameter, for a total of 11 parameters. So again, the model is underidentified: it has –1 *df*.

**Exhibit 7.1** Variance-covariance matrix that includes four variances and six covariances

$$
\begin{pmatrix}
\text{var}_{11} & & & \\
\text{cov}_{21} & \text{var}_{22} & & \\
\text{cov}_{31} & \text{cov}_{23} & \text{var}_{33} & \\
\text{cov}_{41} & \text{cov}_{24} & \text{cov}_{34} & \text{var}_{44}
\end{pmatrix}
$$

Although SEM utilises latent variables (factors) to allow for between-person variability in the slopes and the intercepts, the principle remains the same. With three time points, it is possible to estimate the factor variances for the slope and intercept, the covariance between the factors and the error variance for the observed variables. However, adding an additional factor variance (and covariances) results in an underidentified model. With only three time points, it is not possible to estimate seven unknown parameters (three variances, three covariances and a measurement error variance) with only six unique elements in the variance–covariance matrix. Therefore, in either SEM or MLM, growth models can accommodate one fewer random effect (factor) than there are time points (observations). There are ways to reduce the number of variance–covariance components, either by constraining one or more of the covariances or by constraining the level-1 residual variance in some way. Such approaches require additional assumptions. McCoach et al. (2006) provides an example of constraining the level-1 residual variance.

This rule of thumb indicates the *minimum* number of observations required for identification, *not* the *optimal* number of observations for such a model. Additional time points allow for more thorough examination of the adequacy of the specified functional form. For instance, three time points is adequate for a linear model. However, with only three time points, it is not possible to compare the linear trajectory to a more complex growth trajectory to determine whether the linear model adequately captures the shape of the growth. Allowing the slopes for TVCs to randomly vary also requires additional random effects, necessitating additional time points. Therefore, if possible, we suggest collecting at least two additional time points, over and above the minimal number necessary to fit hypothesised the growth trajectory. For instance, a model with three randomly varying slopes and a randomly varying intercept is identified with only five time points. However, we recommend collecting observations across seven or more time points for such a model.

## The mechanics of coding time for the piecewise model

To capture multiple linear growth slopes, the piecewise linear growth model includes multiple time variables. For example, to examine whether a math intervention actually increases math growth, researchers collected data across six time points. The first three observations occurred prior to implementing the intervention, and the final three time points occurred during the intervention period. If the math intervention is effective, the rate of math growth should be higher (more positive) from time points 3 to 6 than during the baseline period (time points 1–3). If the intervention is ineffective, the rate of math growth should be consistent across the baseline period (times 1–3) and the intervention period (times 3–6). Therefore, our model contains two separate growth rates: a rate of growth for time points one to three, and another rate of growth for time points three to six. In MLM, our two-piece linear growth model contains two *time* variables, as shown in Equation (7.6):

Level 1:

$$y_{ti} = \pi_{0i} + \pi_{1i}(time_{piece1ti}) + \pi_{2i}\left(time_{piece2ti}\right) + e_{ti}$$

Level 2:

$$\pi_{0i} = \beta_{00} + r_{0i}$$
$$\pi_{1i} = \beta_{10} + r_{1i}$$
$$\pi_{2i} = \beta_{20} + r_{2i}$$

(7.6)

Given the discussion in the previous section, the model in Equation (7.6) requires a minimum of four time points to be identified. In SEM, we include an additional latent variable to allow for another growth slope, as shown in Figure 7.5.

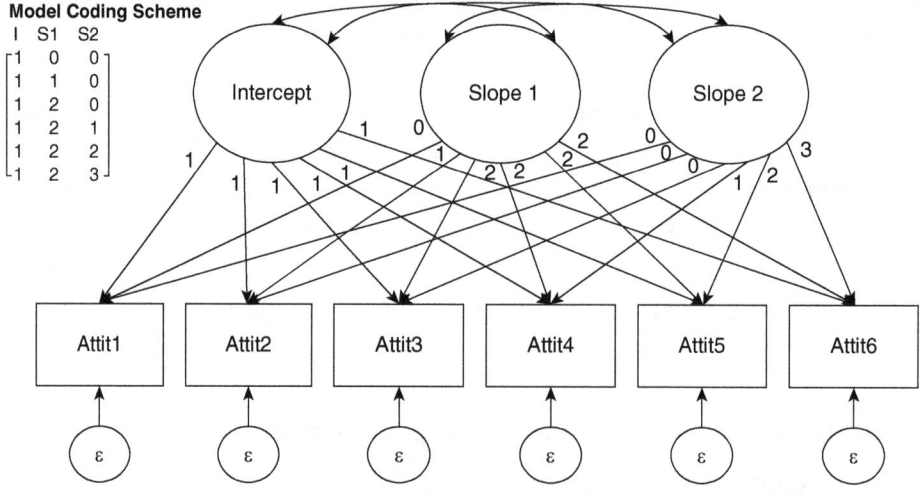

**Figure 7.5** A piecewise growth model with two slopes. The first slope captures growth from waves 1 to 3 and the second slope captures growth from waves 3 to 6

Whether we use MLM or SEM to fit piecewise models, the 'magic' of the piecewise model occurs in the coding of time, and our interpretations of the growth slopes depend on our time codes. In SEM, we code the time information in the factor loadings; in MLM, we create time variables that contain the relevant time information. However, the logic for coding time is identical across the two frameworks.

Moving forward, it may be helpful to think of these systems of time codes as clocks – each 'clock' captures the passage of time and can be turned on or off, as needed. These multiple 'time clocks' produce separate growth rates. Table 7.3 shows the coding scheme that generates two distinct growth rates. The second and third columns contain the two *clocks*. Each clock turns on, captures the growth rate during its respective period and turns off at the end of its period. Therefore, the first clock captures the passage of time in the first, pre-intervention time period and stops at the end of the first time period. The second clock captures the passage of time from the beginning of the intervention period (the second time period) to the end of the study. In Table 7.3, only one clock is 'turned on' at any given time period. Using this coding scheme, the intercept ($\beta_{00}$) is the expected value when both clocks 1 and 2 are 0, at wave 1. The first growth slope ($\beta_{10}$) captures the growth rate during the first time period (from waves 1 to 3) and the second growth slope ($\beta_{20}$) captures the growth rate during the second time period (from waves 3 to 6).

**Table 7.3**  Coding scheme for two distinct growth rates

| Wave | Time 'Clock' 1 | Time 'Clock' 2 |
|---|---|---|
| 1 | 0 | 0 |
| 2 | 1 | 0 |
| 3 | 2 | 0 |
| 4 | 2 | 1 |
| 5 | 2 | 2 |
| 6 | 2 | 3 |

Alternatively, we could model a baseline growth trajectory (slope) and a change in that slope. Think of the change in slope as a *deflection* parameter: it captures the increase or decrease in the slope during the second time period. (To deflect is to change direction or to turn aside from a straight course.) We use the same strategy described above, but we code the first clock differently (See Table 7.4). Under this coding scheme, the first clock is identical to the time coding for a linear growth model: time at wave 1 = 0 and continues through to wave 6 = 5. The coding for the second clock is identical across Tables 7.3 and 7.4: it still clocks the passage of time from the beginning of the intervention period (the second time period) to the end of the study. Although the second clock is identical across the two coding schemes, the meaning of the second growth slope changes. Using the coding scheme in Table 7.4, the second growth slope $(\beta_{20})$ now captures the *change* in the growth rate between the first and the second time periods (from waves 3 to 6). The interpretation of the intercept $(\beta_{00})$ and the first growth slope $(\beta_{10})$ remain constant across the two coding schemes. The intercept $(\beta_{00})$ is still the expected value at wave 1 (when time 1 and time 2 are 0). In either case, the first growth slope $(\beta_{10})$ is the baseline growth rate across the first phase, from waves 1 to 3. However, using the coding system in Table 7.3, $\beta_{20}$ captures the total growth during the second phase of the study, whereas the $\beta_{20}$ using the coding system in Table 7.4 captures the change in growth rate between the first and second phases of the study.

**Table 7.4** Coding scheme for two separate growth rates with baseline growth slope

| Wave | Time 'Clock' 1 | Time 'Clock' 2 |
|------|----------------|----------------|
| 1 | 0 | 0 |
| 2 | 1 | 0 |
| 3 | 2 | 0 |
| 4 | 3 | 1 |
| 5 | 4 | 2 |
| 6 | 5 | 3 |

The coding scheme in Table 7.3 produces two separate slopes, but we can compute the deflection (change) parameter by subtracting $\beta_{20}$ from $\beta_{10}$. Similarly, the coding scheme in Table 7.4 produces a change in slope; to compute the total slope, we can sum $\beta_{20}$ and $\beta_{10}$. These two models are statistically equivalent: they produce the same deviance. To test the hypothesis that the second slope is 0, use the coding scheme in Table 7.3. To test the hypothesis that the change in slope between the first and second phases of the study equals 0, use the coding scheme in Table 7.4.

## Polynomial growth models

All models thus far have been linear growth models. A variety of non-linear growth models exist (Grimm et al., 2016). This chapter discusses the most common (and simplest) method for modelling non-linearity: polynomial growth models. Polynomial growth models represent the outcome variable $(y_{ti})$ as an *n*th degree polynomial of time. Simple linear growth models (and **intercept-only models**) are part of the family of polynomial growth models. Below, we briefly describe zero-order (intercept-only), first-order (linear), second-order (quadratic) and third-order (cubic) growth models using MLM notation and provide a pictorial representation of the corresponding model in SEM. Given space constraints, we present polynomial growth models using the MLM framework. However, the coding system that we described above for piecewise models continues to apply: each slope is a latent variable in the SEM framework, and the polynomial coding of time occurs in the factor loadings.

## Intercept-only model

An intercept-only model is a zero-order growth model in which the outcome variable is a function only of $\pi_{0i}$ * $time_{ti}^0$. Because $time^0 = 1$, we generally drop the $time_{ti}^0$ portion of the level-1 intercept term from the equation, yielding

Level 1:

$$y_{ti} = \pi_{0i} + e_{ti}$$

Level 2:

$$\pi_{0i} = \beta_{00} + r_{0i} \tag{7.7}$$

This model specifies that the outcome variable at time $t$, for individual $i$, is a function of the intercept (which randomly varies across people) and a within-person residual, which captures the time-specific deviation of the outcome variable from the individual's intercept. The intercept represents the person's average level on the outcome variable across time. An intercept-only model specifies a mean (average level) on the outcome variable, and variability around that average level, but does not allow for systematic change across time. (See the horizontal light grey line in Figure 7.6.) In SEM, to fit

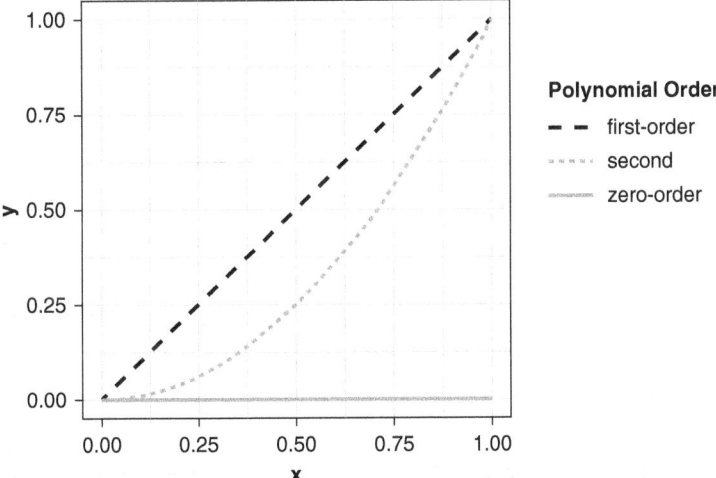

**Figure 7.6**   Polynomial order for polynomial growth models

an intercept-only model, we include a latent variable for the intercept, and we constrain all the factor loadings in the intercept factor to 1.

Although the zero-order growth model may seem overly simplistic, it can be quite useful. Revisiting the anxiety example, imagine that a researcher conducts an observational study of stress and anxiety. The researcher wishes to examine how much average anxiety varies across people (i.e. the between-person variance in anxiety), how much people's anxiety levels vary across time (i.e. the within-person variance in anxiety), whether a person's average level stress predicts their overall (average) anxiety, and whether daily fluctuations in stress levels predict within-person changes in anxiety. The researcher does not believe that people's anxiety levels systematically increase or decrease across time. Instead, she hypothesises that people have a homeostatic[i] level of anxiety: some people are naturally more anxious than others. This homeostasis represents a dynamic state of equilibrium rather than a constant, unchanging state. However, she also hypothesises that stress can knock people out of their equilibrium; higher stress increases anxiety and lower stress decreases anxiety. She could model a zero-order growth trajectory with stress included as a group mean–centred time-varying (level-1) covariate at level 1 and person's average stress level included as a time-invariant (level-2) covariate, as shown in Equation (7.7). $\beta_{00}$ represents expected anxiety (for a person of average stress), $\beta_{01}$ represents the effect of average stress on anxiety, $\beta_{10}$ represents the effect of within-person changes in stress on anxiety for a person with average overall stress levels. In other words, how much do daily changes in stress affect daily anxiety levels (for a person whose stress level is average)? The $\beta_{11}$ parameter represents the effect of overall stress level on the within-person effect of stress on anxiety. How do we interpret these parameters? If $\beta_{10}$ is positive, an average person feels more anxious on more stressful days and less anxious on less stressful days. If $\beta_{11}$ is positive, then $\pi_{1i}$ is more positive for people with higher average stress levels. The effect of daily changes in stress on anxiety is more pronounced for people who report higher overall stress levels. In other words, the higher a person's overall level of stress is, the more changes in their daily stress levels affect their anxiety levels: people with higher average stress levels are more susceptible to fluctuations in (within-person) anxiety levels caused by (within-person) fluctuations in stress levels. In contrast, if $\beta_{11}$ is negative, then $\pi_{1i}$ is more negative for people with higher average stress levels. The anxiety levels for people with higher overall stress levels are less affected by daily fluctuations in stress levels. In other words, the higher a person's overall stress level, the less within-person fluctuations in stress lead to within-person fluctuations in anxiety.

---

[i]Homeostasis refers to stability or equilibrium. Homeostasis represents a dynamic state of equilibrium rather than a constant, unchanging state.

Level 1:

$$Anxiety_{ti} = \pi_{0i} + \pi_{1i}\left(Stress_{.j}\right) + e_t i$$

Level 2:

$$\pi_{0i} = \beta_{00} + \beta_{01}\left(Stress_{..}\right) + r_{0i}$$
$$\pi_{1i} = \beta_{10} + \beta_{11}\left(Stress_{..}\right) + r_{1i} \tag{7.8}$$

## Linear model

Linear growth models are first-order polynomial growth models: the outcome variable $(y_{ti})$ is a function of $time_{ti}^0$ and $time_{ti}^1$. The equations from the linear growth model (Equation 7.1) are:

Level 1:

$$y_{ti} = \pi_{0i} + \pi_{1i}\left(time_{ti}^1\right) + e_{ti}$$

Level 2:

$$\pi_{0i} = \beta_{00} + r_{0i}$$
$$\pi_{1i} = \beta_{10} + r_{1i}$$

In the level-1 equation, $y_{ti}$ is a function of $\pi_{0i}$ * $time_{ti}^0$ and $\pi_{1i}$ * $time_{ti}^1$. The intercept $(\pi_{0i})$ represents individual $i$'s level on the outcome variable when time = 0. The time slope $(\pi_{1i})$ represents the predicted change in the outcome variable $(y_{ti})$ for individual $i$, at time $t$, for every 1-unit change in time. In a linear growth model, although the level of the outcome variable changes across time, the rate of change remains constant over time. (The dashed black line in Figure 7.6 represents a linear trajectory.)

## Quadratic model

Building on linear growth models, in the second-order polynomial model (a.k.a. quadratic growth model), $y_{ti}$ is a function of $\pi_{0i}$ * $time_{ti}^0$, $\pi_{1i}$ * $time_{ti}^1$ and $\pi_{2i}$ * $time_{ti}^2$. The quadratic model is

Level 1:

$$y_{ti} = \pi_{0i} + \pi_{1i}(time_{ti})^1 + \pi_{2i}(time_{ti})^2 + e_{ti}$$

Level 2:

$$\pi_{0i} = \beta_{00} + r_{0i}$$

$$\pi_{1i} = \beta_{10} + r_{1i}$$

(7.9)

$$\pi_{2i} = \beta_{20} + r_{2i}$$

In a linear growth model: as $time_{ti}^1$ increases by 1 unit, the outcome variable $(y_{ti})$ increases by $\pi_{1i}$ units, and this rate of change is constant across the time continuum. However, in a quadratic model, the rate of change is no longer constant across time. Because $time_{ti}^2$ increases more quickly than $time_{ti}^1$ does, the coefficient for the quadratic parameter $(\pi_2)$ has an increasingly large impact on the outcome variable across time.

Quadratic growth models contain three pieces of information: level $(\pi_{0i})$, the instantaneous rate of change when time = 0 $(\pi_{1i})$ and the change in the instantaneous rate of change across time $(\pi_{2i})$. If we code time so that $t = 0$ at initial status, then $\pi_{0i}$ represents the initial level of the outcome variable, and the first-order parameter $(\pi_{1i})$ represents the instantaneous rate of change at initial status. The instantaneous rate of change can be positive (indicating a positive growth slope when $t = 0$), negative (indicating a negative growth slope when $t = 0$) or 0 (indicating a flat trajectory when $t = 0$).

In a quadratic model, the instantaneous rate of change, itself, differs across time; the quadratic parameter $(\pi_{2i})$ describes the change in this rate of change. Other names for this second-order parameter $(\pi_{2i})$ include the acceleration parameter or the curvature parameter. The quadratic parameter essentially describes the degree and direction of the curve in the growth slope over time. (The grey, dotted line in Figure 7.6 is a quadratic trajectory.) The quadratic term can be positive, indicating that the growth slope becomes more positive across time, or the quadratic term can be negative, indicating that the growth slope becomes more negative across time. Notably, negative quadratic terms produce convex curves, whereas positive quadratic terms produce concave curves.

For those readers familiar with calculus, the second derivative of the quadratic equation with respect to time, $2 * \pi_{2i}$, is the rate of change in the first-order (linear) component for a 1-unit change in time (Biesanz et al., 2004): the growth slope changes by $2 * \pi_{2i}$ per unit change in time. For example, in Table 7.5 the first-order growth slope is 0.50 and the second-order growth slope is 0.10. Ignoring the intercept, which remains constant across time, we can compute the total change at each time point. The quadratic component starts at $\pi_{2i}$ (0.10) at time 1. Then, the difference in the quadratic component from time point to time point increases by $2 * \pi_{2i}$ (0.20) per unit change in time. For example, as shown in Table 7.5, the value of the quadratic component at time 0 is 0.00, at time 1 is 0.10, at time 2 is 0.40, at time 3 is 0.90 and at time 4 is 1.60. Therefore, the rate of change increases by .20 units per unit increase in time: the change from time 0 to 1 is .10, from time 1 to 2 is .30, from time 2 to 3 is 0.50 and from time 3 to time 4 is 0.70. Thus, although the rate of change is not constant over time, the change in the rate of change is constant across time: this rate of change increases by 2 * $\pi_{2i}$ (0.20 units) for each additional unit of time (McCoach et al., 2013; McCoach & Yu, 2016).

**Table 7.5** Computation of first-order, quadratic and composite growth parameters

| Time | $\pi_{1i}$ (t) | $\pi_{2i}$ (t²) | First Order | Quadratic | Composite |
|---|---|---|---|---|---|
| 0 | 0.50 (0) | 0.10 (0) | 0.00 | 0.00 | 0.00 |
| 1 | 0.50 (1) | 0.10 (1) | 0.50 | 0.10 | 0.60 |
| 2 | 0.50 (2) | 0.10 (4) | 1.00 | 0.40 | 1.40 |
| 3 | 0.50 (3) | 0.10 (9) | 1.50 | 0.90 | 2.40 |
| 4 | 0.50 (4) | 0.10 (16) | 2.00 | 1.60 | 3.60 |

In addition to the importance of considering each model component separately, it is imperative to remember that it is the *combination* of first-order (linear) and second-order (quadratic) slopes that determines the ultimate shape of the growth trajectory – no single parameter can adequately capture the model for change on its own. The composite column represents the total change from the intercept to the predicted value at time *t*. The model-predicted score when time = 0 is the intercept parameter, and the instantaneous growth rate is .50. At time = 1, the model-predicted score is .60 points higher than the intercept, and the instantaneous growth rate is .60. At time = 2, the model-predicted score is 1.4 points higher than the intercept, and the instantaneous growth rate is .80. At time = 3, the model-predicted score is 2.4 points higher than the intercept, and the instantaneous growth rate is 1.0. At time = 4, the model-predicted score is 3.6 points higher than the intercept, and the instantaneous growth rate is 1.2.

Both the first-order and second-order parameters determine the shape of the quadratic growth trajectory. Figure 7.3 illustrates the interplay of linear and quadratic slope directionality. Figure 7.3 contains four hypothetical quadratic trajectories, each of which is defined by a positive or negative $\pi_{1i}$ and a positive or negative $\pi_{2i}$. The upper-left quadrant depicts a trajectory with positive $\pi_{1i}$ and a positive $\pi_{2i}$ (+/+). When both $\pi_{1i}$ and $\pi_{2i}$ are positive, growth is positive when time = 0 and becomes increasingly positive across time. The upper-right quadrant depicts a trajectory in which $\pi_{1i}$ is positive and $\pi_{2i}$ is negative. This trajectory exhibits positive growth at the start. However, the negative $\pi_{2i}$ means that the growth slope becomes increasingly negative over time. Therefore, the growth slope becomes less and less positive and eventually becomes more and more negative. The bottom-left quadrant illustrates a trajectory with a negative $\pi_{1i}$ and a positive $\pi_{2i}$. This depicts a trajectory that begins as negative but becomes increasingly positive across time. At first, when the slope is negative, the positive $\pi_{2i}$ slows the rate of negative growth. Eventually, the positive $\pi_{2i}$ overtakes the negative $\pi_{1i}$, and the growth rate becomes positive, resulting in a U shape. The bottom-right quadrant (–/–) with a negative $\pi_{1i}$ and a negative $\pi_{2i}$, illustrates initially negative growth (or decline) that becomes increasingly more negative across time.

## Fitting higher order polynomial models

As seen in Figure 7.6, $\pi_{1i}$ can be positive or negative, depending on the centring of time. In polynomial models, the choice of centring does not affect the parameter estimate or interpretation of the highest order coefficient (Biesanz et al., 2004), but it does affect the parameter estimates and interpretations of all lower order coefficients and their variance components. Polynomial models using different centring schemes are statistically equivalent, parameter estimates can be transformed from one centring scheme to another. However, the coefficients for the intercept and the first-order growth slope lower-order parameters may be quite different in magnitude and/or sign. Therefore, centring decisions are especially important in higher-order polynomial models. When fitting polynomial models, it is helpful to generate predicted values across time for prototypical members of the sample and graph these results. Plotting prototypical growth trajectories aids in the interpretation of the results and helps convey the findings to a variety of audiences.

## Cubic model

In a quadratic model, the rate of change is not constant across time. The quadratic parameter captures the change in the rate of change, or the acceleration (deceleration) in the rate of change. However, the change in the rate of change may not be constant across time. Imagine driving a car. If we stay at a constant speed, there is a constant rate of change across time. For example, if I am on the highway driving 60 miles per hour, I travel one mile for every minute I drive. This is akin to a linear growth model. When we accelerate, we change the velocity at which we travel. This is a change in the rate of change, represented by our quadratic growth model. However, the rate at which velocity changes may itself change across time. For instance, if I am on a busy highway, I may accelerate to get to the speed limit (or a comfortable travelling speed; the acceleration parameter is positive), hit the brakes when I hit a traffic jam (the acceleration parameter is negative) and then accelerate again when the traffic jam clears (the acceleration parameter is positive again). In this scenario, the acceleration parameter is not constant; it changes across time. The cubic parameter in a polynomial model captures the change in the acceleration parameter: it represents the rate of *change* in the *change* in the rate of *change*. Many developmental processes follow a cubic trajectory: slow early growth is followed by a period of rapid growth (a growth spurt), which is then followed by another period of slow growth. In Chapter 8, we fit a cubic model to antisocial attitudes in adolescence. Antisocial attitudes grow slowly from ages 11 to 12, very quickly from ages 12 to 14, and then the rate of growth slows down again after age 14.

A quadratic model does not allow for change in the acceleration parameter: the trajectory becomes increasingly positive or increasingly negative: a quadratic model can only have one curve. To model a growth trajectory with more than one curve, it is necessary to fit a higher-order polynomial function. A cubic model does allow for change in the quadratic parameter across time; therefore, cubic models may contain two curves. For example, Figure 7.7 demonstrates two cubic functions where the *a*, *b* or *c* in the function $ax + bx^2 + cx^3$ are either positive or negative. Higher-order polynomials allow for additional numbers of curves (i.e. fourth-order polynomials, but it is fairly rare for polynomial growth models to fit polynomial functions above the cubic function.

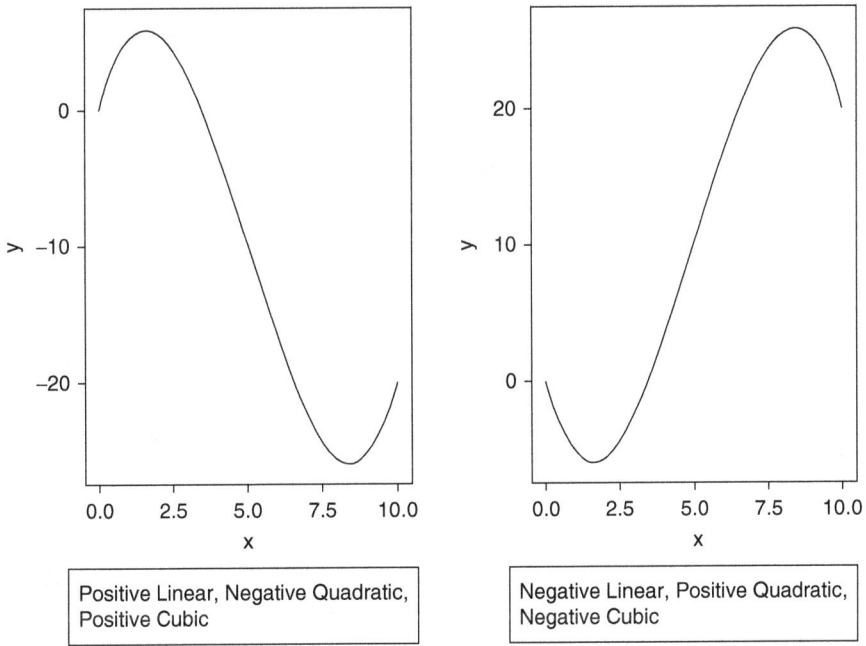

Positive Linear, Negative Quadratic, Positive Cubic

Negative Linear, Positive Quadratic, Negative Cubic

**Figure 7.7** Two cubic functions where the *a*, *b* or *c* in the function $ax + bx^2 + cx^3$ are either positive or negative

To fit a cubic model, we add an additional term to the growth model: time³. The path diagram for a cubic growth model appears in Figure 7.8. The cubic multilevel model appears in Equation (7.9):

Level 1:

$$y_{ti} = \pi_{0i} + \pi_{1i}(time_{ti})^1 + \pi_{2i}(time_{ti})^2 + \pi_{3i}(time_{ti})^3 + e_{ti}$$

Level 2:

$$\pi_{0i} = \beta_{00} + r_{0i}$$

$$\pi_{1i} = \beta_{10} + r_{1i}$$

$$\pi_{2i} = \beta_{20} + r_{2i}$$

(7.10)

$$\pi_{3i} = \beta_{30} + r_{3i}$$

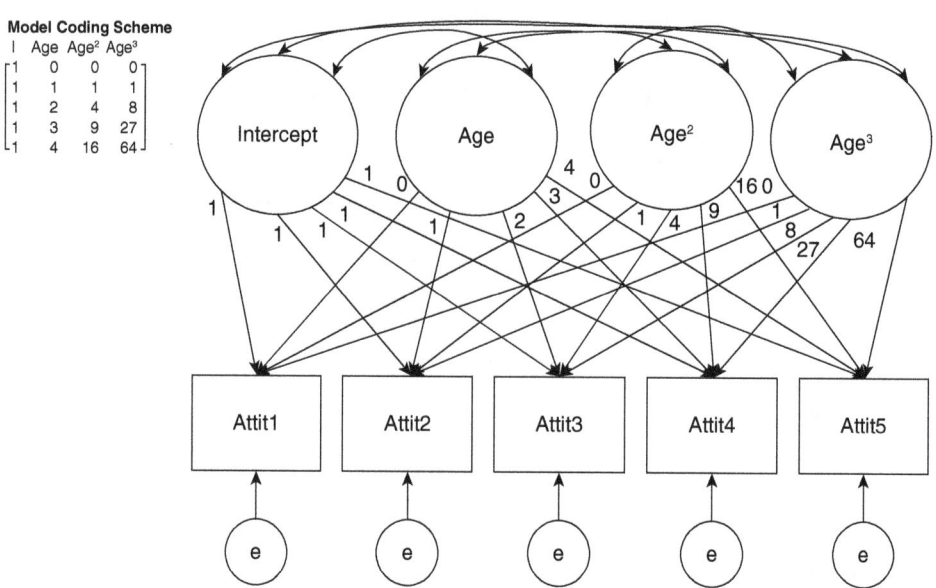

**Figure 7.8** Latent growth model with a cubic growth trajectory in structural equation modelling

This cubic parameter ($\pi_{3i}$) represents the rate of change in the quadratic parameter across time. Therefore, the shape of a cubic model is determined by the parameters for first-, second- and third-order time variables (remember that only the highest order coefficient is invariant to the centring decision). Parameter estimates for the cubic model must be interpreted in the context of the intercept and three growth slopes: they work together as a team to determine the shape of the growth trajectory. Therefore, as was the case for quadratic models, computing predicted values and graphing the growth trajectories is necessary to interpret the results of a cubic model.

To allow all coefficients (the intercept and the three growth slopes) in the cubic model to randomly vary across people requires at least five time points (although more time points are desirable for a model of this complexity). One advantage of a multilevel cubic model with randomly varying intercept and slope coefficients is that every person in the sample can have their own intercept, linear, quadratic and cubic parameters. Thus, in a cubic model with random effects for all four growth parameters, each individual can have a different shaped trajectory, as long as the trajectory has no more than two curves. For example, one person could have a negative linear, negative quadratic and positive cubic parameter, whereas another person could have a positive linear, positive quadratic and negative cubic parameter (recall Figure 7.7). The trajectories of these two people would look very different, but both trajectories could be accommodated by a cubic model that allows all growth parameters to randomly vary across people. Therefore, polynomial models are very flexible, much more so than piecewise models.

Individual growth trajectories are like spaghetti noodles. With a linear growth model, we can build any trajectory that we can form with an uncooked spaghetti noodle. With a piecewise linear model, we can break our spaghetti noodles into a set number of pieces and our individual growth trajectories can take on any shapes we can form by piecing together our uncooked noodles in different shapes. With higher order polynomial models, we cook our spaghetti noodles, allowing us to accommodate individual trajectories with a wide variety of shapes. A quadratic model allows individual growth trajectories of any shape that contains only one curve, cubic models can accommodate any shape with two curves, quartic models can accommodate any shape with three curves and so on. However, the gain in flexibility does come at a cost. Generally, polynomial models are more difficult to interpret. In addition, in some cases, it may be difficult to make a strong theoretical justification for polynomial models.

## Building growth models

Generally, building growth curve models requires a sequential approach. In general, we recommend first focusing on the within-person model of growth. After adequately specifying the shape of the growth trajectory and modelling within-person growth, then add person-level predictors to explain the between-person differences (variability) in within-person growth.

First, build an unconditional growth model that adequately captures the shape of the individual growth trajectories. At this stage, we recommend modelling the growth trajectory without any TVCs except for the time variable(s).

If the a priori assumption is that the quadratic model is likely the best fitting model, then it is sensible to begin with an unconditional level-1 model with an intercept, first-order and quadratic parameters. However, it is wise to compare the quadratic model to both the linear model and the cubic model. To determine whether the higher order components are necessary, we recommend comparing the models using the LRT, AIC or BIC. If the higher order polynomial model does not provide better fit than the lower model (using LRT or AIC/BIC), then the higher order term is probably not necessary, and the data can likely be fit with a simpler model.

After selecting a growth model that adequately reproduces the observed data, introduce TVCs into this model, including any conceptually important interactions among time variables and TVCs. Finally, after selecting the preferred within-person growth model, add time-invariant covariates (TICs) to explain between-person variance in the growth parameters. TICs are predictors that vary across individuals but remain constant across occasions (within a given individual). Examples of TICs include race/ethnicity, gender and other stable, personal characteristics.

Including TICs should help explain individual differences (between-person variability) in the intercept and/or the growth slopes. For instance, when modelling reading growth from ages 6 to 10, person-level predictors such as parents' SES, gender and cognitive ability may help to explain students' initial status and/or growth in reading. **Conditional growth models** estimate residual intercept and slope variances, the variances in the intercept and slope(s) that are not explained by the covariates. Ideally, if person-level covariates help to explain inter-individual variability in the intercept and slope, including those predictors in the model should decrease the (residual) variances for the slope and intercept. Therefore, one way to evaluate the necessity/utility of a given predictor is to determine what proportion of the variance in initial status (intercepts) or growth (slopes) a given predictor can explain.

In a conditional MLM, $\tau_{00}$ represents the residual variance in the intercept after controlling for the modelled covariates. Likewise, $\tau_{11}$ is the between-person residual variance in the time slope after accounting for the individual-level covariates (Raudenbush & Bryk, 2002). The $\tau_{01}$ from the conditional model is the residualised covariance between initial status and growth; standardised $\tau_{01}$ denotes the relationship between initial status and rate of change, after controlling for the other variables in the model.

In SEM, adding between-person covariates to predict the intercept and slope factors transforms the growth factors into endogenous variables. Therefore, instead of estimating exogenous factor variances for the intercept and the slope(s), the model estimates disturbance (residual) variances for the intercept and slope. As in the MLM, the disturbance variances represent unexplained variance in the intercept and slope(s).

**Box 7.2**

### Model Building Steps: Growth Curve Models

1. (Optional). Fit unconditional linear growth model as a baseline model. Often, when we know that the model is non-linear, this model simply serves as a straw man. Even so, fitting the unconditional linear growth model provides a baseline model to compare to other, more realistic models.
2. Model the shape of the growth trajectory without additional covariates. The unconditional growth model should capture shape of the unconditional growth trajectory and should provide reasonable predictions of individual growth data.
3. Add other within-person (time-varying) covariates to the model (if there are any).
4. Add between-person (time-invariant) covariates to the model.

This chapter introduced growth curve models in both MLM and SEM and demonstrated the general equivalence of the two approaches for standard growth curve models. In Chapter 8, we present an applied example which fits a growth curve model using both SEM and MLM.

### Chapter Summary

- This chapter provides a brief introduction to individual growth curve modelling in both the multilevel and structural equation modelling frameworks and discusses the benefits and drawbacks of each framework for growth curve modelling.
- An overview of three of the simplest and most commonly used individual growth models is provided: (1) the linear growth model, (2) the polynomial growth model and (3) the piecewise linear growth model.
- From the multilevel perspective, a simple growth curve model has two levels: (1) the within-individual level (level-1) and (2) the between-individual level (level-2). Observations across time, the level-1 units, are nested within people, the level-2 units. As such, the level-1 model captures the shape of an individual's growth trajectory over time and includes variables that vary across occasions, within individuals. From an SEM perspective, the within-person model is a measurement model.
- We outline model building steps for growth curve models. First (optional), fit an unconditional linear growth model as a baseline model to compare to other, more realistic models. Second, model the shape of the growth trajectory *without* additional covariates. The unconditional growth model should capture shape of the unconditional growth trajectory and should provide reasonable predictions of individual growth data. Third, add other within-person (time-varying) covariates to the model (if there are any). Fourth, add between-person (time-invariant) covariates to the model.

## Further Reading

Bollen, K. A., & Curran, P. J. (2006). *Latent curve models: A structural equation perspective*. Wiley.

This book provides an excellent introduction to latent growth curve models from an SEM perspective.

Grimm, K. J., Ram, N., & Estabrook, R. (2016). *Growth modeling: Structural equation and multilevel modeling approaches*. Guilford Press.

This book provides an extensive, technical review of growth curve modelling from both a multilevel modelling and structural equation modelling framework. Furthermore, the book demonstrates how to implement growth curve models in both frameworks using a variety of commercial and open-source software. This book is an excellent reference for social sciences researchers interested in growth modelling.

Singer, J. D., & Willett, J. B. (2003). *Applied longitudinal data analysis: Modeling change and event occurrence*. Oxford University Press.

The first half of this book provides an excellent introduction to individual growth curve modelling from a multilevel/mixed model framework.

# 8

# AN APPLIED EXAMPLE OF GROWTH CURVE MODELLING IN MLM AND SEM

## Chapter Overview

Model building steps for growth models.................................................. 191

Model 1: the unconditional linear growth model
(age 11 – no covariates) ........................................................................ 192

Model 2: piecewise unconditional linear growth model
(three-piece model)............................................................................... 197

Model 3: piecewise unconditional linear growth model
(two-piece model) ................................................................................. 202

Model 4: unconditional polynomial growth model................................. 206

Model 5: piecewise conditional growth model
(two-piece with gender covariate) ......................................................... 214

Conclusion............................................................................................. 217

Further Reading ..................................................................................... 219

This chapter presents an applied example of growth curve modelling using both SEM and MLM. Pedagogically, this allows us to compare results across the two approaches, and it reinforces the general equivalence of the two frameworks. This chapter serves as a finale of sorts: having integrated SEM and MLM in Chapter 7, we apply both frameworks to build a series of growth models.

The data for this chapter are from the National Youth Survey, a longitudinal study sponsored by the National Institute of Mental Health that explored conventional and deviant behaviour in adolescents (www.icpsr.umich.edu/icpsrweb/ICPSR/series/88).[1] Using these data, we examine adolescents' antisocial attitudes from ages 11 to 15 to answer three main research questions: (1) How antisocial were adolescents' attitudes at age 11? (2) How did adolescents' antisocial attitudes change across time from ages 11 to 15? (What was the shape of the growth trajectory, and how much did they change across that time period?) (3) Did gender moderate either initial antisocial attitudes (at age 11) or growth in antisocial attitudes? In both MLM and SEM, questions 1 and 2 focus on modelling the growth trajectory; neither question requires any person-level covariates. To answer research question 1 (*Where did adolescents start at age 11?*), we estimate the intercept. Research question 2 (*How did adolescents' antisocial attitudes change across time from ages 11 to 15?*) explores the shape of the growth trajectory. In contrast, research question 3 examines inter-individual differences in adolescents' initial status and rate of change as a function of gender, a person-level covariate.

For this example, our primary variable of interest is attitudes towards (i.e. tolerance of) antisocial/deviant behaviour (*ATTIT*) across five time points (from age 11 to 15), for 239 adolescents. Given that *ATTIT* were collected five times for each person, these data include a total of 1079 observations of the *ATTIT* variable. Remember, multilevel growth models require a long data set, containing a row for each observation and a single outcome variable (*ATTIT*). In contrast, fitting a growth model using SEM requires a wide data set: it must contain a separate outcome variable for *ATTIT* at each of the five time points (e.g. *ATTIT*1, *ATTIT*2, . . ., *ATTIT*5).

As shown in Table 8.1, across the entire sample, the mean of *ATTIT* increased at each consecutive time point, from 0.21 at age 11 to 0.45 at age 15; this increase in mean *ATTIT* over time suggests that attitudes towards deviant behaviour increased with age. The variance in *ATTIT* scores also increased across time, indicating greater between-person variability in adolescents' attitudes towards deviant behaviour as they matured (see Figure 8.1). Figure 8.2 plots *ATTIT* scores for males and females. Males had higher levels of attitudes towards deviant behaviour, males had higher *ATTIT* at

[1] Elliott, Delbert S. National Youth Survey [United States]: 1976. Ann Arbor, MI: Inter-university Consortium for Political and Social Research [distributor], 2008-08-01. https://doi.org/10.3886/ICPSR08375.v2

age 11 and across all subsequent measurement occasions and the gap between males and females persisted across time (see Table 8.1 for specific *ATTIT* values by gender).

**Table 8.1** Descriptive statistics for tolerant attitudes towards deviant behaviour (*ATTIT*)

|  | Age 11 | Age 12 | Age 13 | Age 14 | Age 15 |
|---|---|---|---|---|---|
| Overall |  |  |  |  |  |
| N | 202 | 209 | 230 | 220 | 218 |
| Mean (SD) | 0.21 (0.19) | 0.24 (0.21) | 0.33 (0.27) | 0.41 (0.29) | 0.45 (0.30) |
| Min/Max | 0.00/0.80 | 0.00/1.13 | 0.00/1.20 | 0.00/1.24 | 0.00/1.20 |
| Females |  |  |  |  |  |
| N | 96 | 104 | 111 | 105 | 104 |
| Mean (SD) | 0.19 (0.16) | 0.21 (0.21) | 0.30 (0.25) | 0.36 (0.31) | 0.40 (0.32) |
| Min/Max | 0.00/0.69 | 0.00/1.13 | 0.00/0.94 | 0.00/1.24 | 0.00/1.20 |
| Males |  |  |  |  |  |
| N | 106 | 105 | 119 | 115 | 114 |
| Mean (SD) | 0.23 (0.20) | 0.27 (0.21) | 0.37 (0.28) | 0.46 (0.27) | 0.49 (0.27) |
| Min/Max | 0.00/0.80 | 0.00/0.69 | 0.00/1.20 | 0.00/1.13 | 0.00/1.02 |

*Note. SD* = standard deviation; Min = minimum; Max = maximum.

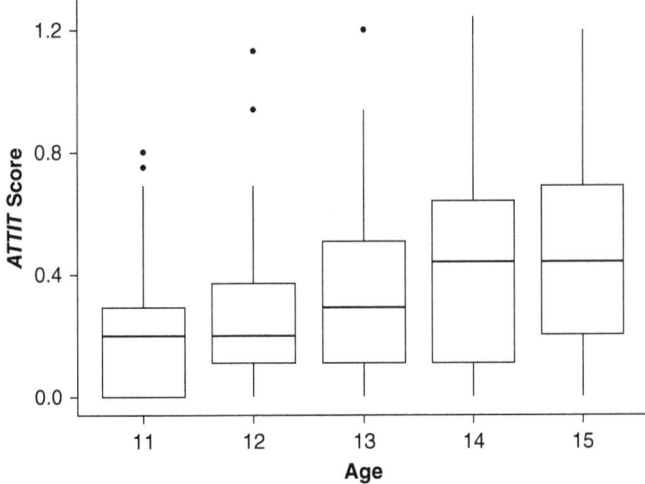

**Figure 8.1** Mean *ATTIT* by age

As seen in Figures 8.1 and 8.2, the growth of attitudes towards antisocial behaviour (*ATTIT*) did not appear to be linear. Instead, *ATTIT* increased slowly from ages 11 to 12, more quickly from ages 12 to 14, then more slowly again from ages 14 to 15. In other words, the growth slope was flatter from ages 11 to 12 and 14 to 15, but steeper from ages 12 to 14. The plots of several individual growth trajectories in Figure 8.3 indicate that within-person change in *ATTIT* was non-linear and somewhat erratic.

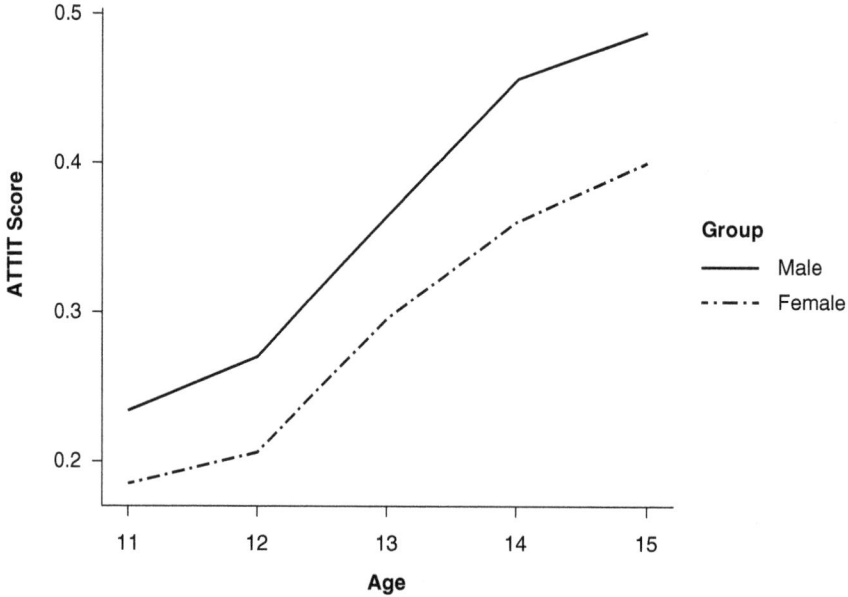

**Figure 8.2** Mean *ATTIT* by age and gender

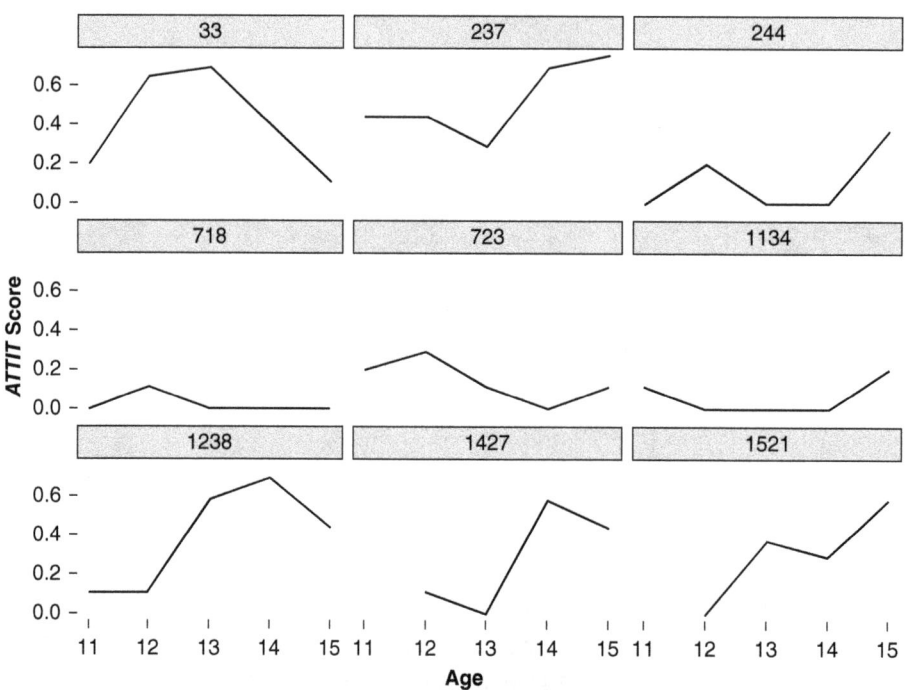

**Figure 8.3** Individual trajectories of ATTIT change for nine individuals in the data. For the purpose of clarity in presentation, not all individuals are shown

## Model building steps for growth models

Figures 8.1 to 8.3 provide preliminary evidence that a linear growth curve model is unlikely to be the best model for change in this sample. Even so, before estimating more complex non-linear models, we first estimate the linear growth curve model and interpret its parameters. This serves two pedagogical purposes: (1) it allows us to present and interpret a simple linear growth model before delving into more complex models and (2) it provides concrete evidence that the linear model does not adequately capture the shape of this growth trajectory.

To demonstrate various growth modelling approaches for evaluating the trajectories of *ATTIT* change, we fit a series of four unconditional growth models to the data using both SEM and MLM: (1) an unconditional linear growth model, (2) an unconditional three-piece growth model, (3) an unconditional two-piece growth model, and (4) an unconditional polynomial growth model. Afterwards, we discuss differences among the four unconditional growth models, compare these models and select our preferred unconditional growth model. Our final growth model includes gender as a predictor of initial status and growth.

As outlined in Chapter 7, to build growth models, first, we model the shape of the growth trajectory without additional covariates. The unconditional growth model should capture shape of the unconditional growth trajectory and should provide reasonable predictions of individual growth data. After selecting the most appropriate unconditional growth model, we add other within-person (time-varying) covariates to the model (if there are any). The final step includes between-person (time-invariant) covariates. These recommended steps underscore the importance of modelling the functional form of change as accurately as possible in the first step in the model building process.

At first glance, our sequence of growth model building steps may appear to deviate from our recommended model building steps for either MLM or SEM. However, that isn't true: we actually follow a similar logic in all three cases. In SEM with latent variables, we recommend fitting the measurement model first, prior to specifying the structural model. After establishing the adequacy of measurement model, we evaluate the structural parameters, commencing with the just-identified structural model. In SEM, misspecifications in the measurement model can lead to errors of inference in terms of interpreting the structural parameters. The same is true in growth modelling. The SEM framework uses CFA to fit growth models; therefore, the unconditional growth model *is* the measurement model. This growth (measurement) model must be adequately specified before considering the effects of structural variables on the growth parameters. Misspecifications in the growth model can lead to errors of inference in interpreting the effects of covariates on the growth parameters. Therefore, we recommend attending to the unconditional growth model first, then adding TVC followed by TIC.

In standard MLM, after fitting the empty model to estimate the ICC, we recommend fitting the full model, which is analogous to our suggestion for fitting path models (structural equation models without latent variables). If there is no measurement model (which is the case in path analysis or MLM), then we recommend fitting and evaluating the full (just-identified) model. In both cases, our strategy incorporates any potential confounders to reduce bias in our parameter estimates. Thus, non-growth MLM and path models use similar model building strategies.

Finally, growth models in MLM use the level-1 model to capture the shape of the growth trajectory. Therefore, in growth models, the level-1 model in MLM serves the same function as the measurement model in SEM. Therefore, it is essential to build the level-1 growth model and assess its adequacy before considering how the conceptual variables of interest moderate the growth parameters. Again, the steps are parallel: in both cases, we attend to the growth trajectory, either as a measurement model or as a level-1 model prior to considering the effects of the structural parameters (level-2 variables) on growth.

Let's return to our example, modelling growth adolescents' antisocial attitudes (*ATTIT*), and begin the model building process.

## Model 1: The unconditional linear growth model (age 11 – no covariates)

The multilevel specification for the unconditional linear growth model is

Level 1:

$$ATTIT_{ti} = \pi_{0i} + \pi_{1i}Age_{ti} + e_{ti}$$

Level 2:

$$\pi_{0i} = \beta_{00} + r_{0i}$$

$$\pi_{1i} = \beta_{10} + r_{1i}$$

(8.1)

In Equation (8.1), subscript $t$ refers to time, and subscript $i$ refers to the individual. In this example, we centre $Age_{ti}$ at age 11 (i.e. Age 11 = 0 or Age = Age – 11), so the intercept $\pi_{0i}$ represents an individual's initial level of *ATTIT* at age 11 (i.e. when time = 0). The time slope ($\pi_{1i}$) describes the change in person $i$'s *ATTIT* score per-unit change in Age (time). *Age* (time) is measured in years, so the slope represents the predicted yearly growth rate in *ATTIT*. In the linear model, *ATTIT*, measures at age or time $t = 0$, 1, . . ., $T$ on individual $i = 1, 2, . . ., n$, increases linearly: as age changes by 1 year,

*ATTIT* changes by $\pi_{1i}$ units. The residual term for time-specific error in the model, $e_{ti}$, represents the time-specific deviation in antisocial attitudes at time *t* from person *i*'s model-predicted score on the *ATTIT* variable at time t. The variance of $e_{ti}$, pooled across both time and people, is $\sigma^2$. Both the intercept and slope coefficients from level 1 ($\pi_{0i}$ and $\pi_{1i}$) become outcome variables in the level-2 equations. The overall intercept, $\beta_{00}$, is expected *ATTIT* when time = 0; $\beta_{10}$ is the expected linear growth slope. $\beta_{00}$ and $\beta_{10}$ coefficients are the same across the *i* individuals in the sample. The level-2 residuals ($r_{0i}$ and $r_{1i}$), the random effects for the intercept and slope, vary across individuals. These person-specific residuals capture the divergence of individual *i*'s personal intercept and time slope from the overall intercept ($\beta_{00}$) and time slope ($\beta_{10}$) (Remember $r_{0i}$ and $r_{1i}$ are normally distributed with means of zero, variances of $\tau_{00}$ and $\tau_{11}$ and a covariance of $\tau_{01}$.) The unconditional linear growth model estimates six parameters: two fixed effects for the growth parameters ($\beta_{00}$ and $\beta_{10}$) and four variance components ($\sigma^2$, $\tau_{00}$, $\tau_{11}$ and $\tau_{01}$).

Figure 8.4 depicts the equivalent SEM path diagram for the unconditional linear growth model. All loadings from the intercept factor to the five *ATTIT* scores are constrained to 1; the fixed loadings for the slope factor (0, 1, 2, 3, 4) represent our time 'clock.' The coding of time is equivalent across the multilevel and structural equation models: 1 unit equals 1 year. In MLM, the variable $Age_{ti}$ clocks the passage of time; in SEM, the fixed factor loadings contain this information. The design matrix in Figure 8.4 depicts coding for the fixed intercept and slope factor loadings. (All remaining figures in this chapter include these design matrices as well.) The structural equation model results in Table 8.2 do not include the constrained factor loadings because they are not freely estimated parameters.

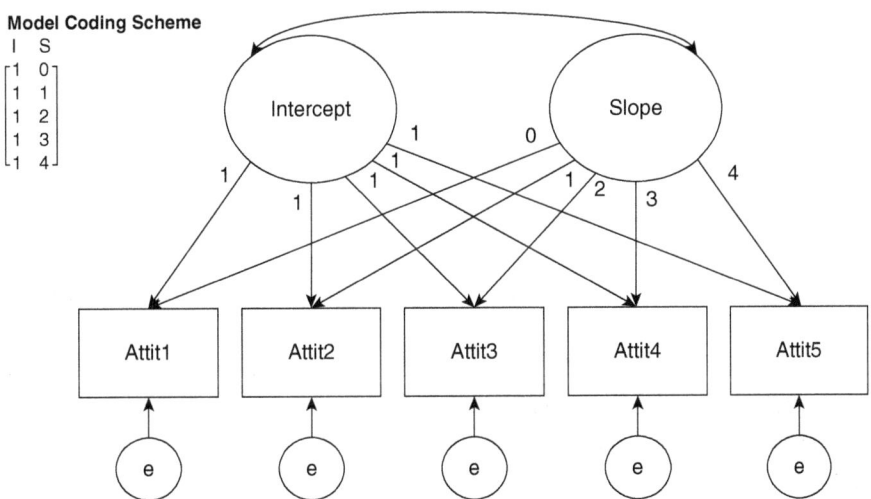

**Figure 8.4** Unconditional linear growth model

**Table 8.2** Model 1 results using the linear growth model in MLM and SEM

|  | MLM | | SEM | |
|---|---|---|---|---|
|  | Estimate | SE | Estimate | SE |
| Fixed effects |  |  |  |  |
| Intercept ($\beta_{00}$) | 0.198*** | 0.012 | 0.198*** | 0.012 |
| Age ($\beta_{10}$) | 0.065*** | 0.005 | 0.065*** | 0.005 |
| Random effects |  |  |  |  |
| Intercept variance ($\tau_{00}$) | 0.013 |  | 0.013 | 0.003 |
| Slope variance ($\tau_{11}$) | 0.003 |  | 0.003 | 0.001 |
| Correlation intercept and slope ($\tau_{01}$) | 0.48 |  | 0.48*** | 0.220 |
| Residual variance ($\sigma^2$) | 0.026 |  | 0.026ª | 0.002 |

*Note.* MLM = multilevel modelling; SEM = structural equation modelling; *SE* = standard error.

ªThe residual variance is the result of fixing the error variance at each respective time point equal (see Figure 8.4). The total number of students in the analysis is 239 with 1079 observations across five time points. The parameter estimates for the multilevel and structural equation models are the full information maximum likelihood estimates.

***$p < .01$. **$p < .05$. *$p < .10$.

Standard **MLM growth models** assume that the time-specific within-person residual variance is constant across time ($\sigma^2$). Therefore, to fit an equivalent structural equation model, we must constrain the five time-specific error variances to be equal. Doing so results in an **SEM growth model** that also estimates six parameters: two latent means (for the intercept and the slope) and four variances and covariances (the factor variance of the intercept, the factor variance of the slope, the covariance of the slope and intercept factors, and 1 error variance). In both SEM and MLM, it is possible to relax the assumption that the residual error variance in *ATTIT* is constant across time and allow individually varying error variances across each of the time points. Doing so increases the number of error variances from 1 to $T$, thereby increasing the number of estimated parameters by $T - 1$ (where $T$ is equal to the number of distinct measurement occasions/waves of data collection in the study). Because this data contains five measurement occasions (at ages 11, 12, 13, 14 and 15), allowing the error variances to vary across time would increase the number of estimated parameters from 6 to 10. Exhibit 8.1 explicitly demonstrates the correspondence between MLM and SEM linear growth.

Table 8.2 displays both MLM and SEM results for the unconditional linear growth model. The chi-square for the SEM linear growth model is 51.39 with 14 *df* ($p < .001$), suggesting that the linear model does not fit the data.

The MLM and SEM parameter estimates are virtually identical. The MLM analyses used the **lme4** package in R (Bates et al., 2015), whereas the structural equation

**Exhibit 8.1**  Comparing the MLM and SEM approaches for the linear growth model

| Multilevel Model | Structural Equation Model With Homogeneous Error Variances |
|---|---|
| Linear Growth Models | |

Level-1 Model:

$$ATTIT_{ti} = \pi_{0i} + \pi_{1i}Age_{ti} + e_{ti}$$

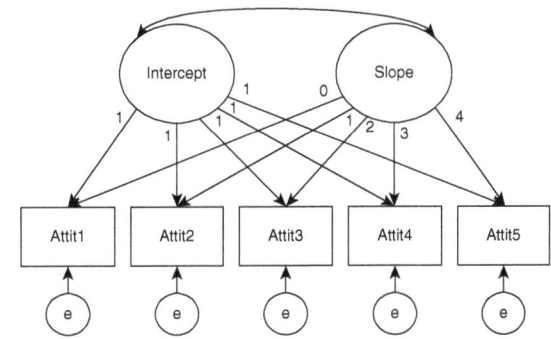

Level-2 Model:

$$\pi_{0i} = \beta_{00} + r_{0i}$$

$$\pi_{1i} = \beta_{10} + r_{1i}$$

<u>Two Fixed Effects:</u>

$\beta_{00}$: expected (mean) initial score

$\beta_{10}$: expected (mean) growth rate

<u>Variance Components:</u>
<u>Level-1:</u>
Variance of $e_{ti}$ ($\sigma^2$): pooled within-person residual variance, assumed to be homogeneous across time points

<u>Level-2:</u>

Variance of $r_{0i}$ ($\tau_{00}$): between-person (residual) variance in initial *ATTIT* scores

Variance of $r_{1i}$ ($\tau_{11}$): between-person (residual) variance in the *ATTIT* growth slope

The covariance of the slope and the intercept is $\tau_{01}$

In the structural equation model, the error variances of for *ATTIT1 to ATTIT5* are constrained to be equal, which corresponds to the level-1 error variance ($\sigma^2$) in the multilevel model. The latent 'Intercept' factor is the initial *ATTIT* level, the mean of 'Intercept' corresponds to the fixed effect, $\beta_{00}$, and the variance of 'Intercept' corresponds to the variance of $r_{0i}$ ($\tau_{00}$). The latent 'Slope' factor is the latent growth rate. The mean of the Slope factor corresponds to the fixed effect, $\beta_{10}$. The variance of 'Slope' corresponds to the variance of $r_{1i}$ ($\tau_{11}$). The covariance of the slope and the intercept is $\tau_{01}$ in multilevel model.

models used the **lavaan** package in R (Rosseel, 2012) and Mplus. (The companion website for this book contains details on model code in different software packages.) SEM produces chi-square tests of model fit for each of our models. One somewhat software specific difference is that lme4 does not report the standard errors for any of the random effects. Therefore, lme4 did not test the statistical significance of $\tau_{01}$, the correlation between the intercept and slope. However, the SEM results indicated that the correlation between the slope and the intercept is statistically significant.

How do we interpret these results? The intercept ($\beta_{00}$) is 0.198, which, rounded to decimal places is 0.20. Because we centred time at age 11, average *ATTIT* scores at age 11 are approximately 0.20. How much do 11-year-olds differ from each other in terms of *ATTIT*? The between-person variance in *ATTIT* at age 11 ($\tau_{00}$) is .013. Taking the square root of $\tau_{00}$ provides the standard deviation of the intercept ($\sqrt{0.013}$ = 0.11). Assuming a normal distribution, approximately 68% of expected (predicted) age 11 *ATTIT* scores fall within 1 *SD* and approximately 95% fall within $2^{ii}$ *SD* of the intercept. Rounding to two decimal places, 68% of predicted age 11 *ATTIT* scores (intercepts) should fall between 0.09 and 0.31 ($0.20 \pm 0.11$); 95% predicted *ATTIT* scores at age 11 should fall between –0.02 and 0.42 ($0.20 \pm (1.96*0.11) = 0.20 \pm 0.22$). Scores on *ATTIT* have a minimum of 0.00, so in reality, 95% of scores at age 11 (intercepts) should be between 0 and 0.42.

The linear slope parameter ($\beta_{00}$) is 0.065, meaning that expected *ATTIT* scores increase by an average of 0.065 units per year. The standard deviation of growth slope provides information about the variability in growth slopes. The slope variance is 0.003, so the standard deviation in slopes is 0.05 ($\sqrt{0.003}$). Thus, approximately 68% of the growth slopes are between 0.015 and 0.115, and about 95% of the growth slopes are between –0.035 and 0.165. The growth rate in *ATTIT* varies considerably across people, and some individuals have flat (or even slightly negative) growth slopes. The correlation between the intercept (age 11 *ATTIT*) and slope (yearly *ATTIT* growth) is .48, indicating a positive relationship between age 11 *ATTIT* and growth in *ATTIT*. Students with higher *ATTIT* at age 11 also tend to have more positive *ATTIT* growth.

The pooled within-person residual variance ($\sigma^2$) is 0.026. At each time point (time $t$), we can compute the difference between a person's model-predicted value and their actual values: $\sigma^2$ pools these residuals both across time and across people, indicating the degree of within-person, time-specific error in our growth model.

To interpret the growth parameters, we can create and graph prototypical predicted values for common types of people across the data collection period. Table 8.3 contains the predicted (model-implied) *ATTIT* values at each age for the unconditional linear growth model. The predicted values of *ATTIT* use the equation for linear growth (ignoring the error term, *e*). Substituting the coefficients from our model (rounding intercept to two decimal places) (i.e. $\beta_{00} = \pi_{0i}$ and $\beta_{10} = \pi_{1i}$), we obtain the equation below:

$$ATTIT_{ti} = \pi_{0i} + \pi_{1i}Age_{ti} \rightarrow ATTIT_{ti} = 0.200 + 0.065Age_{ti}$$

---

[ii.]Technically assuming a normal distribution, 95% of scores should fall within 1.96 *SD* of the mean. We use 2 as a nice approximation here.

**Table 8.3** Predicted values for the linear growth model

| Measurement Occasion | Intercept * | $\beta_{00}$ (0.20) + | Time * | $\beta_{10}$ (0.065) = | Predicted Value | Sample Mean |
|---|---|---|---|---|---|---|
| Age 11 | 1 | 0.20 | 0 | 0.00 | 0.20 | 0.21 |
| Age 12 | 1 | 0.20 | 1 | 0.07 | 0.27 | 0.24 |
| Age 13 | 1 | 0.20 | 2 | 0.13 | 0.33 | 0.33 |
| Age 14 | 1 | 0.20 | 3 | 0.20 | 0.40 | 0.42 |
| Age 15 | 1 | 0.20 | 4 | 0.26 | 0.46 | 0.45 |

Note. 0 = age 11, 1 = age 12, 2 = age 13, 3 = age 14, and 4 = age 15. For the structural equation model, a cell under a given coefficient is the product of the given coefficient value in parentheses and the preceding row for either the time intercept or time. For example, the third cell under $\beta_{00}$ contains the value 0.20, which is the product of 0.20 * 1. Likewise, the third cell under $\beta_{10}$ contains the value 0.13, which is the product of 0.065 * 2. The predicted values can be obtained from summing the coefficient values across a given row (e.g. for row 3, or age 13, the predicted value of ATTIT is 0.20 + 0.13 = 0.33).

The predicted values for each age appear in Table 8.3. The predicted attitude scores are 0.200 at age 11 ($y_{1i} = 0.200 + 0.065 * 0$), 0.265 at age 12 ($y_{2i} = 0.200 + 0.065 * 1$), 0.33 at age 13 ($y_{3i} = 0.200 + 0.065 * 2$), 0.40 at age 14 ($y_{4i} = 0.200 + 0.065 * 3$) and 0.460 ($y_{5i} = 0.200 + 0.065 * 4$) at age 15. The predicted values deviate substantially from the sample means at ages 12 and 14 and deviate somewhat from the sample means at ages 11 and 15, indicating that this linear growth model does not reproduce our observed means very well. Instead, the linear model overpredicts ATTIT at ages 12 and 15 and underpredicts ATTIT at ages 11 and 14.

## Model 2: Piecewise unconditional linear growth model (three-piece model)

Figure 8.2 seems to depict three distinct periods of growth: (1) initial ATTIT growth is modest (slightly positive) from ages 11 to 12, (2) growth becomes faster (more positive) from ages 12 to 14, then (3) growth slows again. From ages 14 to 15, growth looks roughly similar to growth during the first time period. Therefore, we specified the three-piece unconditional linear growth model:

Level 1:

$$ATTIT_{ti} = \pi_{0i} + \pi_{1i}Slope1_{ti} + \pi_{2i}Slope2_{ti} + \pi_{3i}Slope3_{ti} + e_{ti}$$

Level 2:

$$\pi_{0i} = \beta_{00} + r_{0i}$$

$$\pi_{1i} = \beta_{10}$$

$$\pi_{2i} = \beta_{20} + r_{2i} \qquad\qquad (8.2)$$

$$\pi_{3i} = \beta_{30}$$

Table 8.4 contains the coding for the three slopes (pieces) that capture the three different growth rates for the three distinct phases. The first slope (0, 1, 1, 1, 1) measures *ATTIT* change from ages 11 to 12, the second slope (0, 0, 1, 2, 2) captures growth from ages 12 to 14 and the third slope (0, 0, 0, 0, 1) measures *ATTIT* change from ages 14 to 15. Figure 8.5 presents the model diagram for this three-piece piecewise unconditional linear growth model. Table 8.5 reports the MLM and SEM results for the three-piece model.

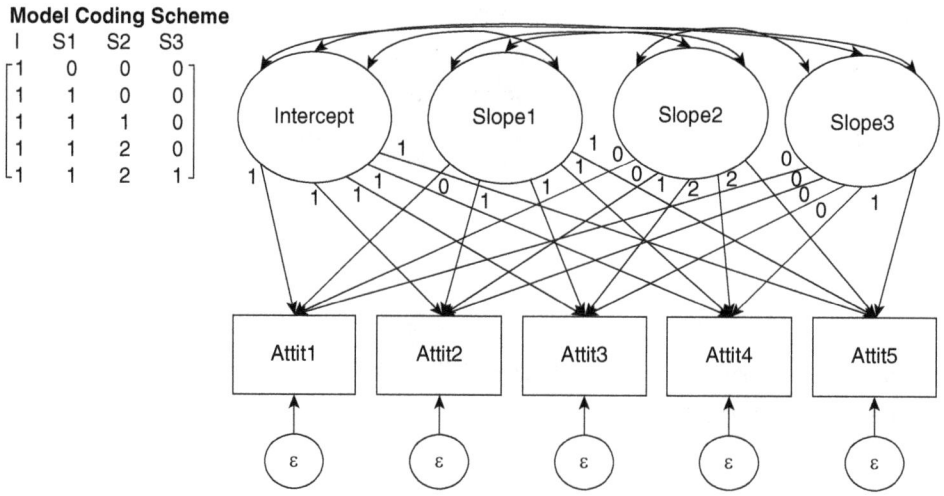

**Figure 8.5**  Three-piece unconditional growth model

**Table 8.4**  Coding for the three-piece linear growth model

| Measurement Occasion | Intercept | Piece 1 | Piece 2 | Piece 3 |
|---|---|---|---|---|
| *ATTIT* (Age 11) | 1 | 0 | 0 | 0 |
| *ATTIT* (Age 12) | 1 | 1 | 0 | 0 |
| *ATTIT* (Age 13) | 1 | 1 | 1 | 0 |
| *ATTIT* (Age 14) | 1 | 1 | 2 | 0 |
| *ATTIT* (Age 15) | 1 | 1 | 2 | 1 |

**Table 8.5**  Model 2 results using the three-piece growth model in MLM and SEM

| | MLM | | SEM | |
|---|---|---|---|---|
| | **Estimate** | *SE* | **Estimate** | *SE* |
| Fixed effects | | | | |
| Intercept ($\beta_{00}$) | 0.213** | 0.014 | 0.213** | 0.014 |
| Slope 1 ($\beta_{10}$) | 0.026 | 0.015 | 0.026 | 0.015 |
| Slope 2 ($\beta_{20}$) | 0.088** | 0.010 | 0.088** | 0.010 |
| Slope 3 ($\beta_{30}$) | 0.031* | 0.014 | 0.031* | 0.014 |
| Random effects | | | | |
| Intercept variance ($\tau_{00}$) | 0.018 | | 0.018 | 0.002 |
| Slope 1 variance ($\tau_{11}$) | — | — | — | — |
| Slope 2 variance ($\tau_{22}$) | 0.008 | | 0.008 | 0.001 |
| Slope 3 variance ($\tau_{33}$) | — | — | — | — |
| Correlation intercept and slope 2 ($\tau_{02}$) | 0.41 | | 0.41** | .153 |
| Residual variance ($\sigma^2$) | 0.025 | | 0.025 | 0.001 |

*Note.* The total number of students in the analysis is 239 with 1079 observations across five time points. The parameter estimates for the multilevel and structural equation models are the maximum likelihood estimates. MLM = multilevel modelling; SEM = structural equation modelling; *SE* = standard error.

**p < .01. *p < .05.

Recall from Chapter 7, given $T$ time points, we can estimate no more than $T - 1$ randomly varying coefficients. Therefore, for five time points, we can estimate a maximum of four random effects, so it is possible to estimate a model with one randomly varying intercept and three randomly varying slopes. The model contains 15 parameters (4 means, 10 taus and $\sigma^2$). However, both MLM and SEM experienced estimation difficulties. The lme4 package failed to converge and produced a warning that one or more of the variance parameters were on the boundary of the feasible parameter space, suggesting that one or more of the variances of the random effects was essentially 0. In HLM, the model converged after more than 7500 iterations; however, the slope variances for pieces 1 and 3 were not statistically significantly different from 0. Such model convergence issues usually indicate there is little to no variance in one or more of the randomly varying slopes (or the randomly varying intercept). The solution is to fix one or more random effects to 0 (McCoach et al., 2018; Raudenbush & Bryk, 2002; Singer & Willett, 2003). Similarly, the SEM analysis produced a warning that the latent variable covariance matrix was not positive definite. The SEM output indicated that the variance for piece 1 factor was 0 and the variance for piece 3 factor was –0.007 (but variances cannot be negative!).

Because both the HLM and Mplus analyses suggested that the variances of the first and third slopes were essentially 0, our next three-piece model included only two random effects (factors). We estimated variances for the intercept and the growth slope for ages 12 to 14, and we constrained the variances for the first and third growth slopes to 0. The period from ages 12 to 14 exhibited the greatest growth; there was also substantial between-person variability in the slope during this growth spurt phase. Constraining the slope variances for the first and third pieces to 0 means everyone has the same predicted growth slope from ages 11 to 12 and from ages 14 to 15. Therefore, any between-person differences in growth rates must occur between ages 12 and 14 (during the second time period). This model estimated two variances ($\tau_{00}$ and $\tau_{22}$) and a covariance ($\tau_{02}$) between the intercept and the age 12 to 14 growth slope. Therefore, this model estimates a total of eight parameters: four means (one mean intercept and three mean slopes), three between-person variances (one intercept variance, one slope variance [for growth from 12 to 14] and one covariance between the ages 12 to 14 growth slope and the intercept) and one pooled within-person error/residual variance. The model that included all variances for the intercept and the three slopes estimates a total of 15 parameters: four means (for the intercept and the three slopes), four between-person variances, six between-person covariances and one within-person error variance. Therefore, fixing the variances of two of the slopes to be 0 eliminates seven parameters: two variances and five covariances. As shown in Table 8.5, the results from our MLM and SEM analyses were again essentially equivalent. In SEM, the chi-square for this model was 27.94 with 12 $df$ ($p < .01$).

The intercept (i.e. the initial starting point in attitudes towards deviant behaviour) is 0.21 (as is the sample mean). The standard deviation of the intercept is 0.13 ($\sqrt{0.018}$), so 68% of predicted scores (intercepts) at age 11 are between 0.08 and 0.34 and 95% of predicted scores at age 11 (intercepts) are between –0.05 and 0.47.

Expected *ATTIT* growth varied across the three phases: (1) the growth from ages 11 to 12 ($\beta_{10}$) was 0.026 points per year, (2) the growth from ages 12 to 14 ($\beta_{20}$) was 0.088 points per year and (3) the growth from ages 14 to 15 was 0.031 points per year. The first ($\beta_{10} = .026$) and third ($\beta_{30} = .031$) slope coefficients are similar: the changes in attitudes towards deviant behaviour between ages 11 and 12 ($\beta_{10}$) and between ages 14 and 15 ($\beta_{30}$) are both approximately 0.03. In contrast, the slope from ages 12 to 14 ($\beta_{20} = 0.088$) is almost three times as large as the other two slopes: *ATTIT* grow almost 0.09 points per year. The standard deviation for the $\beta_{20}$ growth slope is 0.09 ($\sqrt{0.008}$). Rounding to two decimal places, 68% of the estimated yearly age 12 to 14 growth slopes are between 0.00 and 0.18, and 95% of estimated yearly growth slopes are between –0.09 and 0.27. The correlation of .41 between the intercept ($\beta_{00}$) and age 12 to 14 growth slope ($\beta_{20}$) indicates that students with higher *ATTIT* levels at age 11 tended to have more positive growth slopes; their *ATTIT* grew faster from ages 12 to 14. The pooled (across time) within-person residual variance ($\sigma^2 = 0.025$) is slightly lower for this model than it was in the linear growth model ($\sigma^2 = 0.026$).

**Table 8.6** Predicted values for the three-piece piecewise growth model

| Measurement Occasion | $\beta_{00}$ (0.21) | S1 | $\beta_{10}$ * S1(0.03 * S1) | S2 | $\beta_{20}$ * S2(0.09 * S2) | S3 | $\beta_{30}$ * S3(0.03 * S3) | Predicted Values |
|---|---|---|---|---|---|---|---|---|
| Age 11 | 0.21 | 0 | 0.03 | 0 | 0 | 0 | 0.00 | 0.21 |
| Age 12 | 0.21 | 1 | 0.03 | 0 | 0 | 0 | 0.00 | 0.24 |
| Age 13 | 0.21 | 1 | 0.03 | 1 | 0.09 | 0 | 0.00 | 0.33 |
| Age 14 | 0.21 | 1 | 0.03 | 2 | 0.18 | 0 | 0.00 | 0.42 |
| Age 15 | 0.21 | 1 | 0.03 | 2 | 0.18 | 1 | 0.03 | 0.45 |

*Note.* For the structural equation model, a cell under a given coefficient is the product of the given coefficient value in parentheses and the preceding row for either the time intercept or time slope. For example, the third cell under $\beta_{00}$ contains the value 0.21, which is the product of 0.21 * 1. Likewise, the third cell under $\beta_{10}$ contains the value 0.03, which is the product of 0.03 * 3. The predicted values can be obtained from summing the coefficient values across a given row. For example, in row 3, or age 13, the predicted value of *ATT/IT* is 0.21 + 0.03 + 0.09 + 0.00 = 0.33.

We computed model-predicted *ATTIT* values for the unconditional three-piece growth model (see Table 8.6) and compared these predictions to the sample means at each age. The model-predicted values were virtually identical to the sample means in *ATTIT* across all five measurement occasions (see Table 8.1), suggesting that the three-piece growth model adequately reproduces the five *ATTIT* means.

## Model 3: Piecewise unconditional linear growth model (two-piece model)

The slopes for ages 11 to 12 and 14 to 15 are virtually identical, so we can build a more parsimonious growth model which uses only two slope parameters to model the growth across all three developmental periods (and all five time points). The two-slope model constrains the slopes from ages 11 to 12 and from ages 14 to 15 to be equal. Therefore, this model specifies a baseline growth rate from 11 to 12, allows for a second growth rate from ages 12 to 14 and then drops everyone back to their baseline growth rate from ages 14 to 15. Table 8.7 contains the coding for the two-slope piecewise model. Because the baseline growth rate does not randomly vary across people (the random effect/slope variance is constrained to be equal), everyone has same predicted growth rate from 11 to 12 and again from 14 to 15. However, during the growth spurt period, from ages 12 to 14, not only is the growth rate allowed to differ from the baseline growth rate, it is also allowed to vary across people (because the model includes a random effect/factor variance for the age 12 to 14 growth slope). Exhibit 8.2 compares the SEM and MLM specifications.

**Table 8.7** Coding time for the two-piece piecewise model

| Measurement Occasion | Intercept | Piece 1 | Piece 2 |
| --- | --- | --- | --- |
| Age 11 | 1 | 0 | 0 |
| Age 12 | 1 | 1 | 0 |
| Age 13 | 1 | 1 | 1 |
| Age 14 | 1 | 1 | 2 |
| Age 15 | 1 | 2 | 2 |

The two-piece model and the three-piece growth model are nested, so we can compare the fit of the two models using the LRT. The two-piece model is more parsimonious because the age 11 to 12 and age 14 to 15 growth are constrained to be equal. This means that the two-piece model estimates one fewer parameter (and has one more *df*) than the three-piece model does.

**Exhibit 8.2**  Comparing the MLM and SEM approaches for two-piece piecewise growth models

| Multilevel Model | Structural Equation Model With Homogeneous Error Variances |
|---|---|
| Two-Piece Piecewise Growth Models | |

**Multilevel Model**

Level-1 Model:

$$ATTIT_{ti} = \pi_{0i} + \pi_{1i} Slope1_{ti} + \pi_{2i} Slope2_{ti} + e_{ti}$$

Level-2 Model:

$$\pi_{0i} = \beta_{00} + r_{0i}$$

$$\pi_{1i} = \beta_{10}$$

$$\pi_{2i} = \beta_{20} + r_{2i}$$

<u>Three Fixed Effects:</u>

$\beta_{00}$: expected value of the initial score

$\beta_{10}$: expected growth rate from times *ATTIT1* to *ATTIT2* and *ATTIT4* to *ATTIT5*

$\beta_{20}$: expected growth rate from time *ATTIT2* to *ATTIT4*

<u>Variance Components:</u>
<u>Level-1:</u>
Variance of $e_{ti}$ ($\sigma^2$): within-person residual variance, which was assumed to be homogeneous across time points

<u>Level-2:</u>
Variance of $r_{0i}$ ($\tau_{00}$): between-person residual variance in initial *ATTIT* scores

Variance of $r_{2i}$ ($\tau_{22}$): between-person residual variance in the second slope, which represented the growth rate from time *ATTIT2* to *ATTIT4*

**Structural Equation Model With Homogeneous Error Variances**

In the structural equation model, the error variances for each *ATTIT1–ATTIT5* are constrained to be equal. The latent *Intercept* factor is the initial *ATTIT* level at time 1. The mean of the *Intercept* factor corresponds to the fixed effect, $\beta_{00}$. The variance of *Intercept* corresponds to the variance, $\tau_{00}$. The latent slope factors *Slope 1* and *Slope 2* represent the growth rate from times *ATTIT1* to *ATTIT2* and *ATTIT4* to *ATTIT5* and the growth rate from time *ATTIT2* to *ATTIT4*, respectively. The means of *Slope 1* and '*Slope 2*' corresponds to the fixed effects, $\beta_{10}$ and $\beta_{20}$. The variance of '*Slope 2*' corresponds to the variance, $\tau_{22}$. The variance of *Slope 1* is constrained to zero in the structural equation model to conform with the multilevel model.

---

The multilevel equations for two-slope piecewise unconditional growth model are

Level 1:

$$ATTIT_{ti} = \pi_{0i} + \pi_{1i} Slope1_{ti} + \pi_{2i} Slope2_{ti} + e_{ti}$$

Level 2:

$$\pi_{0i} = \beta_{00} + r_{0i}$$

$$\pi_{1i} = \beta_{10}$$

(8.3)

$$\beta_{20}\pi_{2i} = \beta_{20} + r_{2i}$$

Note that the first slope (*Slope*1) combines the coding from slopes 1 and 3 in the three-slope piecewise model. Coding for the second slope (*Slope*2) is identical to that of slope 2 in the three-slope model. Conceptually, this model posits one growth rate from 11 to 12 ($\beta_{10}$), another growth rate from ages 12 to 14 ($\beta_{20}$) and a return to the initial growth rate from ages 14 to 15 ($\beta_{10}$). In addition, Figure 8.6 contains the model diagram for the two-piece unconditional growth model, and Table 8.8 reports the results for this model in both MLM and SEM. We did not estimate a variance for slope 1 (the slope for ages 11–12 and 14–15), but we did model variances for the intercept and the ages 12 to 14 growth slope.

**Table 8.8**  Model 3 results using the two-piece growth model in MLM and SEM

| | MLM | | SEM | |
|---|---|---|---|---|
| | **Estimate** | *SE* | **Estimate** | *SE* |
| Fixed effects | | | | |
| Intercept ($\beta_{00}$) | 0.211** | 0.013 | 0.211** | 0.013 |
| Slope 1 ($\beta_{10}$) | 0.029** | 0.011 | 0.029** | 0.011 |
| Slope 2 ($\beta_{20}$) | 0.088** | 0.010 | 0.088** | 0.010 |
| Random effects | | | | |
| Intercept variance ($\tau_{00}$) | 0.017 | | 0.018 | 0.003 |
| Slope 1 variance ($\tau_{11}$) | — | | — | — |
| Slope 2 variance ($\tau_{22}$) | 0.008 | | 0.008 | 0.001 |
| Correlation intercept and slope 2 ($\tau_{02}$) | 0.41 | | 0.41** | 0.15 |
| Residual variance ($\sigma^2$) | 0.025 | | 0.025 | 0.001 |

*Note.* The total number of students in the analysis is 239 with 1079 observations across five time points. The parameter estimates for the multilevel and structural equation models are the full information maximum likelihood estimates. MLM = multilevel modelling; SEM = structural equation modelling; *SE* = standard error.

**p < .01. *p < .05.

Both the MLM and SEM results are very similar to the results of the three-piece growth model. In the two-piece model, the intercept, predicted *ATTIT* at age 11, is 0.21. The standard deviation of the intercept is 0.13, indicating that 68% of the adolescents have intercepts between 0.08 and 0.34. (This is equivalent to Model 2.) Furthermore, the constrained age 11 to 12 and 14 to 15 slopes are 0.029, indicating

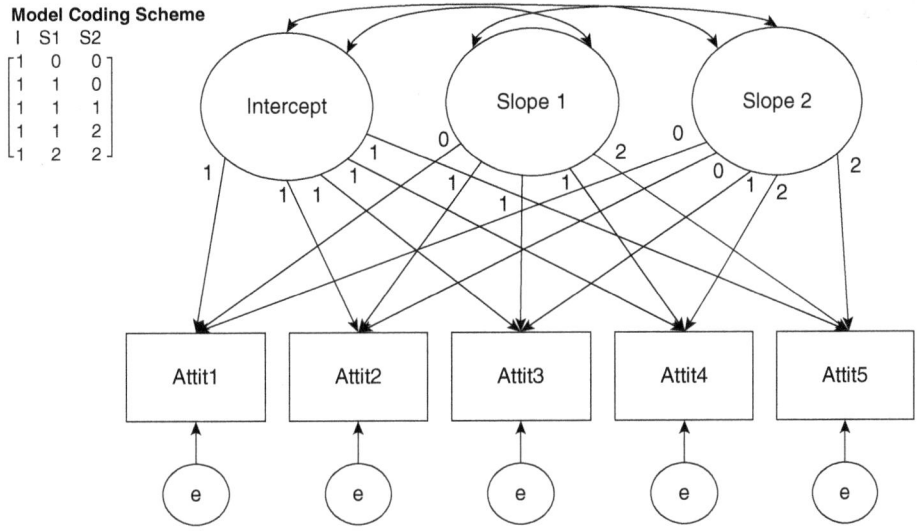

**Figure 8.6** Two-piece unconditional growth model

that growth in antisocial attitudes is 0.029 points per year from ages 11 to 12 and from ages 14 to 15. Because slope 1 is not allowed to randomly vary across people, the model-predicted growth trajectory is 0.029 for everyone. Again, between ages 12 and 14, expected *ATTIT* grow by 0.088 points per year. This slope does vary across people: The standard deviation is approximately 0.09 ($\sqrt{0.008}$). Approximately 68% of the estimated age 12 to 14 slopes are between 0.00 and 0.18, and 95% of the estimated slopes are between –0.09 and 0.27.

Table 8.9 contains predicted values for the unconditional two-piece piecewise growth model. Overall, the two-slope and three-slope piecewise models generate very similar parameter estimates and predicted values, and those predicted values are virtually identical to the sample means at each age. Figure 8.7 graphs the predicted values for each of these models. It is nearly impossible to distinguish between the two-slope piecewise, three-slope piecewise and sample means: they are essentially on top of one another. In contrast, as previously discussed, the predicted values from the unconditional linear growth model do not reproduce or correspond to the sample averages. *ATTIT* growth under the linear model is constant (0.065 points per year), whereas the piecewise models allow differential growth across three defined stages. The piecewise models indicate that growth rate in antisocial attitudes (*ATTIT*) changes across different periods of adolescence. Growth in *ATTIT* between ages 11 and 12 and between ages 14 and 15 is much slower (0.03 units per year), but *ATTIT* grow much more quickly from ages 12 to 14 (nearly 0.09 units per year).

**Table 8.9** Predicted values for the two-piece piecewise growth model

| Measurement Occasion | Time Intercept | $\beta_{00}$ (0.21) | Time Slope 1 | $\beta_{10}$ (0.03) | Time Slope 2 | $\beta_{20}$ (0.09) | Predicted Value |
|---|---|---|---|---|---|---|---|
| Age 11 | 1 | 0.21 | 0 | 0 | 0 | 0 | 0.21 |
| Age 12 | 1 | 0.21 | 1 | 0.03 | 0 | 0 | 0.24 |
| Age 13 | 1 | 0.21 | 1 | 0.03 | 1 | 0.09 | 0.33 |
| Age 14 | 1 | 0.21 | 1 | 0.03 | 2 | 0.18 | 0.42 |
| Age 15 | 1 | 0.21 | 2 | 0.06 | 2 | 0.18 | 0.45 |

*Note.* For the structural equation model, a cell under a given coefficient is the product of the given coefficient value in parentheses and the preceding row for either the time intercept or the time slope. The predicted values can be obtained from summing the coefficient values across a given row (e.g. for row 3, or age 13, the predicted value of *ATTIT* is 0.21 + 0.03 + 0.09 = 0.33).

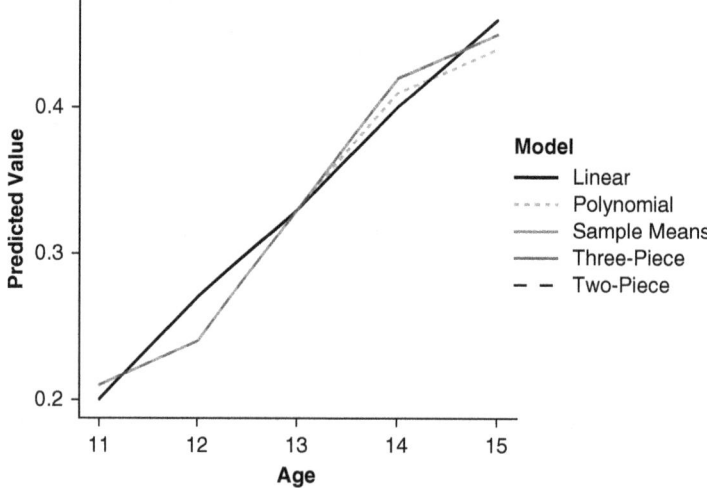

**Figure 8.7** Predicted values by model type. The plot includes the sample means along with the predicted values. Note the sample means are equivalent to the predicted values for the three-piece and two-piece models

## Model 4: Unconditional polynomial growth model

Using separate growth slopes that operate at different times during adolescence, the piecewise model allows for a trajectory where the growth rate increases (becomes more positive) and then slows (becomes less positive/more negative). We could also use a polynomial model to fit such a trajectory. As discussed in Chapter 7, a linear model fits a constant rate of change. A quadratic model allows for a change in the rate of change (an acceleration parameter); however, the change in the rate of change must be constant in a quadratic model. Given that the growth rate increases and then slows down again, we must fit a cubic growth model, which allows the acceleration (curvature) parameter to change.

The multilevel equations for the unconditional cubic growth model are

Level 1:

$$ATTIT_{ti} = \pi_{0i} + \pi_{1i}Age_{ti} + \pi_{2i}(Age_{ti})^2 + \pi_{3i}(Age_{ti})^3 + e_{ti}$$

Level 2:

$$\pi_{0i} = \beta_{00} + r_{0i}$$

$$\pi_{1i} = \beta_{10}$$

$$\pi_{2i} = \beta_{20} + r_{2i} \tag{8.4}$$

$$\pi_{3i} = \beta_{30}$$

The outcome variable ($ATTIT$) is a cubic function of three polynomial terms: Age, $Age^2$ and $Age^3$ (i.e. time, $time^2$ and $time^3$, respectively). Therefore, the model produces separate parameter estimates for the intercept and the linear, quadratic and cubic components of time. Figure 8.8 shows SEM representation of the unconditional cubic growth model, and Exhibit 8.3 compares the MLM and SEM approaches to fitting a cubic model. Exhibit 8.4 explains how to compute the degrees of freedom for these linear, piecewise, and polynomial growth models.

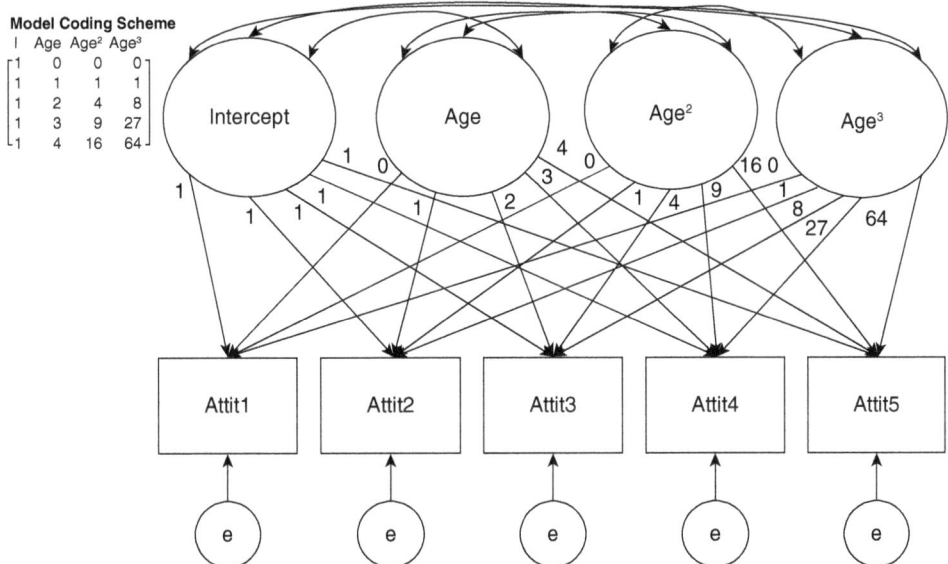

**Figure 8.8** One unconditional cubic growth model

**Exhibit 8.3** Comparing the MLM and SEM approaches for the cubic growth model

| Multilevel Model | Structural Equation Model With Homogeneous Error Variances |
|---|---|
| **Polynomial Growth Models** | |

Level-1 Model:

$$ATTIT_{ti} = \pi_{0i} + \pi_{1i}\, Age_{ti}$$
$$+ \pi_{2i}\, (Age_{ti})^2$$
$$+ \pi_{3i}\, (Age_{ti})^3 + e_{ti}$$

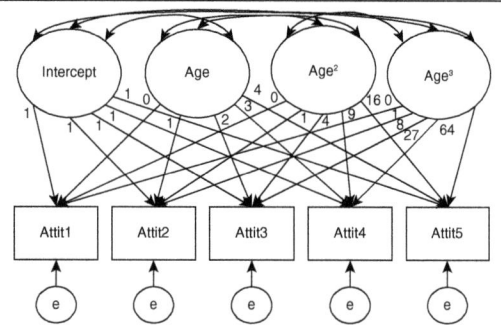

Level-2 Model:

$$\pi_{0i} = \beta_{00} + r_{0i}$$

$$\pi_{1i} = \beta_{10}$$

$$\pi_{2i} = \beta_{20} + r_{2i}$$

$$\pi_{3i} = \beta_{30}$$

Four Fixed Effects:

$\beta_{00}$: expected value of the ATTIT score when time = 0

$\beta_{10}$: expected instantaneous growth rate at time = 0

$\beta_{20}$: curvature (acceleration); rate of change in the (instantaneous) growth rate when time = 0

$\beta_{30}$: rate of change in the acceleration parameter, $\beta_{20}$. The rate at which the change in the rate of change (acceleration) is changing

In the structural equation model, the error variances for ATTIT1 to ATTIT5 are constrained to be equal. The latent Intercept factor was the initial ATTIT level at time 1. The mean of the Intercept factor corresponds to the fixed effect, $\beta_{00}$. The variance of the Intercept corresponds to $\tau_{00}$. The latent slope factor for Age captures the first-order growth rate from time ATTIT1 to ATTIT5. The mean of the age factor corresponds to the fixed effect, $\beta_{10}$. The latent slope factor for $Age^2$ captures the quadratic growth rate from time ATTIT1 to ATTIT5. The mean of $Age^2$ corresponds to the fixed effect, $\beta_{20}$. The variance of $Age^2$ corresponds to $\tau_{22}$. The variances of Age and $Age^3$ were constrained to zero in both the structural equation model and the multilevel model.

**Variance Components:**
Level-1:
Variance of $e_{ti}$ ($\sigma^2$): within-person variance, which was assumed to be homogeneous across time points

Level-2:
Variance of $r_{0i}$ ($\tau_{00}$): between-person residual variance in the initial ATTIT scores

Variance of $r_{2i}$ ($\tau_{22}$): between-person residual variance in the second slope, which represented the growth rate from time ATTIT1 to ATTIT5 squared

---

In a cubic growth model, when time is 0, the intercept ($\beta_{00}$) represents the overall predicted value on the *ATTIT* measure; the linear, or first-order, slope ($\beta_{10}$) is the predicted instantaneous rate of change; the quadratic, or second-order, slope ($\beta_{20}$) characterises change in that instantaneous rate of change; and the cubic, or third-order, slope ($\beta_{30}$) conveys that the change in the rate of change is, itself, changing across time.

**Exhibit 8.4** Calculating degrees of freedom (*df*) for the unconditional linear, piecewise and polynomial growth models for growth in *ATTIT*, measured across five time points

| Model | Parameters to Estimate | *df* |
|---|---|---|
| Linear Growth Model (Latent Factors: Intercept, Slope) | | |
| Constrained Structural Equation Model | No. of error variances ($\sigma^2$): 1; No. of variances in latent factors: 2; No. of covariances among latent factors: 1; No. of means: 2. | $20 - 1 - 2 - 1 - 2 = 14$<br><br>11 *df* (covs)+<br>3 *df* (means) |
| Two-Piece Growth Model (Latent Factors: Intercept, Slope 1, Slope 2) – if all slopes have variances | | |
| Constrained Structural Equation Model | No. of error variances ($\sigma^2$): 1; No. of variances in latent factors: 3; No. covariances among latent factors: 3; No. of means: 3. | $20 - 1 - 3 - 3 - 3 = 10$<br><br>8 *df* (covs)<br>+ 2 *df* (means) |
| Our Two-Piece Growth Model (Latent Factors: Intercept, Slope1, Slope 2) – no slope 1 variance | | |
| Constrained Structural Equation Model | No. of error variances ($\sigma^2$): 1; No. of variances in latent factors: 2; No. covariances among latent factors: 1; No. of means: 3. | $20 - 1 - 2 - 1 - 3 = 13$<br><br>11 *df* (covs)<br>+2 *df* (means) |
| Cubic Growth Model (Latent Factors: Intercept, Age, Age$^2$, Age$^3$) – if all slopes have variances | | |
| Constrained Structural Equation Model | No. of error variances ($\sigma^2$): 1; No. of variances in latent factors: 4; No. covariances among latent factors: 6; No. of means: 4. | $20 - 1 - 2 - 6 - 4 = 5$<br><br>4 *df* (covs)<br>+ 1 *df* (means) |
| Our Cubic Growth Model: Variances for 2 of the latent factors are constrained to 0. | | |
| Constrained Structural Equation Model | No. of error variances ($\sigma^2$): 1; No. of variances in latent factors: 2; No. of covariances among latent factors: 1; No. of means: 4. | $20 - 1 - 2 - 1 - 4 = 12$<br><br>11 *df* (covs)<br>+ 1 *df* (means) |

*Note.* With five time points, the number of unique elements in the variance covariance matrix is 15 and the number of means is 5, so the total number of knowns equals 20.

Like the piecewise model, the cubic model initially had convergence problems when all growth parameters randomly varied across people. Therefore, our cubic model constrains the variances of the linear and cubic terms to 0 but freely estimates variances for the intercept and quadratic terms.

The coding in Table 8.10 demonstrates how the three polynomial slopes work together to modify the rate of change across time. At the first two time points, the coding for all three slopes is the same (all are 0 at the first time point and 1 at the second time point), so each polynomial term exerts a similar degree of influence on the growth curve. However, at the third time point, the quadratic and cubic coding schemes diverge from the linear slope codes, and each polynomial component differentially affects the shape of the growth curve. The quadratic and cubic terms exert ever-increasing influences on the overall growth trajectory: the quadratic parameter is multiplied by time squared and the cubic parameter is multiplied by time cubed. The shape of the trajectory depends on magnitude and direction of all slope coefficients in combination. So, instead of estimating three individual segments of the growth trajectory (as we did in our original piecewise model), the three growth parameters in the cubic model (i.e. $\beta_{10}$, $\beta_{20}$ and $\beta_{30}$) combine to produce the model-predicted values at each time point.

**Table 8.10** Coding time for the cubic polynomial growth model

| Measurement Occasion | Intercept (Time$^0$) | Time$^1$ | Time$^2$ | Time$^3$ |
|---|---|---|---|---|
| Age 11 | 1 | 0 | 0 | 0 |
| Age 12 | 1 | 1 | 1 | 1 |
| Age 13 | 1 | 2 | 4 | 6 |
| Age 14 | 1 | 3 | 9 | 27 |
| Age 15 | 1 | 4 | 16 | 64 |

The intercept in the polynomial model resembles the intercepts from the linear and piecewise models and has a similar interpretation: it represents expected *ATTIT* when time, time$^2$ and time$^3$ = 0 (i.e. at age = 11). However, including the higher order (polynomial) terms in lieu of 'pieces' (as in the piecewise models) dramatically changes the interpretation of the growth slopes. First, $\beta_{10}$ is no longer a constant or linear rate of change. Instead, $\beta_{10}$ represents the instantaneous rate of change when time = time$^2$ = time$^3$ = 0 (at age 11). Notably, this coefficient ($\beta_{10}$ = −0.024) is not statistically different from zero. Therefore, the cubic growth model suggests that at age 11, there is essentially no instantaneous change in *ATTIT*. The quadratic slope ($\beta_{20}$) represents change in the rate of change (the acceleration parameter) at time 0. The positive coefficient for quadratic slope (0.061) indicates that the rate of change is becoming more positive when time = 0. The cubic growth slope captures the change in acceleration parameter across time (Remember: the cubic parameter captures the change in the acceleration parameter). The negative cubic term ($\beta_{30}$ = −0.010) means the acceleration parameter becomes more negative across time; this eventually pulls the composite growth trajectory downward, slowing growth, so the growth curve becomes more negative/less positive (Table 8.11).

**Table 8.11** Model 4 results using the cubic polynomial growth model in MLM and SEM

| | MLM | | SEM | |
|---|---|---|---|---|
| | Estimate | SE | Estimate | SE |
| Fixed effects | | | | |
| Intercept ($\beta_{00}$) | 0.212*** | 0.015 | 0.212*** | 0.015 |
| Age ($\beta_{10}$) | −0.024 | 0.029 | −0.024 | 0.029 |
| Age$^2$ ($\beta_{20}$) | 0.061*** | 0.018 | 0.061*** | 0.018 |
| Age$^3$ ($\beta_{30}$) | −0.010*** | 0.003 | −0.010*** | 0.003 |

|  | MLM | | SEM | |
|---|---|---|---|---|
|  | **Estimate** | *SE* | **Estimate** | *SE* |
| Random effects |  |  |  |  |
| Intercept variance ($\tau_{00}$) | 0.019 |  | 0.019 | 0.003 |
| Age variance ($\tau_{11}$) | — |  | — |  |
| Age$^2$ variance ($\tau_{22}$) | 0.000[a] |  | 0.000[a] | 0.00 |
| Age$^3$ variance ($\tau_{33}$) | — |  | — |  |
| Correlation intercept and age$^2$ ($\tau_{02}$) | 0.58 |  | 0.58*** |  |
| Residual variance ($\sigma^2$) | 0.027 |  | 0.027 |  |

*Note.* The total number of students in the analysis is 239 with 1079 observations across five time points. The parameter estimates for the multilevel and structural equation models are the maximum likelihood estimates. MLM = multilevel modelling; SEM = structural equation modelling; *SE* = standard error.

[a]Not precisely zero. The value is roughly 0.0001.

***$p < .01$. **$p < .05$. *$p < .10$.

Clearly, the slope parameters in the cubic polynomial model are generally more difficult to interpret than those of the piecewise model, and they really cannot be interpreted in isolation. Therefore, the best way to interpret cubic model parameters is to compute and graph model-predicted values (see Table 8.12 and Figure 8.7). In Table 8.12, the predicted values based on the cubic model are virtually identical to the predicted values of *ATTIT* from the piecewise growth model. Like the piecewise model, the cubic model almost perfectly reproduces the observed means across all time points (see Figure 8.7 and Table 8.13 for our comparison of the observed and predicted values of *ATTIT* for each model).

**Table 8.12** Predicted values for the cubic polynomial growth model

| Measurement Occasion | Time | $\beta_{00}$ (0.21) | Time | $\beta_{10}$ (0.21) | Time$^2$ | $\beta_{20}$ (0.06) | Time$^3$ | $\beta_{30}$ (−0.01) | Predicted Value |
|---|---|---|---|---|---|---|---|---|---|
| Age 11 | 1 | 0.21 | 0 | 0 | 0 | 0 | 9 | 0 | 0.21 |
| Age 12 | 1 | 0.21 | 1 | −0.02 | 1 | 0.06 | 1 | −0.01 | 0.24 |
| Age 13 | 1 | 0.21 | 2 | −0.04 | 4 | 0.24 | 8 | −0.08 | 0.33 |
| Age 14 | 1 | 0.21 | 3 | −0.06 | 9 | 0.54 | 27 | −0.27 | 0.41 |
| Age 15 | 1 | 0.21 | 4 | −0.08 | 16 | 0.96 | 64 | −0.64 | 0.44 |

*Note.* For the structural equation model, a cell under a given coefficient is the product of the given coefficient value in parentheses and the preceding row for either the time intercept or time slope. The predicted values can be obtained from summing the coefficient values across a given row (e.g. for row 3 or age 13, the predicted value of *ATTIT* is 0.21 − 0.04 + 0.24 − 0.08 = 0.33).

**Table 8.13** *ATTIT* sample means compared to predicted *ATTIT* values for each longitudinal model

| Variable | Sample Means | Linear | Three-Piece | Two-Piece | Cubic Polynomial |
|----------|:---:|:---:|:---:|:---:|:---:|
| *ATTIT1* | 0.21 | 0.21 | 0.21 | 0.21 | 0.21 |
| *ATTIT2* | 0.24 | 0.27 | 0.24 | 0.24 | 0.24 |
| *ATTIT3* | 0.33 | 0.33 | 0.33 | 0.33 | 0.33 |
| *ATTIT4* | 0.41 | 0.40 | 0.42 | 0.42 | 0.41 |
| *ATTIT5* | 0.45 | 0.46 | 0.45 | 0.45 | 0.44 |

*Note.* The numerical character in each variable name indicates the measurement occasion of the *ATTIT* measure. For example, *ATTIT1* represents antisocial attitudes measured at the first time point. Note the sample means, three-piece and two-piece values are identical.

## Model Comparisons and Selection

Thus far, we have focused on interpreting parameters from each of our estimated growth models and discussing different trajectories for change across these models. The question remains: Which model most accurately and most parsimoniously represents the growth trajectory in attitudes towards deviant behaviour between ages 11 and 15? To answer this question, we can examine a variety of model fit criteria.

Remember that if two models are nested, we can compare the two models directly using the LRT (MLM) or chi-square difference test (SEM). In large samples, the difference between the chi-squares or deviances of two hierarchically nested models is distributed as an approximate chi-square distribution with degrees of freedom equal to the difference in the number of parameters being estimated between the two models (de Leeuw, 2004). When evaluating model fit, we prefer the more parsimonious model, if it does not result in statistically significantly worse fit. In other words, if the model with more parameters (fewer degrees of freedom) fails to reduce the chi-square value or deviance by a substantial amount, we retain the more parsimonious model. Changes in deviance/chi-square that exceed the critical value of chi-square favour the more complex model.

The fit of the two- and three-piece piecewise growth models is virtually identical. The chi-square for two-piece growth model is 28.00 with 13 *df*. The chi-square for the three-piece model is 27.94 with 12 *df*. Eliminating the third growth slope provided additional 1 *df* and increased chi-square by only 0.06. Comparing the difference in chi-square across the two models, the chi-square change of 0.06 is much smaller than the critical value of chi-square with 1 *df* (3.84), favouring the simpler (more parsimonious) two-piece model.

Model fit indices provide an additional model comparison tool, although the model fit indices in MLM and SEM often differ. To use LRT, AIC or BIC to compare the models that differ in terms of their fixed effects, we used FIML. The multilevel model fit indices from lme4 appear in Table 8.14, whereas the model fit indices from lavaan appear in Table 8.15. Information criteria always favour the model with the lowest AIC, BIC and so on. Comparisons of AIC, BIC and SEM fit indices do not require hierarchically nested models. Our cubic model is not nested with our piecewise growth models, so we cannot use chi-square tests to compare all four models. However, we can compare the four models in terms of their AIC, $BIC_1$ and $BIC_2$ (presented in Table 8.14) and their SEM fit indices (presented in Table 8.15). Overall, the multilevel and structural equation model fit information favour the two-piece model (Model 3), which has the best CFI, TLI, RMSEA, $BIC_1$, $BIC_2$ and AIC values.

**Table 8.14**  MLM model fit indices

|         | −2LL    | AIC     | $BIC_1$  | $BIC_2$  |
|---------|---------|---------|----------|----------|
| Model 1 | −338.07 | −326.07 | −296.20  | −305.21  |
| Model 2 | −361.60 | −345.50 | −305.64  | −317.70  |
| Model 3 | −361.45 | −347.40 | −312.56  | −323.11  |
| Model 4 | −328.32 | −312.32 | −272.45  | −284.50  |

Note. Model 1 – unconditional linear growth model, Model 2 – unconditional piecewise growth model (three-piece), Model 3 – unconditional piecewise growth model (two-piece) and Model 4 – polynomial growth model. $BIC_1$ – BIC calculated using $N = 1079$. $BIC_2$ – BIC calculated using $n = 239$. The parameter estimates for the multilevel models are the maximum likelihood estimates. LL = log likelihood; AIC = Akaike information criterion; BIC = Bayesian information criterion.

Both piecewise models (i.e. Models 2 and 3) more accurately depict *ATTIT* growth than the linear model. (The predicted values for the piecewise models were virtually identical to the sample means.) Furthermore, the two-piece model is more parsimonious than the three-piece model and fits equally well. In contrast, even though the unconditional cubic model (Model 4) produced predicted values that were virtually identical to the predicted values from the piecewise model, the model fit for the cubic model was noticeably worse than the fit of either piecewise model.

The high degree of similarity in the parameter estimates and predicted values across the two-piece and three-piece models suggests that we can eliminate the third slope without adversely affecting model-based predictions. The similarity in chi-square across the two models suggests that eliminating the third slope does not adversely affect our ability to reproduce the means and variance–covariance matrix from our set of parameters. Overall, our two-piece model seems to perform just as well as our three-piece model, and it does so with one fewer parameter.

**Table 8.15** SEM model fit indices

|  | $\chi^2$ | df | p | CFI/TLI | RMSEA[90% CI] | SRMR | AIC |
|---|---|---|---|---|---|---|---|
| Model 1 | 51.39 | 14 | <.01 | 0.913/0.938 | 0.106[0.076,0.137] | 0.088 | −326.07 |
| Model 2 | 27.94 | 12 | <.01 | 0.963/0.969 | 0.075[0.038,0.111] | 0.078 | −345.51 |
| Model 3 | 28.00 | 13 | <.01 | 0.965/0.973 | 0.069[0.033,0.105] | 0.078 | −347.45 |
| Model 4 | 61.13 | 12 | <.01 | 0.887/0.906 | 0.130[0.100,0.164] | 0.133 | −312.32 |

*Note.* Model 1 – Unconditional Linear Growth Model, Model 2 – Unconditional Piecewise Growth Model (Three-Piece), Model 3 – Unconditional Piecewise Growth Model (Two-Piece), and Model 4 – Polynomial Growth Model. df – Degrees of freedom. The parameter estimates for the SEM models are the Maximum Likelihood (ML) estimates.

## Model 5: Piecewise conditional growth model (two-piece with gender covariate)

Next, we add gender as a TIC to our piecewise model to explore whether males and females exhibited differences in *ATTIT* levels and differential *ATTIT* growth between ages 11 and 15. Conditional growth models examine how different covariates influence where individuals start and how they change across time, with regard to specified outcomes. Therefore, using a conditional two-piece growth model, we examine three main questions: (1) Does gender influence individuals' initial *ATTIT* levels? (2) Do changes in deviant behaviour from ages 11 to12 and 14 to 15 differ by gender? (3) Does gender moderate the antisocial attitude growth rate from age 12 to 14? To do so, we added gender to our preferred (two-piece) unconditional growth model as a predictor of all three growth parameters. Figure 8.9 contains the two-piece conditional growth model diagram. In equation form, we specified this model as follows:

Level 1:

$$ATTIT_{ti} = \pi_{0i} + \pi_{1i}Slope1_{ti} + \pi_{2i}Slope2_{ti} + e_{ti}$$

Level 2:

$$\pi_{0i} = \beta_{00} + \beta_{01}Female_i + r_{0i}$$

$$\pi_{1i} = \beta_{10} + \beta_{11}Female_i$$

$$\pi_{2i} = \beta_{20} + \beta_{21}Female_i + r_{2i}$$

(8.5)

The $\beta_{01}$, $\beta_{11}$ and/or $\beta_{21}$ parameters estimate the moderating effect of gender on the *ATTIT* intercept and the two separate *ATTIT* growth rates, after controlling for all the other variables in the model: $\beta_{01}$ indicates whether boys and girls differ in terms of their predicted *ATTIT* at age 11, $\beta_{11}$ indicates whether boys and girls grow at different rates during the non–growth spurt periods (from ages 11 to 12 and ages 14 to 15), and $\beta_{21}$ indicates whether boys and girls grow at different rates during the *ATTIT* growth spurt period from ages 12 to 14.

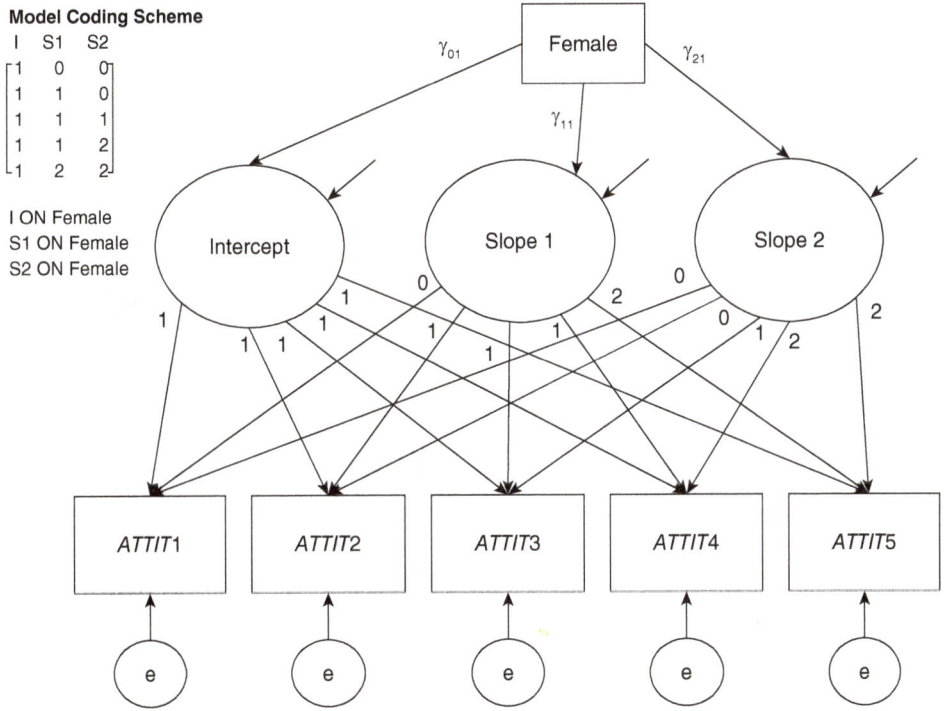

**Figure 8.9** *Two-piece conditional linear growth model*

The results for this conditional model appear in Table 8.16. The estimate for the intercept ($\beta_{00}$) was 0.236, but it no longer represents the overall predicted *ATTIT* score at age 11. Instead, because we coded females as '1' and males as '0', the intercept represents the predicted initial status for males (i.e. when female = 0). The effect of female on the intercept ($\beta_{01}$ = –0.051) indicates that age 11 *ATTIT* scores are 0.051 lower for females than they are for males. The predicted *ATTIT* score at age 11 for females is simply the sum of the intercept and the gender slope: 0.236 – 0.051 = 0.185. Females' expected *ATTIT* score at age 11 is 0.185 and males' expected attitudes score is 0.236, and this difference ($\beta_{01}$ = –0.051) is statistically significantly different from 0. So, gender does moderate *ATTIT* at age 11.

**Table 8.16** Model 5 results using the two-piece piecewise conditional growth model in MLM & SEM

| | MLM | | SEM | |
|---|---|---|---|---|
| | Estimate | SE | Estimate | SE |
| **Fixed Effects** | | | | |
| Intercept ($\beta_{00}$) | 0.236*** | 0.017 | 0.236*** | 0.017 |
| Slope 1 ($\beta_{10}$) | 0.031** | 0.015 | 0.031** | 0.014 |
| Slope 2 ($\beta_{20}$) | 0.096*** | 0.013 | 0.096*** | 0.013 |
| Female ($\beta_{01}$) | −0.051** | 0.025 | −0.051** | 0.025 |
| Slope1xFemale ($\beta_{11}$) | −0.004 | 0.022 | −0.004 | 0.022 |
| Slope2xFemale ($\beta_{21}$) | −0.016 | 0.019 | −0.016 | 0.019 |
| **Random Effects** | | | | |
| Intercept Variance ($\beta_{00}$)) | 0.017 | | 0.017 | 0.003 |
| Slope 1 Variance ($\tau_{11}$)) | – | | – | – |
| Slope 2 Variance ($\tau_{22}$) | 0.008 | | 0.007 | 0.001 |
| Correlation Intercept & Slope 2 ($\tau_{02}$) | 0.40 | | 0.40*** | 0.154 |
| Residual Variance ($\sigma^2$) | 0.025 | | 0.025 | 0.001 |

*Note.* *** $p < .01$, ** $p < .05$, * $p < .10$. The total number of students in the analysis is 239 with 1079 observations across five time points. The parameter estimates for the MLM and SEM models are the Maximum Likelihood (ML) estimates.

Does gender moderate growth in antisocial attitudes from ages 11 to 15? Because our model contains two different growth slopes, gender could moderate either, neither or both of these slopes. For example, perhaps during non–growth spurt periods, the *ATTIT* growth rate is similar for boys and girls, but during growth spurts, one gender grows more quickly than the other. To determine whether gender moderated either the growth spurt (ages 12–14) or the non–growth spurt (ages 11–12 and 14–15) *ATTIT* growth rates, we added gender (i.e. *Female$_i$*) as a level-2 predictor of both growth slopes.

In this conditional model, $\beta_{10}$ (the slope for the change in attitudes between ages 11 and 12 and ages 14 and 15) represents the intercept for the slope. In other words, $\beta_{10}$ is the slope when Female = 0, so $\beta_{10}$ is the model-predicted growth rate (from ages 11 to 12 and 14 to 15) for boys, and $\beta_{11}$ represents the difference in growth between males and females (from ages 11 to 12 and 14 to 15). Males grew at a rate of 0.031 points per year during this period, and females grew 0.004 points more slowly than males ($\beta_{11} = -0.004$); therefore, females' model-predicted growth was 0.027 points per year from ages 11 to 12 and 14 to 15. However, that difference was not statistically significantly different from 0: males and females grew at similar rates during the non–growth spurt periods.

From ages 12 to 14, both males and females grew more quickly in antisocial attitudes. The intercept for the slope ($\beta_{20} = 0.096$) indicates that males grew 0.096 points per year during this period. The effect of female on this slope ($\beta_{21}$ = –0.016) indicates that the yearly growth rate for females was 0.016 lower than the yearly growth rate for males: the model-predicted yearly growth rate for females is 0.08. Again, the gender difference growth rates ($\beta_{21}$ = –0.016) was not statistically significantly different from 0. To recap, males had higher antisocial attitudes than females, but males and females had similar growth rates (this was evident from Figure 8.2). Furthermore, regardless of gender, growth in antisocial attitudes was far faster from ages 12 to 14 than it was from ages 11 to 12 or 14 to 15 (anyone who has raised a middle school student can probably relate to these findings)!

Even after accounting for gender, there was still considerable between-person variability in students' initial antisocial attitudes at age 11 ($\tau_{00}$ = 0.017) and their growth in antisocial attitudes from ages 12 to 14 ($\tau_{22}$ = 0.008; $SD$ = 0.089). On average, students grew at a rate of 0.09 points per year from ages 12 to 14, but 68% of the individual growth slopes were between 0.00 and 0.18 and 95% of the growth slopes were between –0.09 and 0.27. Even though the overall growth rate was positive, certain students actually experienced a decline in antisocial attitudes during that period. Finally, the residual correlation of .40 between the intercept and growth slope from ages 12 to 14 indicates a moderate positive association between initial *ATTIT* and growth in *ATTIT* between the ages of 12 and 14, even after accounting for gender.

In conclusion, the two-piece piecewise model provides the best functional fit for these data. Students' antisocial attitudes grow slowly between ages 11 and 12, rapidly from ages 12 to 14, and then slowly again between ages 14 and 15. Males tend to have more antisocial attitudes at age 11, but the growth rates for males and females are similar, so this gender gap persists across time. Even after controlling for gender, there is substantial variability between adolescents' *ATTIT* at age 11 and growth in *ATTIT* from ages 12 to 14. In contrast, adolescents' *ATTIT* appear to grow at similar rates from ages 11 to 12 and 14 to 15.

## Conclusion

In this example, we demonstrated how to fit and interpret five different growth models: an unconditional linear growth model, an unconditional three-piece growth model, an unconditional two-piece growth model, an unconditional polynomial growth model and a conditional two-piece growth model. Polynomial and piecewise models represent two common and conventional methods for modelling non-linear growth. However, a variety of other non-linear growth models exist, including a large family of exponential growth models and latent basis models. Exponential growth models are far easier to estimate using MLM, whereas latent basis models are much easier to fit in SEM. Grimm, Ram, and Estabrook's (2016) book on growth modelling (*Growth Modeling: Structural Equation and Multilevel Modeling Approaches*) contains an excellent chapter on estimating growth models with non-linear parameters.

This book has covered a great deal of content on modelling. We provided an introduction and overview of both MLM and SEM, provided guidance on how to build and select models in both traditions and then applied both approaches to estimate a variety of growth models. However, this introductory text cannot possibly do justice to all the intricacies of these modelling techniques. We hope that this book has inspired you to learn more, and many wonderful books exist on each of these topics. For example, an excellent introductory book on SEM is *Principles and Practices of Structural Equation Modeling* (fourth edition) by Rex Kline (2015). A classic (and more dense) text is *Structural Equations With Latent Variables* by Ken Bollen (1989). Raudenbush and Bryk's (2002) *Hierarchical Linear Models: Applications and Data Analysis Methods* (second edition) is a classic book in the area of multilevel modelling. Snijders and Bosker's (2012) *Multilevel Analysis: An Introduction to Basic and Advanced Multilevel Modeling* (second edition), Hox's et al.'s (2017) *Multilevel Analysis: Techniques and Applications* (third edition), and Goldstein's (2011) *Multilevel Statistical Models* are also excellent texts. Additionally, there are so many areas of latent variable modelling that we have not had an opportunity to explore in this book. We have discussed basic growth curve models, but there are a wide variety of other longitudinal models, including non-linear mixed effects models, autoregressive models, latent change score models and dynamic SEM (Asparouhov et al., 2018; Grimm et al., 2016; McArdle & Nesselroade, 2014; Stegmann et al., 2018). Additional latent variable modelling techniques include multilevel SEM, mixture modelling and Bayesian SEM (Muthén & Muthén, 1998–2017). We hope that having piqued your interest in these modelling topics, you will delve into more comprehensive and complete treatments of these techniques, and we believe that this book has given you the tools to do so. Happy modelling!

---

### Chapter Summary

- This chapter provides an applied example of fitting linear growth, polynomial growth model and piecewise linear growth models using both SEM and MLM.
- To build growth models in this example, the model building steps of Chapter 7 are used. First, we model the shape of the growth trajectory without additional covariates. The unconditional growth model should capture shape of the unconditional growth trajectory and should provide reasonable predictions of individual growth data. After

selecting the most appropriate unconditional growth model, we add other within-person (time-varying) covariates to the model (if there are any). The final step includes between-person (time-invariant) covariates.

- These recommended steps underscore the importance of modelling the functional form of change as accurately as possible in the first step in the model building process.

## Further Reading

Hoffman, L. (2015). *Longitudinal analysis: Modeling within-person fluctuation and change*. Routledge.

This book introduces and extends individual growth modelling within the mixed/multilevel framework.

McArdle, J. J., & Nesselroade, J. R. (2014). *Longitudinal data analysis using structural equation models*. American Psychological Association.

This book is a tour de force, and it introduces several interesting extensions to latent growth curve models from an SEM framework.

# APPENDIX 1

## BRIEF INTRODUCTION TO MATRICES AND MATRIX ALGEBRA

In Appendices 1 to 4, we provide a brief introduction to matrix algebra and relate this to the specification of a structural equation model using the example model from Chapter 6. Also, we provide the technical details that undergird our discussion of Wright's standardised tracing rules, we discuss Wright's rules for generating the model-implied covariances using the unstandardised path coefficients for confirmatory factor analysis and path models and we demonstrate the equivalence of using either unstandardised tracing rules or *covariance algebra* for generating model-implied covariances. These technical details provide a deeper understanding of the mathematical underpinnings of structural equation modelling (SEM).

To begin our discussion of matrix algebra, it is necessary to define a vector and a matrix. Vectors and matrices (plural of matrix) are rectangular arrays of numbers. A matrix is a set of numbers which are arranged by rows and columns. For example, consider the $3 \times 2$ matrix **A** below (vectors and matrices are displayed using boldface capital letters):

$$\mathbf{A} = \begin{bmatrix} 2 & 3 \\ 4 & 5 \\ 6 & 7 \end{bmatrix}$$

**A** is a $3 \times 2$ matrix because there are three rows and two columns. The rows are typically referenced by the index $m$ and the columns by the index $n$. **A** is an $M \times N$ matrix where $M$ is the total number of rows and $N$ is the total number of columns. (We always indicate the rows first.) The total number of rows and columns together refer to the dimensions of the matrix. The dimension of the matrix **A** is $3 \times 2$: it has three rows and two columns.

A vector is a special type of matrix with either only one row or only one column. Consider the *row* vector **V** below:

$$\mathbf{V} = \begin{bmatrix} 2 & 6 & 1 & 4 \end{bmatrix}$$

**V** is a $1 \times 4$ vector or matrix: there is one row and four columns. We denote the transpose of **V** as **V**′, where

$$\mathbf{V}' = \begin{bmatrix} 2 \\ 6 \\ 1 \\ 4 \end{bmatrix}$$

The transpose of **V** is the $4 \times 1$ *column* vector or matrix **V**′. The (′) symbol is used to refer to a transpose (a T superscript is another symbol which is often used to denote a transpose). All vectors contain either one row or one column (e.g. the transpose of **V** contains one column). A matrix can be transposed as well. For example,

$$\mathbf{A}' = \begin{bmatrix} 2 & 4 & 6 \\ 3 & 5 & 7 \end{bmatrix}$$

$\mathbf{A}'$ is the transpose of $\mathbf{A}$; it is a 2 × 3 matrix. The dimensions of the transposed matrix or vector are switched (i.e. the $m$th row $\mathbf{A}$ becomes the $n$th column of $\mathbf{A}'$).

Other special matrices include a scalar matrix (i.e. a 1 × 1 matrix which is simply a constant), a square matrix (i.e. a matrix with the same number of rows as columns), a symmetric matrix (i.e. a square matrix with equal elements on the upper and lower triangles of the square, in which case the transpose of the matrix will be equal to the original matrix – correlation and covariance matrices are symmetric), a diagonal matrix (i.e. a square matrix with numbers only on the diagonal of the square) and an identity matrix (i.e. a square matrix with 1s on the diagonal of the square). The reader is encouraged to consult Fox (2008) for more detail on matrices and matrix algebra.

We can conduct basic mathematical operations such as addition, subtraction and multiplication on matrices and vectors. To add or subtract two matrices, they must contain the same number of dimensions (i.e. the same number of rows and columns). For example, consider the two 3 × 3 square matrices $\mathbf{B}$ and $\mathbf{C}$:

$$\mathbf{B} = \begin{bmatrix} 1 & 8 & 2 \\ 8 & 4 & 3 \\ 2 & 3 & 1 \end{bmatrix}$$

$$\mathbf{C} = \begin{bmatrix} 1 & 0 & 0 \\ 0 & 1 & 0 \\ 0 & 0 & 1 \end{bmatrix}$$

Notice, in addition to being a square matrix, $\mathbf{B}$ is a symmetric matrix because the elements in the upper triangle of the square matrix are equal to the elements in the lower triangle of the square matrix and we can see that $\mathbf{B} = \mathbf{B}'$. Also, note that $\mathbf{C}$ is a diagonal identity matrix. Let $\mathbf{D}$ be equal to the sum of the two matrices and $\mathbf{E}$ be equal to the difference of the two matrices. Addition and subtraction of matrices is carried out by adding or subtracting the corresponding elements of the matrices. For example,

$$\mathbf{D} = \mathbf{B} + \mathbf{C} = \begin{bmatrix} 1 & 8 & 2 \\ 8 & 4 & 3 \\ 2 & 3 & 1 \end{bmatrix} + \begin{bmatrix} 1 & 0 & 0 \\ 0 & 1 & 0 \\ 0 & 0 & 1 \end{bmatrix} = \begin{bmatrix} 2 & 8 & 2 \\ 8 & 5 & 3 \\ 2 & 3 & 2 \end{bmatrix}$$

and

$$\mathbf{E} = \mathbf{B} - \mathbf{C} = \begin{bmatrix} 1 & 8 & 2 \\ 8 & 4 & 3 \\ 2 & 3 & 1 \end{bmatrix} - \begin{bmatrix} 1 & 0 & 0 \\ 0 & 1 & 0 \\ 0 & 0 & 1 \end{bmatrix} = \begin{bmatrix} 0 & 8 & 2 \\ 8 & 3 & 3 \\ 2 & 3 & 0 \end{bmatrix}$$

Notice, the resulting matrices from the addition and subtraction, respectively, are of the same dimension as the corresponding matrices used in the arithmetic operations. That is, all matrices are 3 × 3.

The multiplication of two vectors or matrices requires that the number of *columns* of the left-hand element (i.e. matrix or vector) be equal to the number of *rows* in the right-hand element. For example, consider the multiplication of the two vectors below:

$$[2 \quad 4 \quad 1] \begin{bmatrix} x_1 \\ x_2 \\ x_3 \end{bmatrix}$$

The number of columns in the left-hand vector is three and the number of rows in the right-hand vector is three. Further, the dimension of the matrix that results from the multiplication of the two vectors will be equal to the number of rows in the left-hand element (here one) by the number of columns in the right-hand element (here one). Therefore, the dimensions of the resulting matrix from multiplying the 1 × 3 row vector with the 3 × 1 column vector is a 1 × 1 scalar.

$$[2x_1 + 4x_2 + x_3]$$

We multiply each corresponding element of the row vector by each corresponding element of the column vector, and adding them up, which results in a scalar (i.e. one number). The resulting 1 × 1 vector may not appear to be a scalar because there are unknown variables in the column vector. However, if the values of these variables were known, we could solve the equation, and the result would be a scalar (i.e. one number). The multiplication of the two vectors is referred to as the **scalar product** because the multiplication of two vectors results in a scalar. The row vector by column vector multiplication shown above is generalised to the case of matrix multiplication. To multiply two matrices, we compute the scalar products for each element of the *m* × *n* matrix, as demonstrated below. For example, consider the multiplication of the matrices **B** and **A**:

$$\mathbf{F} = \mathbf{BA} = \begin{bmatrix} 1 & 8 & 2 \\ 8 & 4 & 3 \\ 2 & 3 & 1 \end{bmatrix} \begin{bmatrix} 2 & 3 \\ 4 & 5 \\ 6 & 7 \end{bmatrix} = \begin{bmatrix} 2*1 + 4*8 + 6*2 & 3*1 + 5*8 + 7*2 \\ 2*8 + 4*4 + 6*3 & 3*8 + 5*4 + 7*3 \\ 2*2 + 4*3 + 6*1 & 3*2 + 5*3 + 7*1 \end{bmatrix}$$

$$= \begin{bmatrix} 46 & 57 \\ 50 & 65 \\ 22 & 28 \end{bmatrix}$$

The result is the 3 × 2 matrix **F**. To obtain the first element of the matrix **F**, multiply the first row vector of **B** with the first column vector of **A**. The individual elements of the resulting **F** matrix are composed of the result of scalar products where the scalar in the *m*th row and *n*th column of **F** is determined by the scalar product of the *m*th row of **B** and *n*th column **A**. For example, the resulting scalar product of the third row of **B** and second column of **A** results in the scalar in the third row and second column of **F** (i.e. 28).

## Systems of equations in matrix notation and path diagrams

Matrix algebra provides a convenient shorthand to represent a system of equations. For instance, consider the following set of three equations:

$$8 = 2x_1 + 4x_2 + x_3$$
$$3 = 6x_1 + 2x_2 - x_3$$
$$5 = x_1 - 8x_2 + 0$$

We can rewrite these three equations in matrix notation as follows:

$$\begin{bmatrix} 8 \\ 3 \\ 5 \end{bmatrix} = \begin{bmatrix} 2 & 4 & 1 \\ 6 & 2 & -1 \\ 1 & -8 & 0 \end{bmatrix} \begin{bmatrix} x_1 \\ x_2 \\ x_3 \end{bmatrix}$$

We can further simplify the matrix notation by defining the following matrices. Let

$$\mathbf{Y} = \begin{bmatrix} 8 \\ 3 \\ 5 \end{bmatrix}, \mathbf{B} = \begin{bmatrix} 2 & 4 & 1 \\ 6 & 2 & -1 \\ 1 & -8 & 0 \end{bmatrix}, \mathbf{X} = \begin{bmatrix} x_1 \\ x_2 \\ x_3 \end{bmatrix}$$

which results in the simplified matrix notation for the system of equations,

**Y = BX**

We encourage the reader to use the matrix algebra rules presented above to verify the equivalence between the matrix notation and the system of equations.

## Chapter 6 example notation in matrix algebra

As an example of using matrix algebra in SEM, consider the simplified matrix notation for the two latent variables in the conceptual model from Chapter 5[i] (see Figure 5.2):

---

[i]For simplicity, here we include only the two latent variables and their indicators.

$$\mathbf{Y} = \Lambda\eta + \varepsilon \tag{A1.1}$$

Below, we have replaced each of the matrices and vectors in Equation (A1.1) with their respective components (note the use of the marker variable strategy):

$$
\begin{bmatrix}
\text{Algebra} \\
\text{Geometry} \\
\text{Measurement} \\
\text{Comprehension} \\
\text{Vocabulary} \\
\text{Fluency}
\end{bmatrix}
=
\begin{bmatrix}
1 & 0 \\
\lambda_2 & 0 \\
\lambda_3 & 0 \\
0 & 1 \\
0 & \lambda_5 \\
0 & \lambda_6
\end{bmatrix}
\begin{bmatrix}
\text{Math} \\
\text{Reading}
\end{bmatrix}
+
\begin{bmatrix}
\varepsilon_1 \\
\varepsilon_2 \\
\varepsilon_3 \\
\varepsilon_4 \\
\varepsilon_5 \\
\varepsilon_6
\end{bmatrix}
$$

Here $\mathbf{Y}$ is the $p \times 1$ vector of the $p$ observed variables, manifest indicators or dependent variables. For the model in Figure 5.2, the six manifest indicators or dependent variables comprising the $\mathbf{Y}$ vector include Algebra, Geometry, Measurement, Comprehension, Vocabulary and Fluency. $\Lambda$ is a $p \times k$ matrix of factor loadings where each component $\lambda_{pk}$ represents the relationships between the $p$ manifest indicators (i.e. the dependent variables in $\mathbf{Y}$) and the $k$ latent variables. The $k$ latent variables are represented with $\eta$ which is a $k \times 1$ vector of latent variables. Here, there are two latent variables: Math and Reading. Finally, $\varepsilon$ is $p \times 1$ vector of the errors of measurement for each manifest indicator or dependent variable. In summary, for Figure 5.2, $\mathbf{Y}$ is a $6 \times 1$ vector, $\Lambda$ is a $6 \times 2$ matrix, $\eta$ is a $2 \times 1$ vector and $\varepsilon$ is a $6 \times 1$ vector.

The 0s in the $\Lambda$ matrix indicate that the path from the latent variable in the column to the observed variable in the row is constrained to be 0. For the Math latent variable, there are three hypothesised structural paths from Math to Algebra, Geometry and Measurement with the values 1, $\lambda_2$ and $\lambda_3$, respectively. These are in the first column of the $\Lambda$ matrix. Moreover, for the Reading latent variable, there are three hypothesised structural paths from Reading to the remaining manifest indicators (i.e. Comprehension, Vocabulary and Fluency) with the values 1, $\lambda_5$ and $\lambda_6$, respectively. These are in the second column of the $\Lambda$ matrix. The lack of a direct effect between the Math (Reading) latent variable and the last three (first three) manifest indicators are depicted by 0s in the first (second) column of the $\Lambda$ matrix, indicating that the direct effects are *constrained* (i.e. fixed) to 0.

# APPENDIX 2

## LINKING PATH DIAGRAMS TO STRUCTURAL EQUATIONS

To introduce Wright's rules, we must first discuss the link between path diagrams and structural equations. As an example, consider the path diagram in Figure A2.1. Using common conventions, we refer to exogenous variables as $X$ variables and endogenous variables as $Y$ variables. In Figure A2.1, we refer to the exogenous variable as $X_1$, the mediator as $Y_1$ and the outcome variable as $Y_2$.

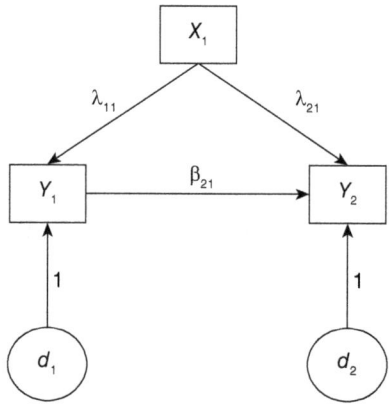

**Figure A2.1**    Diagram of the structural relationships between $X_1$, $Y_1$ and $Y_2$. The coefficient $\lambda_{11}$ is the direct effect of $X_1$ on $Y_1$ and $\lambda_{21}$ the direct effect of $X_1$ on $Y_2$. The coefficient $\beta_{21}$ is the direct effect of $Y_1$ on $Y_2$. The latent variables $d_1$ and $d_2$ are error terms for $Y_1$ and $Y_2$

Figure A2.1 depicts the structural relationships among three observed variables: $X_1$, $Y_1$ and $Y_2$. We can also represent the diagram in Figure A2.1 using a system of equations. To start, let's write an equation for each of the endogenous variables in our path model. The equation for each endogenous variable is analogous to a regression equation. As is the case in regression, the endogenous variable always appears on the left-hand side of the equation. Any (observed or latent) variable with a single-headed arrow leading to the endogenous variable appears on the right-hand side of the equation. Because standard structural equation models are linear models, the terms are additive. Using the path coefficients in the diagram, we construct the system of equations from the path diagram. The system of equations captures all the direct pathways among the variables.[1] Two paths lead to $Y_1$: (1) $X_1$ has a path coefficient of $\lambda_{11}$ and (2) $d_1$ has a path coefficient of 1 (which can be dropped from the equation because $1 * d_1 = d_1$). We use this information to construct the equation for $Y_1$: $Y_1 = \lambda_{11}X_1 + d_1$.

---

[1]The system of equations includes directional paths but not (non-directional) correlations among variables.

As we can see from both the path diagram and the equation for $Y_1$, two separate influences jointly determine the values of $Y_1$. $X_1$ predicts $Y_1$: $\lambda_{11}$ is the regression coefficient that shows the expected change in $Y_1$ per unit change in $X_1$. The disturbance (residual), $d_1$, represents the variance in $Y_1$ that is not explained by $X_1$. In other words, the total variance in $Y_1$ is equal to the sum of $\lambda_{11}X_1$, the variance explained by $X_1$, and $d_1$, the residual variance (the variance in $Y_1$ which is not explained by $X_1$).

Next, we specify the equation for $Y_2$. $Y_2$ is predicted by both $X_1$ and $Y_1$. Again, the endogenous variable, $Y_1$, appears on the left side of the equation. Now, both $X_1$ and $Y_1$ have directional linkages to $Y_2$, so they both appear on the right-hand side of the equation. Finally, $d_2$ represents the residual variance in $Y_2$ that is not explained by $X_1$ or $Y_1$. Thus, our equation for $Y_2$ is $Y_2 = \lambda_{21}X_1 + \beta_{21}Y_1 + d_2$. Because there are multiple predictors, the path coefficient $\lambda_{21}$ is akin to a partial regression coefficient: it represents the change in $Y_2$ per unit change in $X_1$, after controlling for $Y_1$. The path coefficient[ii] $\beta_{21}$ represents the change in $Y_2$ per unit change in $Y_1$, after controlling for $X_1$. Therefore, the two equations for this simple three-variable path model are

$$Y_1 = \lambda_{11}X_1 + d_1 \tag{A2.1}$$

$$Y_2 = \lambda_{21}X_1 + \beta_{21}Y_1 + d_2 \tag{A2.2}$$

Substituting $\lambda_{11}X_1 + d_1$ for $Y_1$ in Equation (A2.2), we obtain

$$Y_2 = \lambda_{21}X_1 + \beta_{21}\left(\lambda_{11}X_1 + d_1\right) + d_2 \tag{A2.3}$$

Distributing across the terms in the parentheses produces

$$Y_2 = \lambda_{21}X_1 + \beta_{21}\lambda_{11}X_1 + \beta_{21}d_1 + d_2 \tag{A2.4}$$

Note that Equation (A2.4) expresses $Y_2$ as a function of exogenous variables and disturbances (i.e. $Y_1$ does not appear in Equation A2.4). $X_1$ predicts $Y_2$ directly: $\lambda_{21}$ is the direct effect of $X_1$ on $Y_2$, after controlling for $Y_1$. $X_1$ also predicts $Y_2$ indirectly, through $Y_1$, and that effect is $\beta_{21} * \lambda_{11}$. The indirect effect of $X_1$ on $Y_2$, $\beta_{21} * \lambda_{11}$, passes through, and is therefore correlated with, $Y_1$. Because $d_1$ represents the variance in $Y_1$ that is not explained by $X_1$, the term $\beta_{21}d_1$ represents the effect of $Y_1$ on $Y_2$ that is not correlated with (predicted by) $X_1$.

---

[ii]Both lambda and beta are path coefficients. Lambda represents a path coefficient from an exogenous variable. Beta represents a path coefficient from an endogenous variable.

# APPENDIX 3
## WRIGHT'S STANDARDISED TRACING RULES

This appendix demonstrates the use of Wright's standardised rules and constructing covariances and correlations using covariance algebra for a path model. In Appendix 4, we introduce Wright's unstandardised rules and provide demonstrations on the use of unstandardised Wright's rules for model-implied variances of endogenous variables as well as the equivalence between Wright's unstandardised rules and covariance algebra.

## Wright's Standardised rules for model-implied correlations of path models

Using the standardised solution, Wright's tracing rules can reproduce the model-implied correlation matrix for the path diagram or model in Figure A2.1. First, we assume that the total variance for each variable in the path diagram is equal to 1. Therefore, all exogenous variables have a variance equal to 1. For endogenous variables, the explained variance and disturbance variance sum to 1.

With Wright's tracing rules, we can generate the model-implied correlations from the standardised parameter estimates. Using Figure A2.1 to construct the model-implied correlation for $Y_2$ and $X_1$ (denoted as cor($Y_2$, $X_1$)), we must sum all the compound paths representing all the unique ways we can trace from $Y_2$ and $X_1$. The Greek letters (e.g. $\lambda$ and $\beta$) represent the parameters for the paths. For example, in Figure A2.1, the model-implied correlation of $Y_2$ with $X_1$ requires summing two compound paths. The first compound path is $X_1 \rightarrow Y_1 \rightarrow Y_2$. The path coefficient for the path from $X_1 \rightarrow Y_1$ is $\lambda_{11}$, and the path coefficient for the path from $Y_1 \rightarrow Y_2$ is $\beta_{21}$. Multiplying these coefficients produces the first compound path for the model-implied correlation of $Y_2$ with $X_1$:

$$\text{cor}(Y_2, X_1) = \lambda_{11}\beta_{21} \ (\textit{Compound Path 1})$$

The second compound path is simply $X_1 \rightarrow Y_2$. The path coefficient encountered along the path is $\lambda_{21}$. Therefore, the second compound path for the model-implied correlation of $Y_2$ with $X_1$ is

$$\text{cor}(Y_2, X_1) = \lambda_{21} \ (\textit{Compound Path 2})$$

Summing the two compound paths produces the model-implied correlation between $Y_2$ and $X_1$:

$$(\textit{Model-Implied}) \ \text{cor}(Y_2, X_1) = \lambda_{11}\beta_{21} + \lambda_{21}$$

The model-implied correlation between $Y_2$ and $X_1$ is shown in Exhibit A3.1. See if you can construct the remaining model-implied correlations for the path diagram in Figure A2.1.

**Exhibit A3.1** The model-implied correlation matrix for the structural equation model in Figure A2.1. The order of variables in this matrix is $Y_1$, $Y_2$ and $X_1$

$$\Sigma(\theta)_{ModelA} = \begin{bmatrix} 1 & & \\ \lambda_{11}\lambda_{21}+\beta_{21} & 1 & \\ \lambda_{11} & \beta_{21}\lambda_{11}+\lambda_{21} & 1 \end{bmatrix}$$

## Constructing covariances and correlations using covariance algebra

We can also use covariance algebra to generate the model-implied variance–covariance matrix. The basic rules of covariance algebra appear in several texts on SEM and path analysis (e.g. Bollen, 1989, p. 21; Kenny, 2004; Paxton et al., 2011, p. 9). We summarise these in Exhibit A3.2.

**Exhibit A3.2** Covariance algebra rules

Defining $c$ as a constant and $X_1$, $X_2$ and $X_3$ as random variables (see *Note*),

1  $cov(c, X_1) = 0$. (The covariance between a constant and a random variable is 0.)
2  $cov(cX_1, X_2) = c*cov(X_1, X_2)$.
3  $cov(X_1 + X_2, X_3) = cov(X_1, X_3) + cov(X_2, X_3)$.
4  $cov(X_1, X_1) = var(X_1)$. (The covariance of a variable with itself is its variance.)

Similarly, four useful variance algebra rules are as follows:

1  $var(aX) = a^2var(X)$.
2  $var(a + X) = var(X)$.
   a  From (1) and (2), we can see that $var(a + bX) = b^2var(X)$.
3  $var(X + Y) = var(X) + var(Y) + 2cov(X + Y)$.
4  If $X$ and $Y$ are independent, $var(X + Y) = var(X) + var(Y)$.

*Note.* A random variable is defined as a function that maps the outcomes of an unpredictable process to numerical quantities, typically real numbers (Wikipedia).

Using covariance and variance algebra rules, we can reconstruct the model-implied covariance matrices and correlation matrices for Figure A2.1 using the system of equations defined by the structural equation model. Note, the population covariance matrix for the structural equation model in Figure A2.1 is

$$\Sigma = \begin{bmatrix} \sigma^2_{Y_1} & & \\ \sigma_{Y_1,Y_2} & \sigma^2_{Y_2} & \\ \sigma_{X_1,Y_1} & \sigma_{X_1,Y_2} & \sigma^2_{X_1} \end{bmatrix}$$

or,

$$\Sigma = \begin{bmatrix} cov(Y_1,Y_1) & & \\ cov(Y_2,Y_1) & cov(Y_2,Y_2) & \\ cov(X_1,Y_1) & cov(X_1,Y_2) & cov(X_1,X_1) \end{bmatrix}$$

The elements on the diagonal of the population covariance matrix are the variances of the three different variables in the model (i.e. $Y_1$, $Y_2$ and $X_1$), whereas the off-diagonal elements of the matrix represent the covariances between each unique pair of the three variables in the model. We use covariance algebra notation, that is, $\sigma_{Y_1}^2 = \text{cov}(Y_1, Y_1)$, to call attention to the fact that a covariance of variable with itself is equivalent to the variance of that variable: $\text{cov}(Y_1, Y_1) = \text{var}(Y_1)$. To further clarify the variances and covariances in Figure A2.1, an extended version of Figure A2.1, Figure A3.1, includes coefficients for the variance parameters in the structural equation model ($\phi$s and $\psi$s). Again, recall that our system of equations for the model in Figure A2.1 and A3.1 is comprised of

*Equation 1:* $Y_1 = \lambda_{11}X_1 + d_1$ and

*Equation 2:* $Y_2 = \lambda_{21}X_1 + \beta_{21}Y_1 + d_2$ .

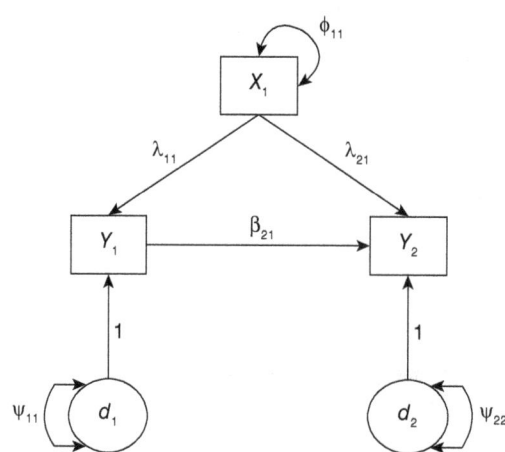

**Figure A3.1** The SEM from Figure A2.1 with depictions of the variance parameters

## Constructing the model-implied covariance of $X_1$ and $Y_1$ with covariance algebra

As our first example, we construct the model-implied $\text{cov}(X_1, Y_1)$ for the model in Figure A3.1. First, we write the $\text{cov}(X_1, Y_1)$ as

$$\text{cov}(X_1, Y_1) = \text{cov}(X_1, \lambda_{11}X_1 + d_1)$$

Here, we have replaced $Y_1$ with $\lambda_{11}X_1 + d_1$. Using covariance algebra rule (3), we can then rewrite the equation as

$$\text{cov}(X_1, Y_1) = \text{cov}(X_1, \lambda_{11}X_1) + \text{cov}(X_1, d_1)$$

We assume that the residual $(d_1)$ is uncorrelated with the predictor variable, $X_1$. Therefore, $\text{cov}(X_1, d_1) = 0$. Using covariance algebra rule (2), we remove $\lambda_{11}$ from within the parentheses of $\text{cov}(X_1, \lambda_{11}X_1)$. With these simplifications, we can now reduce the model-implied $\text{cov}(X_1, Y_1)$ to $\lambda_{11}\phi_{11}$ as follows:

$$(\textit{Model Implied}) \; \text{cov}(X_1, Y_1) = \lambda_{11}\text{cov}(X_1, X_1) = \lambda_{11}\text{var}(X_1) = \lambda_{11}\phi_{11}$$

This model-implied covariance is in Exhibit A3.3 (row 3, column 1).[i]

**Exhibit A3.3**  The model-implied covariance matrix for the structural equation model in Figures A2.1 and A3.1. The order of variables in this matrix is $Y_1$, $Y_2$ and $X_1$

$$\Sigma(\theta)_{\text{Model A}} = \begin{bmatrix} \lambda_{11}^2\phi_{11} + \psi_{11} & & \\ (\lambda_{11}^2\beta_{21} + \lambda_{11}\lambda_{21})\phi_{11} + \beta_{21}\psi_{11} & (\beta_{21}^2\lambda_{11}^2 + 2\beta_{21}\lambda_{11}\lambda_{21} + \lambda_{21}^2)\phi_{11} + \beta_{21}^2\psi_{11} + \psi_{11} & \\ \lambda_{11}\phi_{11} & (\beta_{21}\lambda_{11} + \lambda_{21})\phi_{11} & \phi_{11} \end{bmatrix}$$

## Constructing the model-implied covariance of $Y_2$ and $Y_1$ with covariance algebra

Likewise, we can use covariance algebra to solve for the covariance between $Y_2$ and $Y_1$. First, we substitute the equations for $Y_1$ and $Y_2$, and also substitute the equation for $Y_1$ within the equation for $Y_2$:

$$\text{cov}(Y_1, Y_2) = \text{cov}(\lambda_{11}X_1 + d_1, \beta_{21}(\lambda_{11}X_1 + d_1) + \lambda_{21}X_1 + d_2)$$

Then we distribute $\beta_{21}$ and eliminate the parentheses.

$$\text{cov}(Y_1, Y_2) = \text{cov}(\lambda_{11}X_1 + d_1, \beta_{21}\lambda_{11}X_1 + \beta_{21}d_1 + \lambda_{21}X_1 + d_2)$$

Next, we use covariance algebra rule (3), which gives us

$$\text{cov}(Y_1, Y_2) = \text{cov}(\lambda_{11}X_1, \beta_{21}\lambda_{11}X_1) + \text{cov}(\lambda_{11}X_1, \beta_{21}d_1) + \text{cov}(\lambda_{11}X_1, \lambda_{21}X_1) + \text{cov}(\lambda_{11}X_1, d_2) + \text{cov}(d_1, \beta_{21}\lambda_{11}X_1) + \text{cov}(d_1, \beta_{21}d_1) + \text{cov}(d_1, \lambda_{21}X_1) + \text{cov}(d_1, d_2)$$

Before simplifying the equation, we can drop terms based on the assumption that the disturbances, $d_1$ and $d_2$, are uncorrelated with $X_1$. Therefore, $\text{cov}(\lambda_{11}X_1, \beta_{21}d_1) = 0$,

---

[i]Note that the example for this covariance matrix demonstration is from Paxton et al. (2011). However, we note here that their depiction of this covariance matrix on p. 11 contains a simple error in factoring out the parameter $\phi_{11}$ from the $\text{cov}(Y_1, Y_2)$. See Exhibit C.3 for clarification and the exhibit on the covariance algebra rules.

$\text{cov}(\lambda_{11}X_1, d_2) = 0$, $\text{cov}(d_1, \beta_{21}\lambda_{11}X_1) = 0$ and $\text{cov}(d_1, \lambda_{21}X_1) = 0$. We assume that the $\text{cov}(d_1, d_2) = 0$ in Figure A3.1 (there is no structural relationship between $d_1$ and $d_2$). This leaves us with $\text{cov}(Y_2, Y_1) = \text{cov}(\lambda_{11}X_1, \beta_{21}\lambda_{11}X_1) + \text{cov}(\lambda_{11}X_1, \lambda_{21}X_1) + \text{cov}(d_1, \beta_{21}d_1)$.

Using a combination of the covariance algebra rules above, we can rewrite the equation as $\text{cov}(Y_2, Y_1) = \lambda_{11}^2\beta_{21}\text{var}(X_1) + \lambda_{11}\lambda_{21}\text{var}(X_1) + \beta_{21}\text{var}(d_1)$. Therefore, the model-implied covariance of $Y_2$ and $Y_1$ can be written as

$$(\textit{Model Implied})\ \text{cov}(Y_2, Y_1) = \lambda_{11}^2\beta_{21}\phi_{11} + \lambda_{11}\lambda_{21}\phi_{11} + \beta_{21}\psi_{11}$$

## Constructing the model-implied covariance of $Y_2$ and $Y_2$ with covariance algebra

Next, we construct the model-implied $\text{cov}(Y_2, Y_2)$ for the model in Figure A3.1. Substituting the prediction equation for $Y_2$, the model-implied $\text{cov}(Y_2, Y_2)$ for the model in Figure A3.1 is

$$\text{cov}(Y_2, Y_2) = \text{cov}(\beta_{21}Y_1 + \lambda_{21}X_1 + d_2, \beta_{21}Y_1 + \lambda_{21}X_1 + d_2)$$

Then we can substitute the prediction equation for $Y_1$, resulting in

$$\text{cov}(Y_2, Y_2) = \text{cov}(\beta_{21}(\lambda_{11}X_1 + d_1) + \lambda_{21}X_1 + d_2, \beta_{21}(\lambda_{11}X_1 + d_1) + \lambda_{21}X_1 + d_2)$$

We distribute terms, resulting in

$$\text{cov}(Y_2, Y_2) = \text{cov}(\beta_{21}\lambda_{11}X_1 + \beta_{21}d_1 + \lambda_{21}X_1 + d_2, \beta_{21}\lambda_{11}X_1 + \beta_{21}d_1 + \lambda_{21}X_1 + d_2)$$

Then using covariance algebra rule (3), we can re-express the equation above as

$$\begin{aligned}
\text{cov}(Y_2, Y_2) = {} & \text{cov}(\beta_{21}\lambda_{11}X_1, \beta_{21}\lambda_{11}X_1) + \text{cov}(\beta_{21}\lambda_{11}X_1, \beta_{21}d_1) + \text{cov}(\beta_{21}\lambda_{11}X_1, \lambda_{21}X_1) + \\
& \text{cov}(\beta_{21}\lambda_{11}X_1, d_2) + \text{cov}(\beta_{21}d_1, \beta_{21}\lambda_{11}X_1) + \text{cov}(\beta_{21}d_1, \beta_{21}d_1) + \text{cov}(\beta_{21}d_1, \\
& \lambda_{21}X_1) + \text{cov}(\beta_{21}d_1, d_2) + \text{cov}(\lambda_{21}X_1, \beta_{21}\lambda_{11}X_1) + \text{cov}(\lambda_{21}X_1, \beta_{21}d_1) + \\
& \text{cov}(\lambda_{21}X_1, \lambda_{21}X_1) + \text{cov}(\lambda_{21}X_1, d_2) + \text{cov}(d_2, \beta_{21}\lambda_{11}X_1) + \text{cov}(d_2, \beta_{21}d_1) + \\
& \text{cov}(d_2, \lambda_{21}X_1) + \text{cov}(d_2, d_2)
\end{aligned}$$

We assume the disturbances do not correlate with the predictor variables or with each other ($\text{cov}(X_1, d_1) = 0$ and $\text{cov}(d_1, d_2) = 0$), so we can further simplify to the following:

$$\begin{aligned}
\text{cov}(Y_2, Y_2) = {} & \text{cov}(\beta_{21}\lambda_{11}X_1, \beta_{21}\lambda_{11}X_1) + \text{cov}(\beta_{21}\lambda_{11}X_1, \lambda_{21}X_1) + \text{cov}(\beta_{21}d_1, \beta_{21}d_1) + \\
& \text{cov}(\lambda_{21}X_1, \beta_{21}\lambda_{11}X_1) + \text{cov}(\lambda_{21}X_1, \lambda_{21}X_1) + \text{cov}(d_2, d_2)
\end{aligned}$$

Using covariance algebra rule (2), we can move the constants outside the covariance expressions, producing

$$\text{cov}(Y_2, Y_2) = \beta_{21}^2 \lambda_{11}^2 \text{cov}(X_1, X_1) + \beta_{21}\lambda_{11}\lambda_{21}\text{cov}(X_1, X_1) + \beta_{21}^2 \text{cov}(d_1, d_1) + \lambda_{21}\beta_{21}\lambda_{11}$$
$$\text{cov}(X_1, X_1) + \lambda_{21}^2 \text{cov}(X_1, X_1) + \text{cov}(d_2, d_2)$$

Using covariance algebra rule (4), we can re-express the covariance of a variable with itself as a variance and rewrite the above as

$$\text{cov}(Y_2, Y_2) = \beta_{21}^2 \lambda_{11}^2 \text{var}(X_1) + \beta_{21}\lambda_{11}\lambda_{21}\text{var}(X_1) + \beta_{21}^2 \text{var}(d_1) + \lambda_{21}\beta_{21}\lambda_{11}\text{var}(X_1) +$$
$$\lambda_{21}^2 \text{var}(X_1) + \text{var}(d_2)$$

Lastly, when we substitute the variances from Figure A3.1 and rearrange the terms, we see that the model-implied covariance of $Y_2$ with $Y_2$ or the variance of $Y_2$ is

$$\text{cov}(Y_2, Y_2) = \beta_{21}^2 \lambda_{11}^2 \phi_{11} + 2\beta_{21}\lambda_{11}\lambda_{21}\phi_{11} + \lambda_{21}^2 \phi_{11} + \beta_{21}^2 \psi_{11} + \psi_{22}$$

or,

$$\text{cov}(Y_2, Y_2) = (\beta_{21}^2 \lambda_{11}^2 + 2\beta_{21}\lambda_{11}\lambda_{21} + \lambda_{21}^2)\phi_{11} + \beta_{21}^2 \psi_{11} + \psi_{22}$$

which is equal to the model-implied covariance in Exhibit A3.3 and is equivalent to the model-implied covariance that we derived using the Wright's rules for the unstandardised solution. We encourge the reader to determine the remaining model-implied covariance matrix elements. The entire set of covariances can be seen in Exhibit A3.3. These are the same covariances that can be constructed using Wright's rules for the unstandardised solution (see Appendix 4).

It is also possible to compute the model-implied correlations from the model-implied covariances. We can construct the model-implied correlations for the path diagram in Figure A3.1 from the model-implied covariances in Exhibit A3.3. To start, let's compute the correlation for $Y_1$ with itself, which we know is 1. The formula for the correlation between two random variables $V_1$ and $V_2$ is

$$\text{cor}(V_1, V_2) = \frac{\text{cov}(V_1, V_2)}{\sqrt{\text{var}(V_1)} \sqrt{\text{var}(V_2)}}$$

Therefore, to find the model-implied correlation for $Y_1$ with itself, we use the following equation:

$$\text{cor}(Y_1, Y_1) = \frac{\text{cov}(Y_1, Y_1)}{\sqrt{\text{var}(Y_1)}\sqrt{\text{var}(Y_1)}}$$

Using the cov($Y_1$, $Y_1$) from Exhibit A3.3, and the fact that cov($Y_1$, $Y_1$) = var($Y_1$), we can rewrite the above formula as follows:

$$\text{cor}(Y_1, Y_1) = \frac{\lambda_{11}^2 \phi_{11} + \psi_{11}}{\sqrt{\lambda_{11}^2 \phi_{11} + \psi_{11}} \sqrt{\lambda_{11}^2 \phi_{11} + \psi_{11}}}$$

Using basic algebra, we can rewrite the above equation as

$$\text{cor}(Y_1, Y_1) = \frac{\lambda_{11}^2 \phi_{11} + \psi_{11}}{\lambda_{11}^2 \phi_{11} + \psi_{11}}$$

From the above, we have shown how the equation for the correlation of $Y_1$ reduces to the following:

$$\text{cor}(Y_1, Y_1) = 1$$

With the same algebraic manipulations, we can easily see how a correlation for any variable with itself is 1.

To demonstrate how the algebra of model-implied correlations works for off-diagonal correlations, we need to again assume that the variance of each variable in the structural equation model is equal to 1 or standardised. Using the same correlation formula as above, we can construct the correlation between $Y_2$ and $X_1$ as follows:

$$\text{cor}(Y_2, X_1) = \frac{\text{cov}(Y_2, X_1)}{\sqrt{\text{var}(Y_2)} \sqrt{\text{var}(X_1)}}$$

Using the appropriate elements from the covariance matrix in Exhibit A3.3, we arrive at the following:

$$\text{cor}(Y_2, X_1) = \frac{\beta_{21} \lambda_{11} \phi_{11} + \lambda_{21} \phi_{11}}{\sqrt{1} \sqrt{1}}$$

Simplifying the denominator, and using the fact that $\phi_{11} = 1$ we can rewrite the above as

$$\text{cor}(Y_2, X_1) = \beta_{21} \lambda_{11} + \lambda_{21}$$

which is identical to the model-implied cor($Y_2$, $X_1$) computed using Wright's tracing rules.

Lastly, we construct the model-implied correlation between $Y_2$ and $Y_1$, which is slightly more complicated. Again, using the formula for the correlation we have

$$\text{cor}(Y_2, Y_1) = \frac{\text{cov}(Y_2, Y_1)}{\sqrt{\text{var}(Y_2)} \sqrt{\text{var}(Y_1)}}$$

Recall, the cov($Y_2$, $Y_1$) between $Y_2$ and $Y_1$ can be written as

$$\text{cov}(Y_2, Y_1) = \lambda_{11}^2 \beta_{21} \text{var}(X_1) + \lambda_{11}\lambda_{21}\text{var}(X_1) + \beta_{21}\text{var}(d_1) + \text{cov}(d_1, d_2)$$

Noting that the cov($d_1$, $d_2$) = 0, and that the var($X_1$) = 1, we can simplify the above covariance further to

$$\text{cov}(Y_2, Y_1) = \lambda_{11}^2 \beta_{21} + \lambda_{11}\lambda_{21} + \beta_{21}\text{var}(d_1)$$

But what is var($d_1$)? Recall that var($d_1$) refers to the disturbance variance for $Y_1$. For an endogenous variable, we know that the explained variance and disturbance variance sum to 1. Therefore, we can rewrite the variance of $Y_1$, which is an endogenous variable, as

$$\text{var}(Y_1) = \text{Explained variance} + \text{Disturbance variance} = 1$$

Or, using covariance algebra,

$$\text{var}(Y_1) = \text{var}(\lambda_{11}X_1 + d_1) = \text{var}(\lambda_{11}X_1) + \text{var}(d_1) = 1$$

Therefore, the disturbance variance is var($d_1$) = 1 – the explained variance because var($Y_1$) = 1. Because the explained variance in $Y_1$ is $\text{var}(\lambda_{11}X_1) = \lambda_{11}^2$, the var($d_1$) = $1 - \lambda_{11}^2$.

Substituting into the cov($Y_2$, $Y_1$) produces

$$\text{cov}(Y_2, Y_1) = \lambda_{11}^2 \beta_{21} + \lambda_{11}\lambda_{21} + \beta_{21}(1 - \lambda_{11}^2)$$
$$= \lambda_{11}^2 \beta_{21} + \lambda_{11}\lambda_{21} + \beta_{21} - \beta_{21}\lambda_{11}^2$$

The cancellation of $\lambda_{11}^2 \beta_{21}$ and $\beta_{21} \lambda_{11}^2$ leaves

$$\text{cov}(Y_2, Y_1) = \lambda_{11}\lambda_{21} + \beta_{21}$$

Finally, substituting the cov($Y_2$, $Y_1$) back into the formula for the correlation produces

$$\text{cov}(Y_2, Y_1) = \frac{\lambda_{11}\lambda_{21} + \beta_{21}}{\sqrt{1}\sqrt{1}}$$

Or a model-implied correlation of

$$\text{cov}(Y_2, Y_1) = \lambda_{11}\lambda_{21} + \beta_{21}$$

The result is equivalent to what we found using the standardised tracing rules above. Try to determine the remaining model-implied correlation matrix elements. Recall, the implied correlation matrix for the Figure in A3.1 is in Exhibit A3.1.

Using Wright's rules or covariance algebra enables us to determine the model-implied variance–covariance matrix from the parameter estimates. If the specified model is correct, then the model-implied variance–covariance matrix should be able to reproduce the population variance–covariance matrix. Given that we are using a sample variance–covariance matrix as an estimate of the population variance–covariance matrix, we would not expect the model-implied variance–covariance matrix to be identical to the observed variance–covariance matrix. However, we would expect the model-implied variance–covariance matrix to be similar to the observed variance–covariance matrix.

# APPENDIX 4

## WRIGHT'S UNSTANDARDISED TRACING RULES AND COVARIANCE ALGEBRA

Although it is generally easier to calculate and interpret correlations and standard-ised coefficients, SEM uses the variance–covariance matrix and produces unstand-ardised (as well as standardised) parameter estimates. To demonstrate how we can construct the model-implied covariance matrix, often depicted as $\Sigma(\theta)$, we can use two methods: (1) Wright's (unstandardised) tracing rules and (2) rules of covariance algebra. Here, we demonstrate Wright's unstandardised tracing rules to construct the model-implied covariances.

Heise (1975) extended Wright's tracing rules to derive expected covariances in addition to expected correlations. Although the unstandardised tracing rules are a bit more complex and a bit more tedious than the standardised rules, the unstand-ardised tracing rules provide a way to more intuitively generate the model-implied variances and covariances without using covariance algebra. First, we demonstrate Wright's rules for the unstandardised solution, using the path model in Figure A3.1. After presenting Wright's rules for generating the model-implied variance–covariance matrix from the unstandardised parameter estimates, we demonstrate the equivalence of Wright's (1918, 1934) tracing rules and the rules of covariance algebra (Bollen, 1989; Jöreskog & Sörbom, 1993), for constructing the model-implied covariance matrix (McArdle & Boker, 1990). The equivalence is demonstrated with an example using endogenous variables. Understanding the relationship between Wright's rules and the rules of covariance algebra provides deep insight into the underpinnings of SEM.

## Wright's Unstandardised rules for determining model-implied covariances

Recall, the population covariance matrix for the structural equation model in Figure A3.1 is

$$\Sigma = \begin{bmatrix} \sigma^2_{Y_1} & & \\ \sigma_{Y_1,Y_2} & \sigma^2_{Y_2} & \\ \sigma_{X_1,Y_1} & \sigma_{X_1,Y_2} & \sigma^2_{X_1} \end{bmatrix}$$

The elements on the diagonal of the population covariance matrix are the variances of the three different variables in the model (i.e. $Y_1$, $Y_2$ and $X_1$), whereas the off-diagonal elements of the matrix represent the covariances between each unique pair of the three variables in the model. Using the unstandardised tracing rules with the unstandardised parameter estimates, it is possible to reproduce the model-implied variance–covariance matrix (see Exhibit A3.3).

Neale and Cardon (1992) describe Wright's tracing rules for unstandardised variables. First, ensure that all the variances (including disturbance variances) are included in the SEM diagram. To incorporate variances and residual variances in the SEM diagram, we use double-headed arrows in the shape of a circle (see Figure A3.1). Generating the covariance between two variables follows the same basic logic as the standardised tracing rule. However, to account for the variances in the computation of the model-implied covariances, *we must include the variance of the predictor variable in the compound path anytime we change direction in our trace.* The Wright's standardised rules specify that a trace can go backward and then forward, but not forward and then backward. To form the tracing chain between two variables using the unstandardised tracing rules, *we always start at the endogenous variable and trace backward.* Anytime we change direction and switch from backward to forward, we include the variance of the variable in the trace. For example, in Figure A3.1, we trace backward, change direction at the two-headed arrow (representing the variance of the predictor variable) and then trace forward (Neale & Maes, 1998, p. 92). If the trace does not contain a change in direction, we include the double-headed arrow at the end of the trace to account for the variance (this is required in the first model-implied covariance between $X_1$ and $Y_2$). Finally, when incorporating a covariance, or a double-headed curved arrow into the trace, the covariance is included in the compound path, but the variances of those two variables are not (Heise, 1975, p. 120).

As with the standardised coefficients, to find the expected covariance, we compute the sum of the compound paths. We multiply each of the elements in a compound path to obtain the complete compound path and then we sum over all possible compound paths. To derive the variance for an endogenous variable, we sum all the compound paths from the endogenous variable back to itself. We demonstrate how to use Wright's rules for unstandardised variables to construct the implied covariance matrix from a path diagram, using Figure A3.1. There are three observed variables. Therefore, we can express the population covariance matrix, $\Sigma$, as

$$\Sigma = \begin{bmatrix} \text{cov}(Y_1,Y_1) & & \\ \text{cov}(Y_2,Y_1) & \text{cov}(Y_2,Y_2) & \\ \text{cov}(X_1,Y_1) & \text{cov}(X_1,Y_2) & \text{cov}(X_1,X_1) \end{bmatrix}$$

We can now use Wright's tracing rules for determining the model-implied covariances for the path diagram in Figure A3.1. We apply Wright's tracing rules for unstandardised variables (Heise, 1975; Neale & Cardon, 1992) to determine each of the six elements, or covariances, in the model-implied covariance matrix. First, let's compute the model-implied covariance between $X_1$ and $Y_2$: $\text{cov}(X_1, Y_2)$. To do so, for each unique trace from $X_1$ to $Y_2$, we compute the compound path, adhering to the unstandardised tracing rules, and then we sum all the compound paths computed from each of the unique traces, just as we did earlier. The sum of those compound paths is the model-implied covariance between $X_1$ and $Y_2$.

The first compound path from $X_1$ to $Y_2$ is through the *direct* path (or arrow) from $X_1$ to $Y_2$, which has the path coefficient $\lambda_{21}$. For the unstandardised tracing rules, we always start at the endogenous variable and therefore we can represent our trace backwards as $Y_2 \leftarrow X_1$. Also, this compound paths' trace ends at $X_1$, and we do not have a change of direction, so we must include the double-headed arrow at the end of the trace to account for the exogenous variance of $X_1$, $\phi_{11}$, in the compound path. Therefore, the first compound path then for the model-implied covariance between $X_1$ and $Y_2$ is $\phi_{11}\lambda_{21}$.

$$\text{cov}(X_1, Y_2) = \phi_{11}\lambda_{21} \ (Compound\ Path\ 1)$$

Again, we multiply the parameters in the model to determine the compound path. Specifically, we multiply the estimate of the variance of $X_1$ ($\phi_{11}$) by the estimate of the direct effect of $Y_2$ on $X_1$ ($\lambda_{21}$) as these were the two coefficients we encountered. Including the exogenous variance in the unstandardised traces may make more sense if you recall that to compute the regression coefficient from a variance–covariance matrix in a simple linear regression, the unstandardised regression coefficient of $Y$ regressed on $X$, $b_{yx}$, is $\text{cov}_{xy}/\text{var}_x$. Therefore, to go from $b_{yx}$ to $\text{cov}_{xy}$, we must multiply $b_{yx}$ by $\text{var}_x$.

The second compound path from $X_1 \rightarrow Y_2$ is the indirect path from $X_1 \rightarrow Y_1 \rightarrow Y_2$. Again, we begin at $Y_2$ and head backward from $Y_2$ to $Y_1$ through $\beta_{21}$ and then travel from $Y_1$ to $X_1$ through $\lambda_{11}$, or $Y_2 \leftarrow Y_1 \leftarrow X_1$. Again, this compound paths' trace ends at $X_1$ without a change of direction. Therefore, we must include the double-headed arrow at the end of the trace to account for the exogenous variance of $X_1$, $\phi_{11}$, in the compound path. The second compound path for the model-implied covariance between $X_1$ and $Y_2$ is $\phi_{11}\lambda_{11}\beta_{21}$.

$$\text{cov}(X_1, Y_2) = \phi_{11}\lambda_{11}\beta_{21} \ (Compound\ Path\ 2)$$

Lastly, the implied covariance between $X_1$ and $Y_2$ is the sum of the two compound paths: $\phi_{11}\lambda_{21} + \phi_{11}\lambda_{11}\beta_{21} = (\lambda_{21}+\lambda_{11}\beta_{21})\,\phi_{11}$

$$(Model\ Implied)\ \text{cov}(X_1, Y_2) = \phi_{11}\lambda_{21} + \phi_{11}\lambda_{11}\beta_{21} = (\lambda_{21}+\lambda_{11}\beta_{21})\phi_{11}$$

The model-implied covariance between $X_1$ and $Y_2$ is depicted in Exhibit A3.3 (row 3 and column 2). The exhibit also contains all the other model-implied covariances between the remaining variables. See if you can recreate the model-implied covariance between $Y_1$ and $Y_2$ using the Wright's unstandardised rules.

## Wright's Unstandardised rules for model-implied variances of endogenous variables

To further demonstrate the unstandardised tracing rules, and the equivalence to covariance algebra, we reproduce the model-implied variance of $Y_1$ which is the covariance of $Y_1$ with itself, that is, $\text{cov}(Y_1, Y_1)$. The total variance in $Y_1$ has been partitioned into two parts: the part that is explained by $X_1$ and the part that is unexplained by $X_1$, the disturbance of $Y_1$ $(\psi_{11})$. The path between $Y_1$ and $d_1$ is fixed to one; we estimate the disturbance variance $(\psi_{11})$. To compute the first compound path, we first trace back from $Y_1$ to $d_1$, circle through the double-headed error representing the disturbance variance of $Y_1$, $\psi_{11}$, thereby changing direction, and then trace back to $Y_1$. This trace is $Y_1 \leftarrow d_1 \rightarrow Y_1$. Note that the path between the covariance is 1. Therefore, the first compound path for the model-implied variance of $Y_1$ is $\psi_{11}$.

$$\text{cov}(Y_1, Y_1) = 1 * \psi_{11} * 1 = \psi_{11} \quad (Compound\ Path\ 1)$$

Therefore, we see that the first compound path for the model-implied $\text{cov}(Y_1, Y_1)$ is simply the disturbance variance $\psi_{11}$.

To construct the second compound path for the variance of $Y_1$, we need to account for the variance that is explained by $X_1$. Therefore, we travel backward from $Y_1$ to $X_1$, circle through the variance of $X_1$ $(\phi_{11})$, and then we travel forward through the arrow from $X_1$ back to $Y_1$. This trace, $Y_1 \leftarrow X_1 \rightarrow Y_1$, contains three elements: $\lambda_{11}$, $\phi_{11}$ and $\lambda_{11}$. Thus, the product of the path coefficients and the second compound path for the variance of $Y_1$ is $\lambda_{11}\phi_{11}\lambda_{11}$, which equals $\lambda_{11}^2\phi_{11}$.

$$\text{cov}(Y_1, Y_1) = \lambda_{11}\phi_{11}\lambda_{11} = \lambda_{11}^2\phi_{11} \quad (Compound\ Path\ 2)$$

Therefore, the model-implied $\text{cov}(Y_1, Y_1)$ or variance of $Y_1$ is $\lambda_{11}^2\phi_{11} + \psi_{11}$.

$$(Model\ Implied)\ \text{cov}(Y_1, Y_1) = \lambda_{11}^2\phi_{11} + \psi_{11}$$

The model-implied $\text{cov}(Y_1, Y_1)$, or variance of $Y_1$ implied by the model, consists of two pieces: (1) the variance of $Y_1$ which is explained by $X_1$ or the explained variance of $Y_1$, $\lambda_{11}^2\phi_{11}$, and (2) the disturbance variance $\psi_{11}$ (i.e. the variance of $Y_1$ which is not explained by $X_1$).

In large models, using the unstandardised tracing rule to generate the variance of an endogenous variable within a complex set of variables can be tedious and somewhat confusing. We generate the model-implied variance of $Y_2$ using the unstandardised version of Wright's rules. The greatest danger in using the tracing rule to reproduce model-implied covariances (or correlations) is inadvertently excluding one of the necessary compound paths. Generating the model-implied variance of $Y_2$ requires summing six compound paths. We outline the six compound paths below, in no particular order:

1   Leave $Y_2$, travel backward to $Y_1$ and then backward to $X_1$, change direction at $X_1$, so include the variance of $X_1$ in the compound path, travel forward to $Y_2$. The trace is then $Y_2 \leftarrow Y_1 \leftarrow X_1 \rightarrow Y_2 = \beta_{21}\lambda_{11}\phi_{11}\lambda_{21}$.

2   Leave $Y_2$, travel backward to $X_1$, change direction at $X_1$, so include the variance of $X_1$ in the compound path, travel forward to $Y_1$ and then to $Y_2$. The trace is then $Y_2 \leftarrow X_1 \rightarrow Y_1 \rightarrow Y_2 = \lambda_{21}\phi_{11}\lambda_{11}\beta_{21}$.

3   Travel backward from $Y_2$ to the disturbance, include the variance and circle back around to $Y_2$. The trace is then $Y_2 \leftarrow d_2 \rightarrow Y_2 = 1\psi_{22}1 = \psi_{22}$.

4   Leave $Y_2$, travel backward to $Y_1$, change direction at $Y_1$, so include the disturbance variance of $Y_1$ in the compound path, then return from $Y_1$ to $Y_2$. The trace is then $Y_2 \leftarrow Y_1 \leftarrow d_1 \rightarrow Y_1 \rightarrow Y_2 = \beta_{21}1\psi_{11}1\beta_{21} = \beta_{21}^2\psi_{11}$.

5   Travel backward from $Y_2$ to $X_1$, include the variance of $X_1$ and then circle back and return to $Y_2$. The trace is then $Y_2 \leftarrow X_1 \rightarrow Y_2 = \lambda_{21}\phi_{11}\lambda_{21} = \lambda_{21}^2\phi_{11}$.

6   Travel backward from $Y_2$ to $Y_1$, then from $Y_1$ to $X_1$, include the variance of $X_1$ and then circle back from $X_1$ to $Y_1$ and back to $Y_2$. The trace is then $Y_2 \leftarrow Y_1 \leftarrow X_1 \rightarrow Y_1 \rightarrow Y_2 = \beta_{21}\lambda_{11}\phi_{11}\lambda_{11}\beta_{21} = \beta_{21}^2\lambda_{11}^2\phi_{11}$.

The sum of these compound paths is $\beta_{21}^2\lambda_{11}^2\phi_{11} + 2\beta_{21}\lambda_{11}\lambda_{21}\phi_{11} + \lambda_{21}^2\phi_{11} + \beta_{21}^2\psi_{11} + \psi_{22}$ or $(\beta_{21}^2\lambda_{11}^2 + 2\beta_{21}\lambda_{11}\lambda_{21} + \lambda_{21}^2)\phi_{11} + \beta_{21}^2\psi_{11} + \psi_{22}$. These steps are succinctly outlined in Exhibit A4.1. From the above examples, we can reconstruct the model-implied covariance matrix for the remaining bivariate relations using the unstandardised tracing rules. In most cases, using the tracing rules is quicker than working out the covariance algebra associated with the variables (see Appendix 3). However, with very large and complex models, using covariance algebra may be more expedient. We leave it to the reader to compute the remaining model-implied covariances for the structural equation model in Figure A3.1. Each of the covariances is depicted in Exhibit A3.3.

**Exhibit A4.1**   Wright's unstandardised tracing rules cov($Y_2$, $Y_2$)

---

Compound Path 1:
$Y_2 \leftarrow Y_1 \leftarrow X_1 \rightarrow Y_2 = \beta_{21}\lambda_{11}\phi_{11}\lambda_{21}$

Compound Path 2:
$Y_2 \leftarrow X_1 \rightarrow Y_1 \rightarrow Y_2 = \lambda_{21}\phi_{11}\lambda_{11}\beta_{21}$

Compound Path 3:
$Y_2 \leftarrow d_2 \rightarrow Y_2 = 1\psi_{22}1 = \psi_{22}$

Compound Path 4:
$Y_2 \leftarrow Y_1 \leftarrow d_1 \rightarrow Y_1 \rightarrow Y_2 = \beta_{21}1\psi_{11}1\beta_{21} = \beta_{21}^2\psi_{11}$

Compound Path 5:
$Y_2 \leftarrow X_1 \rightarrow Y_2 = \lambda_{21}\phi_{11}\lambda_{21} = \lambda_{21}^2\phi_{11}$

---

Compound Path 6:

$Y_2 \leftarrow Y_1 \leftarrow X_1 \rightarrow Y_1 \rightarrow Y_2 = \beta_{21}\lambda_{11}\phi_{11}\lambda_{11}\beta_{21} = \beta_{21}^2\lambda_{11}^2\phi_{11}$

Adding the compound paths together, we are left with

$\text{cov}(Y_2, Y_2) = \beta_{21}\lambda_{11}\phi_{11}\lambda_{21} + \lambda_{21}\phi_{11}\lambda_{11}\beta_{21} + \psi_{22} + \beta_{21}^2\psi_{11} + \lambda_{21}^2\phi_{11} + \beta_{21}^2\lambda_{11}^2\phi_{11}$

Or, after rearranging terms and factoring out $\phi_{11}$

$\text{cov}(Y_2, Y_2) = (\beta_{21}^2\lambda_{11}^2 + 2\beta_{21}\lambda_{11}\lambda_{21} + \lambda_{21}^2)\phi_{11} + \beta_{21}^2\psi_{11} + \psi_{22}$

which is equal to the model-implied covariance in Exhibit A3.3 (row 2, column 2).

# GLOSSARY

**Absolute fit indices:** Absolute fit indices evaluate the degree to which the specified model reproduces the sample data. Common absolute fit indices include the root mean square error of approximation (RMSEA) and the standardised root mean square residual (SRMR).

**Between-cluster variance:** Variability that is explained by the cluster; between-cluster differences.

**Between-individual level:** In a simple growth model, the between-individual level is level 2 of a multilevel model. Observations across time (level 1) are nested within individuals or persons.

**Cluster mean:** The mean of a variable at the level of the cluster. The cluster mean is often used to group mean level-1 variables. When doing so, generally, the cluster mean should be included in the model at level 2.

**Clustered data:** Observations can be hierarchically arranged into clusters, and observations within clusters exhibit some degree of dependency (non-independence).

**Combined model:** A model that combines level-1 and level-2 equations.

**Compound path:** A compound path, or trace, is a pathway connecting the two variables following Wright's rules and is the product of its constituent paths. There can be multiple compound paths between the same set of two variables.

**Conceptual structural model:** A conceptual model depicts the researcher's hypothesis about the interrelations of variables in a path diagram. The variables can be latent or observed.

**Conceptually omitted (deleted) paths:** Paths that are not hypothesised in a researcher's conceptual structural equation model.

**Conditional growth model:** A growth model that includes additional time-invariant or time-varying covariates at the within (level 1) or between (level 2) level.

**Conditional ICC:** The proportion of (residual) between-cluster variance after accounting for all the variables in the model.

**Confirmatory factor analysis (CFA):** A specific form of factor analysis that is used to test a researcher's hypothesised nature of the relation between observed and latent variables.

**Construct:** A concept, model or schematic idea.

**Contextual variable:** A variable that exists at the cluster level or is aggregated to the cluster level. These variables describe characteristics of the clusters.

**Correlation:** An unstandardised measure of the linear dependency between two variables.

**Covariance:** A standardised measure of the linear dependency between two variables.

**Covariance algebra:** A general set of rules for computing the covariance between random variables. In SEM, the use of covariance algebra is for determining the model-implied correlations or covariances in a path diagram (see Exhibit A3.2 in Appendix 3, and Appendix 4).

**Covariance/correlation matrix:** A matrix containing the covariances or correlations between variables in an SEM. The covariance is central to SEM.

**Cross-level interaction:** Interactions across levels of a multilevel model, for example, a level-2 contextual variable is a predictor of a level-1 slope.

**Design effect (DEFF):** The ratio of the sampling variability for the study design compared to the sampling variability that would be expected if the study used a simple random sample.

**Deviance:** Deviance is a measure of the badness of fit of a given model, it describes how much worse the specified model is than the best possible model.

**Direct effect:** An effect of one variable directly on another that contains no intermediate variables.

**Disturbance:** Unexplained variance in an endogenous variable; variance in the variable that is not explained by the model.

**Effective sample size:** The effective sample size is the actual sample size divided by the design effect (see *Design effect*). In MLM, the effective sample size may be smaller than the actual sample size due to non-independence. Using the effective sample size, we can adjust our standard errors to account for non-independence.

**Endogenous variable:** Variables that are predicted by one or more variables in the model.

**Equivalent models:** Equivalent models have different causal structures but produce identical fit to the data.

**Exogenous variable:** Variables that predict other variables, but they are not predicted by any other variables in the model.

**Fixed effect:** A fixed effect is a parameter that does not vary across clusters.

**Fixed factor variance strategy:** Constrains the variance of each factor to 1.0. In standard CFA models, the factor's mean is constrained to 0. Therefore, the fixed factor variance strategy results in a standardised solution: each latent variable in the model has a mean of 0 and a variance of 1.

**Full contextual/theoretical model:** The model that describes the complete theory of a researcher. Includes all within- and between-cluster variables, all hypothesised cross-level (and same-level) interactions, and all theoretically relevant random effects.

**Grand mean centring:** A method to centre covariates. In grand mean centring, the overall mean of the independent variable is subtracted from all scores.

**Group mean centring:** A method to centre covariates. In group mean centring, the cluster mean is subtracted from each score.

**Growth curve model:** Models that allow for the estimation of systematic growth or decline over time.

**Growth slope:** Describes the rate of change or growth in the model. The variance in the growth slope can be measured across individuals or persons in the model. See also *Latent slope factor*.

**Heywood case:** Negative variances or correlations above 1. Heywood cases result in inadmissible solutions.

**Hierarchical linear modelling (HLM)/multilevel modelling (MLM):** A modelling framework that allows researchers to adjust for and model non-independent clustered data. This non-independent data generally results from the clustering or nesting of observations.

**Hybrid model:** A latent variable path model; combines a measurement model and structural model.

**Hybrid structural equation:** An SEM that combines a measurement model (see *Measurement model*) and a structural model (see *Structural model*).

**Identification:** Identification involves ensuring that all the parameters in a model are uniquely identified. If all the parameters in the model are uniquely identified, then the model itself is identified.

**Inadmissible solution:** A solution that is not mathematically possible. These include, for example, negative error variances and standardised regression weights above 1 (Heywood cases).

**Incremental fit indices:** Incremental fit indices measure the proportionate amount of improvement in fit when the specified model is compared with a nested baseline model. Common incremental fit indices include the Tucker–Lewis index (TLI) and the comparative fit index (CFI).

**Indirect effect:** An effect of one variable on another that works through intermediate variables. Models with indirect effects are often referred to as mediational models.

**Individual level:** The level of the individual. This is level 1 in organisational models, but level 2 in longitudinal models. Predictors at the individual level characterise aspects about the individual (e.g. the test scores of an individual).

**Initial status:** In a growth model, the initial status describes where people start in their growth trajectory. If time is centred at initial status, then this is the latent intercept factor.

**Integrative framework of $R^2$:** Framework proposed by Right and Sterba in 2019. The integrative $R^2$ framework allows for the computation of proportion of variance explained at various levels and for the full contextual model.

**Intercept-only model:** Specifies that the outcome variable is a function of the intercept (which randomly varies across people) and a within-person residual, which captures the time-specific deviation of the outcome variable from the individual's intercept.

**Intra-class correlation coefficient (ICC):** Quantifies the degree of dependence among observations from the same cluster. The ICC is a measure of how similar, or homogeneous, individuals are within clusters and how much they vary across clusters.

**Just-identified structural equation model:** A just-identified structural model contains one linkage between each of the structural variables, even if some of these linkages are not part of the conceptual model. That is, the just-identified model may contain conceptually omitted (deleted) paths (see *Paths*).

**Latent construct:** A non-observable construct (see *Construct*), such as cognitive ability, self-concept or optimism. Latent variables are not directly observed, but their presence and influence can be inferred based on variables that are directly observed.

**Latent intercept factor:** A latent variable for the intercept in an SEM growth model. If time is centred at initial status, then the parameter estimate for latent intercept factor is the expected initial status (score at the initial time point).

**Latent slope factor:** A latent variable for the growth slope in an SEM growth model.

**Latent variable:** A variables that is not directly observable within the sample data.

**lavaan:** An R package for fitting and analysing structural equation modelling and latent variable models.

**Level-1 equation:** A regression equation that is specified as a function of level-1 predictors.

**Level-2 equation:** A regression equation that is specified as a function of level-2 predictors.

**Linear growth model:** A growth model that assumes a constant rate of growth.

**lme4:** An R package for fitting and analysing mixed effects models that can be used to fit multilevel and hierarchical linear models.

**Local independence:** The correlations among the observed variables or indicators is solely a function of their mutual relationship with the latent factor(s). There are no relationships among the unexplained variances of the observed variables or indicators.

**Marker variable strategy:** The marker variable strategy constrains one unstandardised path coefficient for each factor 1 and freely estimates the variance of the latent factor. The variable whose unstandardised regression coefficient is fixed to 1 becomes the marker variable for the factor, and the factor's freely estimated variance is scaled in the same metric as the marker variable.

**Measurement error:** The variance in an observed variable that is not attributable to true score variance (or the latent variable). Error variance that attenuates the regression weight from the predictor variable to the dependent variable biasing estimates of the paths. One of the advantages of SEM is that the modelling framework accounts for potential measurement error.

**Measurement model:** Depicts the relationships between the observed variables (also called indicators) and their underlying latent variables.

**Measurement weights/pattern coefficients:** Represent the direct effects of the factor on the indicator variable.

**Mediator:** A 'middle man'; an intervening variable between a predictor variable and a dependent variable.

**MLM growth model:** A growth model that is fit in a multilevel modelling framework. MLM growth models handle time-unstructured data more easily than SEM growth models (see *SEM growth models*), can easily incorporate nested data and can more easily handle fully non-linear growth models. MLM growth models require data be stored in a long format.

**Model fit:** Model fit describes how well a model reproduces the data. Model fit measures include, for example, deviance, the Akaike information criterion (AIC) and the Bayesian information criterion (BIC).

**Model-implied correlation/covariance:** The variance–covariance matrix that is reproduced, using the model-estimated parameters.

**Model prediction:** The adequacy with which a model predicts a criterion variable of interest. The proportion of variance in the criterion variable that is explained by the model provides a measure of the adequacy of model prediction (see *R-squared*).

**Multidimensional:** The statistical dependence among observed variables or indicators is captured by multiple latent constructs. More than one latent variable is predictive of responses on observed indicators.

**Multilevel modelling:** Multilevel modelling of data repeatedly measured across time. In this case, observations are nested within individuals (units) across time.

**Multiple group SEM (MG-SEM):** Enables between-group comparisons of any model parameters, including latent means, in SEM models (see SEM definition).

**Non-independence:** Observations are non-independent if they are correlated or related in some manner. The observations exhibit dependence due to clustering or as a result of being repeatedly measured across time.

**Non-linear growth model:** A growth model that allows for non-linear growth (see e.g. **polynomial growth model**).

**Non-recursive structural equation model:** An SEM is non-recursive if there are 'loops' or 'cycles' that imply feedback loops or reciprocal relationships. These loops occur when, in a path diagram, a causal pathway leaves one variable and then returns to that variable. An SEM can also be non-recursive if there are covariances among disturbances.

**Observed variable:** A variable that is directly observed/measured.

**Organisational models:** multilevel models where clustering occurs within organisations or groups such as schools, companies or families.

**Parsimonious model:** The simpler, less complex, less parameterised model; the model with fewer free parameters.

**Partial regression coefficient:** A regression coefficient or slope that represents a change in the dependent variable when one independent variable is increased by 1 unit and the other independent variables in the model are held constant.

**Path diagram:** A diagram used to represent structural equation models and a system of equations.

**Piecewise growth model:** A growth model that allows for slopes to be estimated in multiple linear pieces; sometimes referred to as a spline model.

**Polynomial growth model:** A growth model that allows for curvilinear change or change with a possible point of inflection. Linear, quadratic and cubic models are all examples of polynomial models.

**Preregistration:** Preregistration involves (a) committing to a series of analytic steps prior to engaging with the data, (b) providing detailed descriptions of the proposed analytic techniques and (c) specifying exactly how the authors will conduct all data analyses, without advance knowledge of the research outcomes.

**Proportion of variance unexplained by a factor:** Represents the proportion of variance in the observed variable that is not explained by a factor but rather by error, noise or other variables not in the model.

**Random coefficients model:** A statistical model that allows for random intercepts and slopes. MLM allows for random intercepts and slopes.

**Random effect:** A random effect varies across clusters.

**Random intercept:** In an MLM, a random intercept allows the intercept or mean of the dependent variable to vary across clusters.

**Random slope:** In an MLM, a random slope allows the effect of a predictor variable on the dependent variable to vary across clusters.

**Recursive structural equation model:** A path diagram where there are no 'loops' or 'cycles' that imply feedback loops or reciprocal relationships. Also, the path diagram has no covariances among the disturbance terms.

**Residual variance:** Variance that is not explained by the model. Often referred to as unexplained variance.

**Respecification:** The amendment of certain aspects of a model of interest; often occurs when the model does not fit the data.

**R-squared:** The R-squared characterises the proportion of the variance in a dependent that is predictable from the independent variables in a model.

**Sampling variability:** Uncertainty or variance introduced by sampling from a larger population.

**SEM growth model:** A growth model that is in a structural equation framework. Compared to MLM growth models, SEM growth models more easily incorporates measurement models and allow for more flexibility in terms of the modelling of the residual variance–covariance structure. SEM growth models require data be stored in a wide format.

**Simple random sample:** A subset of observations from a larger population. In a simple random sample, observations are chosen at random such that each observation has the same probability of being selected.

**Standard error:** The square root of the sampling variance. It provides a statistical measure of the accuracy of an estimate. Larger standard errors are less accurate.

**Standardised path coefficient:** The expected unit of change in a dependent variable given a 1-unit change in a predictor variable, holding the other variables in the model constant, analogous to a path coefficient when all the variables are in $z$-scores (the mean = 0 and $SD$ = 1) units.

**Standardised tracing rules:** A set of rules for determining the model-implied correlation between any two variables in a path diagram (see Appendix 3).

**Structural equation modelling (SEM):** Refers to a family of techniques that makes use of the variance–covariance matrix to examine the structural relations among a set of variables; path analysis and confirmatory factor analysis are two specific types of SEM.

**Structural model:** The structural model depicts the predictive paths and consists of the structural paths between and among the latent variables and any observed variables that are not indicators of an underlying variable.

**Sufficient statistic:** A sufficient statistic provides all the necessary information to estimate model parameters. Under normal circumstances, the variance–covariance matrix serves as the sufficient statistic for an SEM analysis.

**Systems of equations:** A set of equations for the variables and parameters in a path diagram.

**Three steps of structural model specification:** A model building approach for SEM analysis. The first step involves fitting a just-identified structural model (see *Just-identified structural model*). The second step involves deleting any of the conceptually omitted (deleted) paths that were not found to be statistically significant in the first step and rerunning the model. The third (optional) step includes trimming non-statistically significant, non-necessary conceptual paths.

**Time-structured data:** All participants' data is collected on the same schedule; the interval between data collection is the same across all sample participants.

**Time-unstructured data:** Data are time-unstructured if time intervals can vary both within and across people. With time-unstructured data, the data collection schedule can be completely different for every person in the sample. Forcing time-unstructured data to follow a time-structured pattern can decrease precision and increase bias in the parameter estimates.

**Time-varying covariate:** Time-varying covariates are level-1 variables that can vary across time as well as across people. Thus, TVC can have different values for the same person at different time points throughout the study period. Although the time variable itself is technically a time-varying covariate (TVC), the term *TVC* usually refers to a variable other than time that varies both within and across people.

**Total effect:** The total effect of one variable on another; the total of direct and indirect effects of a system of variables in an SEM analysis.

**Unconditional growth model:** A growth model that does not include any time-invariant or time-varying covariates at the within (level 1) or between (level 2) level.

**Unconditional ICC:** The proportion of (residual) between-cluster variance before accounting for all the variables in the model.

**Unconditional random effects ANOVA model:** The null model or random effects analysis of variance model that includes no predictors at level 1 or level 2. The unconditional random effects model is used to compute the unconditional ICC.

**Unidimensional:** The statistical dependence among observed variables or indicators is captured by a single latent construct.

**Unstandardised path coefficient:** The expected unit of change in a dependent variable given a 1-unit change in a predictor variable, holding the other variables in the model constant. Unstandardised path coefficients reflect the scales of measurement of the predictor and dependent variables and must be interpreted accordingly.

**Unstandardised tracing rules:** A set of rules for determining the model-implied covariances between any two variables in a path diagram (see Appendix 4).

**Variance components:** The variance components are the variances and covariances of the random effects (see **random effect**, **random intercept** and **random slope**) in a model. The variance components allow us to partition the variance in the dependent or outcome variable into within- and between-cluster variance components.

**Vertical scaling:** Vertical scaling equates assessment scores across age or grade levels, rendering the scores directly comparable across time, even though examinees of different ages, grades or ability levels may complete different assessments.

**Within-cluster variance:** The variability among observations within the same cluster.

**Within-individual level:** Level 1 in a growth model; includes observations across time.

**Wright's tracing rules:** Sewall Wright's basic rules and principles for path analysis that provide a method to generate model-implied covariances or correlations for a model from the path diagram.

---

**Check out the next title in the collection: *Statistical Approaches to Causal Analysis*, for guidance on Causal Inference in Quantitative Research.**

# REFERENCES

Aiken, L. S., & West, S. G. (1991). *Multiple regression: Testing and interpreting interactions*. Sage.

Airoldi, E. M., Blei, D. M., Erosheva, E. A., & Feinberg, S. E. (Eds.). (2015). *Handbook of mixed membership models and their applications*. CRC Press. https://doi.org/10.1201/b17520

Anderson, J. C., & Gerbing, D. W. (1982). Some methods for respecifying measurement models to obtain unidimensional construct measurement. *Journal of Marketing Research*, *19*(4), 453–460. https://doi.org/10.2307/3151719

Anderson, J. C., & Gerbing, D. W. (1988). Structural equation modeling in practice: A review and recommended two-step approach. *Psychological Bulletin*, *103*(3), 411–423. https://doi.org/10.1037/0033-2909.103.3.411

Asparouhov, T., Hamaker, E. L., & Muthén, B. (2018). Dynamic structural equation models. *Structural Equation Modeling*, *25*(3), 359–388. https://doi.org/10.1080/10705511.2017.1406803

Baltes, P. B., & Nesselroade, J. R. (1979). History and rationale of longitudinal research. In J. R. Nesselroade & P. B. Baltes (Eds.), *Longitudinal research in the study of behavior and development* (pp. 1–39). Academic Press.

Bandalos, D. L. (2002). The effects of item parceling on goodness-of-fit and parameter estimate bias in structural equation modeling. *Structural Equation Modeling*, *9*(1), 78–102. https://doi.org/10.1207/S15328007SEM0901_5

Baron, R. M., & Kenny, D. A. (1986). The moderator–mediator variable distinction in social psychological research: Conceptual, strategic, and statistical considerations. *Journal of Personality and Social Psychology*, *51*(6), 1173–1182. https://doi.org/10.1037/0022-3514.51.6.1173

Bast, J., & Reitsma, P. (1997). Matthew effects in reading: A comparison of latent growth curve models and simplex models with structured means. *Multivariate Behavioral Research*, *32*(2), 135–167. https://doi.org/10.1207/s15327906mbr3202_3

Bates, D., Machler, M., Bolker, B., & Walker, S. (2015). Fitting linear mixed effects models using lme4. *Journal of Statistical Software, 67*(1), 1–48. https://doi. org/10.18637/jss.v067.i01

Bauer, D. J., Preacher, K. J., & Gil, K. M. (2006). Conceptualizing and testing random indirect effects and moderated mediation in multilevel models: New procedures and recommendations [Supplemental material]. *Psychological Methods, 11*(2), 142–163. https://doi.org/10.1037/1082-989X.11.2.142.supp

Bentler, P. M. (1990). Comparative fit indexes in structural models. *Psychological Bulletin, 107*(2), 238–246. https://doi.org/10.1037/0033-2909.107.2.238

Bentler, P. M., & Bonett, D. G. (1980). Significance tests and goodness of fit in the analysis of covariance structures. *Psychological Bulletin, 88*(3), 588–606. https:// doi.org/10.1037/0033-2909.88.3.588

Bentler, P. M., & Raykov, T. (2000). On measures of explained variance in nonrecursive structural equation models. *Journal of Applied Psychology, 85*(1), 125–131. https://doi.org/10.1037/0021-9010.85.1.125

Beretvas, S. N. (2008). Cross-classified random effects model. In A. A. O'Connell & D. B. McCoach (Eds.), *Multilevel modeling of educational data* (pp. 161–197). Information Age.

Berry, W. D. (1984). *Nonrecursive causal models*. Sage. https://doi. org/10.4135/9781412985321

Biesanz, J. C. (2012). Autoregressive longitudinal models. In R. H. Hoyle (Ed.), *Handbook of structural equation modeling* (pp. 459–471). Guilford Press.

Biesanz, J. C., Deeb-Sossa, N., Papadakis, A. A., Bollen, K. A., & Curran, P. J. (2004). The role of coding time in estimating and interpreting growth curve models. *Psychological Methods, 9*(1), 30–52. https://doi.org/10.1037/1082-989X.9.1.30

Bollen, K. A. (1989). *Structural equations with latent variables*. Wiley. https://doi. org/10.1002/9781118619179

Bollen, K. A. (2002). Latent variables in psychology and the social sciences. *Annual Review of Psychology, 53*, 605–634. https://doi.org/10.1146/annurev. psych.53.100901.135239

Bollen, K. A., & Bauldry, S. (2011). Three Cs in measurement models: Causal indicators, composite indicators, and covariates. *Psychological Methods, 16*(3), 265–284. https://doi.org/10.1037/a0024448

Bollen, K. A., & Curran, P. J. (2004). Autoregressive latent trajectory (ALT) models: A synthesis of two traditions. *Sociological Methods & Research, 32*(3), 336–383. https://doi.org/10.1177/0049124103260222

Bollen, K. A., & Curran, P. J. (2006). *Latent curve models: A structural equation perspective*. Wiley. https://doi.org/10.1002/0471746096

Bollen, K. A., & Diamantopoulos, A. (2017). In defense of causal-formative indicators: A minority report. *Psychological Methods*, *22*(3), 581–596. https://doi.org/10.1037/met0000056

Bollen, K., & Long, J. S. (Eds.). (1993). *Testing structural equation models*. Sage.

Borsboom, D. (2005). *Measuring the mind: Conceptual issues in contemporary psychometrics*. Cambridge University Press. https://doi.org/10.1017/CBO9780511490026

Borsboom, D., Mellenbergh, G. J., & Van Heerden, J. (2003). The theoretical status of latent variables. *Psychological Review*, *110*(2), 203–219. https://doi.org/10.1037/0033-295X.110.2.203

Bozdogan, H. (1987). Model selection and Akaike's information criterion (AIC): The general theory and its analytical extensions. *Psychometrika*, *52*(3), 345–370. https://doi.org/10.1007/BF02294361

Brannick, M. T. (1995). Critical comments on applying covariance structure modeling. *Journal of Organizational Behavior*, *16*(3), 201–213. https://doi.org/10.1002/job.4030160303

Breckler, S. (1990). Applications of covariance structure modelling in psychology: Cause for concern? *Psychological Bulletin*, *107*(2), 260–273. https://doi.org/10.1037/0033-2909.107.2.260

Brown, T. A. (2015). *Confirmatory factor analysis for applied research* (2nd ed.). Guilford Press.

Browne, M. W., MacCallum, R. C., Kim, C., Anderson, B. L., & Glaser, R. (2002). When fit indices and residuals are incompatible. *Psychological Methods*, *7*(4), 403–421. https://doi.org/10.1037/1082-989X.7.4.403

Burnham, K. P., & Anderson, D. R. (2004). Multimodel inference: Understanding AIC and BIC in model selection. *Sociological Methods & Research*, *33*(2), 261–304. https://doi.org/10.1177/0049124104268644

Burt, R. S. (1976). Interpretational confounding of unobserved variables in structural equation models. *Sociological Methods & Research*, *5*(1), 3–52. https://doi.org/10.1177/004912417600500101

Campbell, D. T., & Fiske, D. W. (1959). Convergent and discriminant validity by the multitrait-multimethod matrix. *Psychological Bulletin*, *56*(2), 81–105. https://doi.org/10.1037/h0046016

Campbell, D. T., & Kenny, D. A. (1999). *Regression artifacts*. Guilford Press.

Cole, D. A., Perkins, C. E., & Zelkowitz, R. L. (2016). Impact of homogeneous and heterogeneous parceling strategies when latent variables represent multidimensional constructs. *Psychological Methods*, *21*(2), 164–174. https://doi.org/10.1037/met0000047

Cole, D. A., & Preacher, K. J. (2014). Manifest variable path analysis: Potentially serious and misleading consequences due to uncorrected measurement error. *Psychological Methods, 19*(2), 300–315. https://doi.org/10.1037/a0033805

Collins, L. (2006). Analysis of longitudinal data: The integration of theoretical model, temporal design, and statistical model. *Annual Review of Psychology, 57,* 505–528. https://doi.org/10.1146/annurev.psych.57.102904.190146

Cook, T. D., & Campbell, D. T. (1979). *Quasi-experimentation: Design and analysis issues for field settings.* Rand McNally.

Coyne, M. D., McCoach, D. B., Ware, S., Austin, C., Loftus, S., & Baker, D. (2019). Racing against the vocabulary gap: Matthew effects in early vocabulary instruction and intervention. *Exceptional Children, 85,* 163–179. https://ies. ed.gov/funding/grantsearch/details.asp?ID=1086

Crocker, L., & Algina, J. (1986). *Introduction to classical and modern test theory.* Holt, Rinehart & Winston.

Curran, P. J., & Bauer, D. J. (2011). The disaggregation of within-person and between-person effects in longitudinal models of change. *Annual Review of Psychology, 62,* 583–619. https://doi.org/10.1146/annurev.psych.093008.100356

Curran, P. J., Howard, A. L., Bainter, S. A., Lane, S. T., & McGinley, J. S. (2014). The separation of between-person and within-person components of individual change over time: A latent curve model with structured residuals. *Journal of Consulting and Clinical Psychology, 82,* 879–894. https://doi.org/10.1037/a0035297

Curran, P. J., Lee, T., Howard, A. L., Lane, S., & MacCallum, R. (2012). Disaggregating within-person and between-person effects in multilevel and structural equation growth models. In J. R. Harring & G. R. Hancock (Eds.), *Advances in longitudinal methods in the social and behavioral sciences* (pp. 217–253). Information Age.

Curran, P. J., West, S. G., & Finch, J. F. (1996). The robustness of test statistics to nonnormality and specification error in confirmatory factor analysis. *Psychological Methods, 1*(1), 16–29. https://doi.org/10.1037/1082-989X.1.1.16

de Leeuw, J. (2004). [Review of the book Multilevel analysis: Techniques and applications, by J. J. Hox]. *Journal of Educational Measurement, 41*(1), 73–77. https://doi.org/10.1111/j.1745-3984.2004.tb01160.x

Duncan, T. E., Duncan, S. C., Stryker, L. A., Li, F., & Alpert, A. (1999). *An introduction to latent variable growth curve modeling.* Lawrence Erlbaum.

Dweck, C. S., Walton, G. M., & Cohen, G. L. (2014). *Academic tenacity: Mindsets and skills that promote long-term learning.* Bill & Melinda Gates Foundation. http://k12education.gatesfoundation.org/download/?Num=2807&filename=30-Academic-Tenacity.pdf

Edwards, J., & Bagozzi, R. (2000). On the nature and direction of relationships between constructs and measures. *Psychological Methods, 5*(2), 155–174. https://doi.org/10.1037/1082-989X.5.2.155

Enders, C. K., & Tofighi, D. (2007). Centering predictor variables in cross-sectional multilevel models: A new look at an old issue. *Psychological Methods, 12*(2), 121–138. https://doi.org/10.1037/1082-989X.12.2.121

Fabrigar, L. R., & Wegener, D. T. (2012). *Exploratory factor analysis.* Oxford University Press. https://doi.org/10.1093/acprof:osobl/9780199734177.001.0001

Fielding, A., & Goldstein, H. (2006). *Cross classified and multiple membership structures in multilevel modelling: An introduction and a review* (Research Report No. 791). Department for Education and Skills. https://dera.ioe.ac.uk/6469/1/RR791.pdf

Finney, S., & DiStefano, C. (2006). Nonnormal and categorical data in structural equation modeling. In G. Hancock & R. Mueller (Eds.), *Structural equation modeling: A second course* (pp. 439–492). Information Age.

Fox, J. (2008). *Applied regression analysis and generalized linear models* (2nd ed.). Sage.

Gerbing, D. W., & Anderson, J. C. (1993). Monte Carlo evaluations of goodness-of-fit indices for structural equation models. In K. A. Bollen & J. S. Long (Eds.), *Testing structural equation models* (pp. 40–65). Sage.

Glass, G. V., & Hopkins, K. D. (1996). *Statistical methods in education and psychology* (3rd ed.). Allyn & Bacon.

Goldstein, H. (2011). *Multilevel statistical models* (4th ed.). Wiley. https://doi.org/10.1002/9780470973394

Gorsuch, R. L. (1997). Exploratory factor analysis: Its role in item analysis. *Journal of Personality Assessment, 68*(3), 532–560. https://doi.org/10.1207/s15327752jpa6803_5

Gribbons, B. C., & Hocevar, D. (1998). Levels of aggregation in higher level confirmatory factor analysis: Application for academic self-concept. *Structural Equation Modeling, 5*(4), 377–390. https://doi.org/10.1080/10705519809540113

Grimm, K. J., Ram, N., & Estabrook, R. (2016). *Growth modeling: Structural equation and multilevel modeling approaches.* Guilford Press.

Gully, S. M., & Phillips, J. M. (2019). On finding your level. In S. E. Humphrey & J. M. LeBreton (Eds.), *The handbook of multilevel theory, measurement, and analysis* (pp. 11–38). American Psychological Association. https://doi.org/10.1037/0000115-002

Gurka, M. J. (2006). Selecting the best linear mixed model under REML. *The American Statistician, 60*(1), 19–26. https://doi.org/10.1198/000313006X90396

Hancock, G. R., & Mueller, R. O. (Eds.). (2006). *Structural equation modeling: A second course.* Information Age.

Hayakawa, K. (2019). Corrected goodness-of-fit test in covariance structure analysis. *Psychological Methods, 24*(3), 371–389. https://doi.org/10.1037/met0000180

Heck, R. H., & Thomas, S. L. (2000). *An introduction to multilevel modeling techniques.* Lawrence Erlbaum.

Heck, R. H., & Thomas, S. L. (2015). *An introduction to multilevel modelling techniques: MLM and SEM approaches using Mplus* (3rd ed.). Routledge. https://doi.org/10.4324/9781315746494

Heise, D. R. (1975). *Causal analysis.* Wiley.

Hershberger, S. L. (2006). The problem of equivalent structural models. In G. R. Hancock & R. O. Mueller (Eds.), *Structural equation modeling: A second course* (pp. 13–41). Information Age.

Hoffman, L. (2015). *Longitudinal analysis: Modeling within-person fluctuation and change.* Routledge. https://doi.org/10.4324/9781315744094

Hox, J. J. (2010). *Multilevel analysis: Techniques and applications* (2nd ed.). Routledge. https://doi.org/10.4324/9780203852279

Hox, J. J., Moerbeek, M., & Van de Schoot, R. (2017). *Multilevel analysis: Techniques and applications* (3rd ed.). Routledge. https://doi.org/10.4324/9781315650982

Hoyle, R. H. (1995). The structural equation modeling approach: Basic concepts and fundamental issues. In R. H. Hoyle (Ed.), *Structural equation modeling: Concepts, issues and application* (pp. 158–176). Sage.

Hoyle, R. H. (Ed.). (2012). *Handbook of structural equation modeling.* Guilford Press.

Hu, L.-T., & Bentler, P. M. (1995). Evaluating model fit. In R. H. Hoyle (Ed.), *Structural equation modelling: Concepts, issues and application* (pp. 76–99). Sage.

Hu, L.-T., & Bentler, P. M. (1999). Cutoff criteria for fit indexes in covariance structure analysis: Conventional criteria versus new alternatives. *Structural Equation Modeling, 6*(1), 1–55. https://doi.org/10.1080/10705519909540118

Jöreskog, K. G., & Sörbom, D. (1993). *LISREL 8: Structural equation modeling with the SIMPLIS command language.* Scientific Software.

Kaplan, D. (2000). *Structural equation modeling: Foundation and extensions.* Sage.

Kazantzis, N., Ronan, K. R., & Deane, F. P. (2001). Concluding causation from correlation: Comment on Burns and Spangler (2000). *Journal of Consulting and Clinical Psychology, 69*(6), 1079–1083. https://doi.org/10.1037/0022-006X.69.6.1079

Kelloway, E. K. (1995). Structural equation modeling in perspective. *Journal of Organizational Behavior, 16*(3), 215–224. https://doi.org/10.1002/job.4030160304

Kenny, D. A. (2004). *Correlation and causality* (Rev. ed.). Wiley-Interscience. http://davidakenny.net/doc/cc_v1.pdf

Kenny, D. A. (2019). Enhancing validity in psychological research. *American Psychologist, 74*(9), 1018–1028. https://doi.org/10.1037/amp0000531

Kenny, D. A., & Campbell, D. T. (1989). On the measurement of stability in over-time data. *Journal of Personality*, *57*(2), 445–481. https://doi.org/10.1111/j.1467-6494.1989.tb00489.x

Kenny, D. A., Kaniskan, B., & McCoach, D. B. (2015). How small is too small? The performance of RMSEA in models with small degrees of freedom. *Sociological Methods & Research*, *44*(3), 486–507. https://doi.org/10.1177/0049124114543236

Kenny, D. A., & Kashy, D. A. (1992). Analysis of the multitrait-multimethod matrix by confirmatory factor analysis. *Psychological Bulletin*, *112*(1), 165–172. https://doi.org/10.1037/0033-2909.112.1.165

Kenny, D. A., Kashy, D., & Bolger, N. (1998). Data Analysis in Social Psychology. In D. Gilbert, S. Fiske, & G. Lindsay (Eds.), *Handbook of Social Psychology* (4th ed. pp. 233–265). New York: McGraw Hill.

Kenny, D. A., & McCoach, D. B. (2003). Effect of the number of variables on measures of fit in structural equation modeling. *Structural Equation Modeling*, *10*(3), 333–351. https://doi.org/10.1207/S15328007SEM1003_1

Kenny, D. A., & Milan, S. (2012). Identification: A nontechnical discussion of a technical issue. In R. H. Hoyle (Ed.), *Handbook of structural equation modeling* (pp. 145–163). Guilford Press.

Kish, L. (1965). *Survey sampling*. Wiley.

Kline, R. B. (1998). *Structural equation modeling*. Guilford Press.

Kline, R. B. (2015). *Principles and practices of structural equation modeling* (4th ed.). Guilford Press.

Kreft, I. G., & de Leeuw, J. (1998). *Introducing multilevel modeling*. Sage. https://doi.org/10.4135/9781849209366

LaHuis, D. M., & Ferguson, M. W. (2009). The accuracy of significance tests for slope variance components in multilevel random coefficient models. *Organizational Research Methods*, *12*(3), 418–435. doi:10.1177/1094428107308984

Lee, S., & Hershberger, S. (1990). A simple rule for generating equivalent models in covariance structure modeling. *Multivariate Behavioral Research*, *25*(3), 313–334. https://doi.org/10.1207/s15327906mbr2503_4

Little, T. D. (2013). *Longitudinal structural equation modeling*. Guilford Press.

Little, T. D., Cunningham, W. A., Shahar, G., & Widaman, K. F. (2002). To parcel or not to parcel: Exploring the question, weighing the merits. *Structural Equation Modeling*, *9*(2), 151–173. https://doi.org/10.1207/S15328007SEM0902_1

Little, T. D., Rhemtulla, M., Gibson, K., & Schoemann, A. M. (2013). Why the items versus parcels controversy needn't be one. *Psychological Methods*, *18*(3), 285–300. https://doi.org/10.1037/a0033266

Loehlin, J. C. (2004). *Latent variable models: An introduction to factor, path, and structural equation analysis* (4th ed.). Lawrence Erlbaum. https://doi.org/10.4324/9781410609823

MacCallum, R. C. (1986). Specification searches in covariance structure modelling. *Psychological Bulletin, 100*, 107–20.

MacCallum, R. C. (2001). 2001 Presidential address: Working with imperfect models. *Multivariate Behavioral Research, 38*(1), 113–139. https://doi.org/10.1207/S15327906MBR3801_5

MacCallum, R. C., Browne, M. W., & Sugawara, H. M. (1996). Power analysis and determination of sample size for covariance structure modeling. *Psychological Methods, 1*(2), 130–149. https://doi.org/10.1037/1082-989X.1.2.130

MacCallum, R. C., Wegener, D. T., Uchino, B. N., & Fabrigar, L. R. (1993). The problem of equivalent models in applications of covariance structure analysis. *Psychological Bulletin, 114*(1), 185–199. https://doi.org/10.1037/0033-2909.114.1.185

MacKinnon, D. P. (2008). *Introduction to statistical mediation analysis.* Lawrence Erlbaum.

MacKinnon, D. P., & Fairchild, A. J. (2009). Current directions in mediation analysis. *Current Directions in Psychological Science, 18*(1), 16–20. https://doi.org/10.1111/j.1467-8721.2009.01598.x

MacKinnon, D. P., Valente, M. J., & Gonzalez, O. (2020). The correspondence between causal and traditional mediation analysis: The link is the mediator by treatment interaction. *Prevention Science, 21*(2), 147–157. https://doi.org/10.1007/s11121-019-01076-4

Marcoulides, G. A., & Schumacker, R. E. (Eds.). (2001). *New developments and techniques in structural equation modeling.* Lawrence Erlbaum. https://doi.org/10.4324/9781410601858

Marsh, H. W. (1987). The factorial invariance of responses by males and females to a multidimensional self-concept instrument: Substantive and methodological issues. *Multivariate Behavioral Research, 22*(4), 457–480. https://doi.org/10.1207/s15327906mbr2204_5

Marsh, H. W. (1993). Stability of individual differences in multiwave panel studies: Comparison of simplex models and one-factor models. *Journal of Educational Measurement, 30*(2), 157–183. https://doi.org/10.1111/j.1745-3984.1993.tb01072.x

Marsh, H. W., & Grayson, D. (1995). Latent variable models of multitrait-multimethod data. In R. H. Hoyle (Ed.), *Structural equation modeling: Concepts, issues, and applications* (pp. 177–198). Sage.

Marsh, H. W., & Hocevar, D. (1983). Confirmatory factor analysis of multitrait-multimethod matrices. *Journal of Education Measurement, 20*(3), 231–248. https://doi.org/10.1111/j.1745-3984.1983.tb00202.x

Marsh, H. W., Lüdtke, O., Nagengast, B., Morin, A. J. S., & von Davier, M. (2013). Why item parcels are (almost) never appropriate: Two wrongs do not make a right—camouflaging misspecification with item parcels in CFA models [Supplemental material]. *Psychological Methods, 18*(3), 257–284. https://doi.org/10.1037/a0032773.supp

Maslowsky, J., Jager, J., & Hemken, D. (2015). Estimating and interpreting latent variable interactions: A tutorial for applying the latent moderated structural equations method. *International Journal of Behavioral Development, 39*(1), 87–96. https://doi.org/10.1177/0165025414552301

Mayer, A., Thoemmes, F., Rose, N., Steyer, R., & West, S. G. (2014). Theory and analysis of total, direct, and indirect causal effects. *Multivariate Behavioral Research, 49*(5), 425–442. https://doi.org/10.1080/00273171.2014.931797

McArdle, J. J. (2009). Latent variable modeling of differences and changes with longitudinal data. *Annual Review of Psychology, 60*, 577–605. https://doi.org/10.1146/annurev.psych.60.110707.163612

McArdle, J. J., & Boker, S. M. (1990). *RAMpath: A computer program for automatic path diagrams.* Lawrence Erlbaum.

McArdle, J. J., & Nesselroade, J. R. (2014). *Longitudinal data analysis using structural equation models.* American Psychological Association. https://doi.org/10.1037/14440-000

McCoach, D. B. (2018). Multilevel modeling. In G. R. Hancock, L. M. Stapleton, & R. O. Mueller (Eds.), *The reviewer's guide to quantitative methods in the social sciences* (2nd ed.). Routledge. https://doi.org/10.4324/9781315755649-22

McCoach, D. B., & Black, A. C. (2008). Evaluation of model fit and adequacy. In A. A. O'Connell & D. B. McCoach (Eds.), *Multilevel modeling of educational data* (pp. 245–272). Information Age.

McCoach, D. B., & Black, A. C. (2012). Introduction to estimation issues in multilevel modeling. *New Directions for Institutional Research, 2012*(154), 23–39. https://doi.org/10.1002/ir.20012

McCoach, D. B., Black, A. C., & O'Connell, A. A. (2007). Errors of inference in structural equation modeling. *Psychology in the Schools, 44*(5), 461–470. https://doi.org/10.1002/pits.20238

McCoach, D. B., Gable, R. K., & Madura, J. P. (2013). *Instrument development in the affective domain: School and corporate applications* (3rd ed.). Springer. https://doi.org/10.1007/978-1-4614-7135-6

McCoach, D. B., Gubbins, E. J., Foreman, J. L., Rubenstein, L. D., & Rambo-Hernandez, K. E. (2014). Evaluating the efficacy of using pre-differentiated and enriched mathematics curricula for grade 3 students: A multi-site cluster-randomized trial. *Gifted Child Quarterly*, *58*(4), 272–286. https://doi. org/10.1177/0016986214547631

McCoach, D. B., & Kaniskan, B. (2010). Using time-varying covariates in multilevel growth models. *Frontiers in Psychology*, *1*, Article 17. https://doi.org/10.3389/ fpsyg.2010.00017

McCoach, D. B., Madura, J., Rambo, K., O'Connell, A. A., & Welsh, M. (2013). Longitudinal data analysis. In T. Teo (Ed.), *Handbook of quantitative methods for educational research* (pp. 199–230). Sense. https://doi.org/10.1007/978-94-6209-404-8_10

McCoach, D. B., O'Connell, A. A., Reis, S. M., & Levitt, H. A. (2006). Growing readers: A hierarchical linear model of children's reading growth during the first 2 years of school. *Journal of Educational Psychology*, *98*(1), 14–28. https://doi. org/10.1037/0022-0663.98.1.14

McCoach, D. B., Rambo, K. E., & Welsh, M. (2013). Assessing the growth of gifted students. *Gifted Child Quarterly*, *57*(1), 56–67. https://doi. org/10.1177/0016986212463873

McCoach, D. B., & Rambo-Hernandez, K. E. (2018). Issues in the analysis of change. In C. Secolsky & D. B. Denison (Eds.), *Handbook of measurement, assessment, and evaluation in higher education* (2nd ed.). Routledge. https://doi. org/10.4324/9781315709307-31

McCoach, D. B., Rifenbark, G., Newton, S. D., Li, X., Kooken, J., Yomtov, D., Gambino, A., & Bellara, A. (2018). Does the package matter? A comparison of five common multilevel modeling software packages. *Journal of Educational and Behavioral Statistics*, *43*(5), 594–627. https://doi.org/10.3102/1076998618776348

McCoach, D. B., & Yu, H. H. (2016). Using individual growth curve models to understand reading fluency development. In K. Cummings & Y. Petscher (Eds.), *The fluency construct* (pp. 269–308). Springer. https://doi. org/10.1007/978-1-4939-2803-3_10

Meehl, P. E., & Waller, N. G. (2002). The path analysis controversy: A new statistical approach to strong appraisal of verisimilitude. *Psychological Methods*, *7*(3), 283–300. https://doi.org/10.1037/1082-989X.7.3.283

Mulaik, S. A. (2009). *Linear causal modeling with structural equations*. CRC Press. https://doi.org/10.1201/9781439800393

Muthén, L. K., & Muthén, B. O. (1998–2017). *Mplus user's guide* (8th ed.). Muthén & Muthén. https://www.statmodel.com/download/usersguide/ MplusUserGuideVer_8.pdf

Myung, I. J. (2003). Tutorial on maximum likelihood estimation. *Journal of Mathematical Psychology*, *47*(1), 90–100. https://doi.org/10.1016/S0022-2496(02)00028-7

Nagase, M., & Kano, Y. (2017). Identifiability of nonrecursive structural equation models. *Statistics & Probability Letters*, *122*, 109–117. https://doi.org/10.1016/j.spl.2016.11.010

Neale, M. C., & Cardon, L. R. (1992). *Methodology for genetic studies of twins and families*. Kluwer Academic. https://doi.org/10.1007/978-94-015-8018-2

Neale, M. C., & Maes, H. M. (1998). *Methodology for genetic studies of twins and families*. Kluwer Academic.

Nesselroade, J. R., & Baltes, P. B. (1979). *Longitudinal research in the study of behavior and development*. Academic Press.

Nosek, B. A., Ebersole, C. R., DeHaven, A. C., & Mellor, D. T. (2018). The preregistration revolution. *Proceedings of the National Academy of Sciences of the United States of America*, *115*(11), 2600–2606. https://doi.org/10.1073/pnas.1708274114

Nunnally, J. C., & Bernstein, I. H. (1994). *Psychometric theory* (3rd ed.). McGraw-Hill.

O'Connell, A. A., & McCoach, D. B. (Eds.). (2008). *Multilevel modeling of educational data*. Information Age.

Paxton, P., Hipp, J. R., & Marquart-Pyatt, S. T. (2011). *Nonrecursive models: Endogeneity, reciprocal relationships, and feedback loops* (Vol. *168*). Sage. https://doi.org/10.4135/9781452226514

Preacher, K. J. (2015). Advances in mediation analysis: A survey and synthesis of new developments. *Annual Review of Psychology*, *66*(1), 825–852. https://doi.org/10.1146/annurev-psych-010814-015258

Preacher, K. J., & Hayes, A. F. (2008). Contemporary approaches to assessing mediation in communication research. In A. F. Hayes, M. D. Slater, & L. B. Snyder (Eds.), *The SAGE sourcebook of advanced data analysis methods for communication research* (pp. 1–54). Sage. https://doi.org/10.4135/9781452272054.n2

Preacher, K. J., Zhang, Z., & Zyphur, M. J. (2011). Alternative methods for assessing mediation in multilevel data: The advantages of multilevel SEM. *Structural Equation Modeling*, *18*(2), 161–182. https://doi.org/10.1080/10705511.2011.557329

Preacher, K. J., Zhang, Z., & Zyphur, M. J. (2016). Multilevel structural equation models for assessing moderation within and across levels of analysis [Supplemental material]. *Psychological Methods*, *21*(2), 189–205. https://doi.org/10.1037/met0000052.supp

Preacher, K. J., Zyphur, M. J., & Zhang, Z. (2010). A general multilevel SEM framework for assessing multilevel mediation [Supplemental material]. *Psychological Methods*, *15*(3), 209–233. https://doi.org/10.1037/a0020141.supp

Raftery, A. E. (1995). Bayesian model selection in social research. *Sociological Methodology*, *25*, 111–163. https://doi.org/10.2307/271063

Raudenbush, S. W., & Bryk, A. S. (2002). *Hierarchical linear models: Applications and data analysis methods* (2nd ed.). Sage.

Raudenbush, S. W., Bryk, A., Cheong, Y. F., & Congdon, R. (2004). *HLM 6: Hierarchical linear and nonlinear modeling*. Scientific Software.

Raykov, T., & Marcoulides, G. A. (2000). *A first course in structural equation modeling*. Lawrence Erlbaum.

Raykov, T., & Penov, S. (1999). On structural equation model equivalence. *Multivariate Behavioral Research, 34*(2), 199–244. https://doi.org/10.1207/S15327906Mb340204

Rhemtulla, M. (2016). Population performance of SEM parceling strategies under measurement and structural model misspecification. *Psychological Methods, 21*(3), 348–368. https://doi.org/10.1037/met0000072

Rhemtulla, M., van Bork, R., & Borsboom, D. (2020). Worse than measurement error: Consequences of inappropriate latent variable models. *Psychological Methods, 25*(1), 30–45. https://doi.org/10.1037/met0000220

Rigdon, E. E. (1995). A necessary and sufficient identification rule for structural models estimated in practice. *Multivariate Behavioral Research, 30*(3), 359–383. https://doi.org/10.1207/s15327906mbr3003_4

Rights, J. D., & Sterba, S. K. (2019). Quantifying explained variance in multilevel models: An integrative framework for defining R-squared measures. *Psychological Methods, 24*(3), 309–338. https://doi.org/10.1037/met0000184

Rosseel, Y. (2012). lavaan: An R package for structural equation modeling. *Journal of Statistical Software, 48*(2), 1–36. https://doi.org/10.18637/jss.v048.i02

Sass, D. A., & Smith, P. L. (2006). The effects of parceling unidimensional scales on structural parameter estimates in structural equation modeling. *Structural Equation Modeling, 13*(4), 566–586. https://doi.org/10.1207/s15328007sem1304_4

Schumacker, R. E., & Lomax, R. G. (1996). *A beginner's guide to structural equation modeling*. Lawrence Erlbaum.

Schwarz, G. (1978). Estimating the dimension of a model. *Annals of Statistics, 6*(2), 461–464. https://doi.org/10.1214/aos/1176344136

Self, S. G., & Liang, K. Y. (1987). Asymptotic properties of maximum likelihood estimators and likelihood ratio tests under nonstandard conditions. *Journal of the American Statistical Association, 82*(398), 605–610. https://doi.org/10.1080/01621459.1987.10478472

Shaw, M., Rights, J. D., Sterba, S. K., & Flake, J. K. (2020, October 16). r2mlm: R-Squared Measures for Multilevel Models. https://doi.org/10.31234/osf.io/xc4sv

Shrout, P. E., & Bolger, N. (2002). Mediation in experimental and nonexperimental studies: New procedures and recommendations. *Psychological Methods, 7*(4), 422–445. https://doi.org/10.1037/1082-989X.7.4.422

Singer, J. D., & Willett, J. B. (2003). *Applied longitudinal data analysis: Modeling change and event occurrence*. Oxford University Press. https://doi.org/10.1093/acprof: oso/9780195152968.001.0001

Skrondal, A., & Rabe-Hesketh, S. (2008). Multilevel and related models for longitudinal data. In J. de Leeuw & E. Meijer (Eds.), *Handbook of multilevel analysis* (pp. 275–299). Springer. https://doi.org/10.1007/978-0-387-73186-5_7

Snijders, T. A. B., & Bosker, R. J. (2012). *Multilevel analysis: An introduction to basic and advanced multilevel modeling* (2nd ed.). Sage.

Spybrook, J., Bloom, H., Congdon, R., Hill, C., Martinez, A., & Raudenbush, S. W. (2011). Optimal Design Plus Empirical Evidence 3.0 Manual. http://hlmsoft.net/od/od-manual-20111016-v300.pdf

Stegmann, G., Jacobucci, R., Harring, J. R., & Grimm, K. J. (2018). Nonlinear mixed-effects modeling programs in R. *Structural Equation Modeling, 25*(1), 160–165. https://doi.org/10.1080/10705511.2017.1396187

Steiger, J. H. (2002). When constraints interact: A caution about reference variables, identification constraints, and scale dependencies in structural equation modeling. *Psychological Methods, 7*(2), 210–227. https://doi.org/10.1037/1082-989X.7.2.210

Stelzl, I. (1986). Changing a causal hypothesis without changing the fit: Some rules for generating equivalent path models. *Multivariate Behavioral Research, 21*(3), 309–331. https://doi.org/10.1207/s15327906mbr2103_3

Stoel, R. D., & Garre, F. G. (2011). Growth curve analysis using multilevel regression and structural equation modeling. In J. J. Hox & J. K. Roberts (Eds.), *Handbook of advanced multilevel analysis* (pp. 97–112). Taylor & Francis.

Stram, D. O., & Lee, J. W. (1994). Variance components testing in the longitudinal mixed effects model. *Biometrics, 50*(4), 1171–1177. https://doi.org/10.2307/2533455

Thomas, S., & Heck, R. (2001). Analysis of large-scale secondary data in higher education research: Potential perils associated with complex sampling designs. *Research in Higher Education, 42*(5), 517–540. https://doi.org/10.1023/A:1011098109834

Tomarken, A. J., & Waller, N. G. (2003). Potential problems with well-fitting models. *Journal of Abnormal Psychology, 112*(4), 578–598. https://doi.org/10.1037/0021-843X.112.4.578

Tomarken, A. J., & Waller, N. G. (2005). Structural equation modeling: Strengths, limitations, and misconceptions. *Annual Review of Clinical Psychology, 1*, 31–65. https://doi.org/10.1146/annurev.clinpsy.1.102803.144239

VanderWeele, T. (2015). *Explanation in causal inference: Methods for mediation and interaction*. Oxford University Press.

Verbeke, G., & Molenberghs, G. (2000). *Linear mixed models for longitudinal data.* Springer-Verlag. https://doi.org/10.1007/978-1-4419-0300-6

Wagenmakers, E., & Farrell, S. (2004). AIC model selection using Akaike weights. *Psychonomic Bulletin & Review, 11*(1), 192–196. https://doi.org/10.3758/BF03206482

Weaklim, D. L. (2004). Introduction to the special issue on model selection. *Sociological Methods & Research, 33*(2), 167–187. https://doi.org/10.1177/0049124104268642

Weaklim, D. L. (2016). *Hypothesis testing and model selection.* Guilford Press.

West, B. T., Welch, K. B., & Galecki, A. T. (2015). Three-level models for clustered data: The classroom example. In B. T. West, K. B. Welch, A. Galecki, & B. Gillespie (Eds.), *Linear mixed models: A practical guide using statistical software* (2nd ed., pp. 135–198). Taylor & Francis.

West, S. G., Finch, J. F., & Curran, P. J. (1995). Structural equation models with non-normal variables: Problems and remedies. In R. H. Hoyle (Ed.), *Structural equation modeling: Issues, concepts, and applications* (pp. 56–75). Sage.

Willett, J. B. (1988). Questions and answers in the measurement of change. *Review of Research in Education, 15*, 345–422. https://doi.org/10.2307/1167368

Willett, J. B. (1997). Measuring change: What individual growth modeling buys you. In E. Amsel & K. A. Renninger (Eds.), *Change and development: Issues of theory, method, and application* (pp. 213–243). Lawrence Erlbaum.

Wright, S. (1918). On the nature of size factors. *Genetics, 3*(4), 367–374.

Wright, S. (1920). The relative importance of heredity and environment in determining the piebald pattern of guinea-pigs. *Proceedings of the National Academy of Sciences of the United States of America, 6*(6), 320–332. https://doi.org/10.1073/pnas.6.6.320

Wright, S. (1923). The theory of path coefficients: A reply to Niles's criticism. *Genetics, 8*(3), 239–255.

Wright, S. (1934). The method of path coefficients. *Annals of Mathematical Statistics, 5*(3), 161–215. https://doi.org/10.1214/aoms/1177732676

Yzerbyt, V., Muller, D., Batailler, C., & Judd, C. M. (2018). New recommendations for testing indirect effects in mediational models: The need to report and test component paths [Supplemental material]. *Journal of Personality and Social Psychology, 115*(6), 929–943. https://doi.org/10.1037/pspa0000132.supp

Zucchini, W. (2000). An introduction to model selection. *Journal of Mathematical Psychology, 44*(1), 41–61. https://doi.org/10.1006/jmps.1999.1276

# INDEX

Page numbers in *italic* indicate figures and exhibits, and in **bold** indicate tables.

absolute fit indices, 86
Akaike information criterion (AIC),
    37–8, 39–41, **39**, **40**, 53, 54, 184
ANOVA, unconditional random-effects
    model, 32, 34–5, 50, 56
autoregressive models, 146–7

Baltes, P. B., 146
Bayesian information criterion (BIC), 37,
    38–41, **39**, **40**, 53, 184
Bentler, P. M., 86
between-cluster variance, 4, 7–8, 19
Bollen, K. A., 79, 85, 89, 96
bootstrapping, 77–8
Brown, T. A., 84
Bryk, A. S., 20, 25, 31, 32, 33, 35, 47–9
Burt, R. S., 119

Cardon, L. R., 243
causality, 78, 90–1
centring, 26–9, 50, 156, 157–8, **157**,
    165–6, 167, 180
chi-square difference test, 35–7, 85–6,
    87, 88, 121
cluster means, 7, 8, 17–18, 19, 26–7,
    28, 30, 32–3, 34–5
clustered data, 2–6
    effective sample size, 8–12, *11*, *12*
    intraclass correlation coefficient (ICC),
        6–8, 30, 33
    *see also* multilevel modelling (MLM)
Cohen's *d*, 43, 52
Cole, D. A., 69
combined models, 18, 21
comparative fit index (CFI), 86
compound paths, 109, 111, 232, 243–4,
    245–6, *246–7*
conceptually omitted paths, 120, 122–3,
    *122*, **137**, 138, *139*

conditional growth models, 184
    example model, 214–17, *215*, **216**
conditional ICC, 52, 61
confirmatory factor analysis (CFA), 68, 81
    *see also* measurement/CFA models
covariance algebra, 107, 233–40, *233*,
    *235*, 242–6
cross-classified random-effects models, 2
cross-level interactions, 23, 29, 50, 51, 52
cubic growth models, 180–3, *181*, *182*
    example model, 206–11, *207*, *208*, *209*,
        **210**, **211**, **212**

dependence, 2, 3–6
    effective sample size, 8–12, *11*, *12*
    intraclass correlation coefficient (ICC),
        6–8, 30, 33
    *see also* multilevel modelling (MLM)
design effect (DEFF), 10–11, *12*
direct effects, 69, 75–6, *75*, 77,
    140–2, **140–1**
disturbances, 74, 83–4

effect sizes, 43, 52, 60
effective sample size, 8–12, *11*, *12*
empirical Bayes estimation, 32, 33–5
empirical underidentification, 90, 99
endogenous variables, 74–5, 83, 245–6
estimation *see* multilevel modelling (MLM)
        estimation; structural equation
        modelling (SEM) estimation
exogenous variables, 74–5, 82
exploratory factor analysis (EFA), 81
exploratory research, 46

factor analysis, 68, 71, 80–1
    *see also* measurement/CFA models
first-order polynomial growth models, *175*,
    177, **179**

fixed effects, in multilevel models, 19, 23, 31, 36, 46
fixed factor variance strategy, 102–3, 104, 105, 160
full contextual/theoretical models, 23
full information maximum likelihood (FIML), 31, 35, 36, 39, 177

grand mean centring, 26–9, 165–6
group mean centring, 26–9, 50, 165–6
growth curve modelling, 5–6, 146–8
  conceptual introduction, 148–53, *150*
  MLM versus SEM frameworks, 153–4
  model building steps, 183–5, 191–2
  popularity of, 147, *147*
  requirements, 150–2, *151*
  time-structured versus time-unstructured data, 152–3, 161
  time-varying covariates (TVCs), 164–9
  *see also* linear growth models; piecewise linear growth models; polynomial growth models
growth curve modelling example, 188–218
  data, 188–9, **189**, *189*, *190*
  model building steps, 191–2
  model comparisons and selection, 212–13, **213**, **214**
  three-piece unconditional growth model, 197–202, **198**, *198*, **199**, **201**, *206*, **212**
  two-piece conditional growth model, 214–17, *215*, **216**
  two-piece unconditional growth model, 202–5, **202**, *203*, **204**, *205*, **206**, *206*, *209*, **212**
  unconditional linear growth model, 192–7, *193*, **194**, *195*, *197*, *208*, **212**
  unconditional polynomial growth model, 206–11, *207*, *208*, *209*, **210**, **211**, **212**

Heise, D. R., 242, 243
Heywood cases, 90
hierarchical linear models, 2
  *see also* multilevel modelling (MLM)
Hu, L.-T., 86
hybrid structural equation models, 118
  degrees of freedom, 105–6, *106*
  measurement model, 118–19, 120–1
  model building steps, 120–6, *125*
  structural model, 118, 119–20, 121–6, *122*
hybrid structural equation models example, 126–42
  data, 128, **128–9**
  direct, indirect and total effects, 140–2, **140–1**
  measurement model, 129–35, *131*, **132–4**
  structural model, *127*, **132**, 135–8, *136*, **137**, *139*

incremental fit indices, 86
independence, 70, 102
  *see also* non-independence
indirect effects, 69, *75*, 76–8, 140–2, **140–1**
information criteria (ICs), 37–41, **39**, **40**, 53, 54, 184
integrative framework of R², 42–3, 49, 50–4, 61
intercept-only models, 175–7, *175*
intraclass correlation coefficient (ICC), 6–8, 30, 33
  conditional, 52, 61
  unconditional, 47, 50
  *see also* effective sample size

just-identified models, 99, 118, 120, 122–4, **132**, 135–7, *136*, **137**

latent constructs, 68, 79–80, *80*, 81–2, *82*
latent growth models, 69, 153–4
  degrees of freedom, 161–2
  linear growth models, 159, *159*, 161–4, **164**
  parameter interpretation, 163–4, **164**
  piecewise growth models, 169–74, *172*, **173**, **174**
  time-varying covariates (TVCs), 168–9
  *see also* growth curve modelling example
latent intercept factor, 162
latent slope factor, 162
latent variable models
  *see* measurement/CFA models
latent variable structural models, 118
  degrees of freedom, 105–6, *106*
  measurement model, 118–19, 120–1
  model building steps, 120–6, *125*
  structural model, 118, 119–20, 121–6, *122*
latent variables, 68–9, 78–9
  scaling, 102–3, 104, 105
lavaan (R package), 194–5
level-1 equations, 17
level-2 equations, 17
likelihood ratio test (LRT), 35–7, 38, 39–41, **39**, **40**, 54, 184
linear growth models
  example model, 192–7, *193*, **194**, *195*, **197**, *208*, **212**
  as first-order polynomial growth models, *175*, 177, **179**
  specification in MLM, 154–8, **157**
  specification in SEM, 159, *159*, 161–4, **164**
  time-varying covariates (TVCs), 164–9
  *see also* piecewise linear growth models
linearity assumption, 70
lme4 (R package), 194, 195, 199
local independence, 102
longitudinal modelling, 5–6, 146–7
  *see also* growth curve modelling

McArdle, Jack, 146
MacCallum, R. C., 89
marker variable strategy, 102–3,
    104, 159, 160–1
matrix algebra, 222–6
maximum likelihood estimation
    multilevel modelling, 30–1, 35, 36, 177
    structural equation modelling, 70, 84
measurement errors, 68–9, 78, 83–4,
    101–2, 104, 113–15
measurement weights, 83
measurement/CFA models, 80, 81–3,
    82, 118, 159
    degrees of freedom, 104
    disturbances, 83–4
    latent variable scaling, 102–3, 104, 105
    means and intercepts, 160–1
    measurement errors, 83–4, 101–2,
        104, 113–15
    model identification, 102–5
    model specification, 100–2, 101
    quantifying unexplained variance, 83–4
    standardised tracing rules for, 111–15,
        112, 113, 114
    system of equations for, 106, 107
    see also hybrid structural equation models;
        latent growth models
mediational models, 75–8, 75
    see also hybrid structural equation
        models example
mixed-effects models, 2
    see also multilevel modelling (MLM)
model fit
    multilevel modelling, 35–41, 39, 40
    structural equation modelling, 84–8, 89
model modification indices, 88–9, 121
model trimming, 88
model-implied correlation/covariance matrix
    see Wright's tracing rules
modification indices, 88–9, 121
multidimensionality, 102
multilevel modelling (MLM), 2–3
    centring, 26–9, 50
    combined models, 18, 21
    conceptual introduction, 16–29
    effective sample size, 8–12, 11, 12
    full contextual models, 23
    intraclass correlation coefficient (ICC),
        6–8, 30, 33
    model building steps, 46–54
    model with level-1 predictors, 20–3, 22
    model with no predictors, 16–18
    nested data and non-independence, 2, 3–6
    quantifying explained variance, 41–3,
        49, 50–4
    randomly varying intercepts, 17, 20–5,
        22, 26–9, 52, 53–4

randomly varying slopes, 20–6, 22, 51, 53–4
residual variances, 5–6, 19
multilevel modelling (MLM) estimation,
    29–30
    deviances, 35–41
    effect sizes, 43, 52, 60
    empirical Bayes estimation, 32, 33–5
    maximum likelihood estimation, 30–1,
        35, 36, 177
    model fit criteria, 35–41, 39, 40
    model selection, 38, 39–41, 39, 40
    quantifying explained variance, 41–3,
        49, 50–4
    randomly varying level-1 coefficients, 32–3
    reliability, 32–3, 34–5
multilevel modelling (MLM) example, 54–64
    evaluating variance components, 61–3
    random coefficients models, 57, 58–64
    random intercept-only model, 56–8, 57
multilevel modelling (MLM)
        growth models, 153–4
    centring, 156, 157–8, 157, 165–6, 167, 180
    linear growth models, 154–8, 157
    piecewise growth models, 169–74, 173, 174
    time-varying covariates (TVCs), 164–8
    variance components, 157–8, 166–8, 170–1
    see also growth curve modelling example;
        polynomial growth models
multilevel structural equation modelling, 70
multiple group structural equation modelling
    (MG-SEM), 69
multiple-membership models, 2

Neale, M. C., 243
Nesselroade, J. R., 146
nested data, 2, 3–6
    see also multilevel modelling (MLM)
non-independence, 2, 3–6
    effective sample size, 8–12, 11, 12
    intraclass correlation coefficient (ICC),
        6–8, 30, 33
    see also multilevel modelling (MLM)
non-normed fit index (NNFI), 86
non-recursive structural equation models,
    96, 99, 119
normality assumption, 70
Nosek, B. A., 46–7

organisational models, 2–3, 4–5
overidentified models, 100, 104, 105

parsimonious models, 36, 38, 40, 53,
    62–3, 87, 88, 100
path analysis/models, 68, 71, 72, 118
    bootstrapping, 77–8
    causality, 78, 90–1
    degrees of freedom, 98–9, 98

direct, indirect and total effects,
69, 75–8, *75*
disturbances, 74
exogenous and endogenous variables, 74–5
linking path diagrams to structural
equations, 228–9, *228*
with mediation, 75–8, *75*
path coefficients defined, 73
path diagrams overview, 72–4, *72*, *73*
*see also* hybrid structural equation models;
Wright's tracing rules
pattern coefficients, 83
*p*-hacking, 46, 120
piecewise linear growth models, 169–74, *172*,
**173**, **174**
example three-piece unconditional
growth model, 197–202, **198**, *198*,
**199**, **201**, *206*, **212**
example two-piece conditional growth
model, 214–17, *215*, **216**
example two-piece unconditional growth
model, 202–5, **202**, *203*, **204**, *205*,
**206**, *206*, *209*, **212**
polynomial growth models, 174–5
centring, 180
cubic models, 180–3, *181*, *182*
example cubic model, 206–11, *207*, *208*,
*209*, **210**, **211**, **212**
intercept-only models, 175–7, *175*
linear models, *175*, 177, **179**
quadratic models, *175*, 177–80, **179**
Preacher, K. J., 69
preregistration, 46–7
proportional reduction in variance, 48–9

quadratic growth models, *175*, 177–80, **179**

R packages
lavaan, 194–5
lme4, 194, 195, 199
r2mlm, 43
r2mlm (R package), 43
random effects
defined, 19
*see also* variance components
random-effects models, 2
*see also* multilevel modelling (MLM)
randomly varying intercepts, 17, 20–5, *22*,
26–9, 52, 53–4
in cubic growth models, 182–3
in example models, 55, 56–63
in piecewise growth models, 169–71
randomly varying slopes, 20–6, *22*, 51, 53–4
in cubic growth models, 182–3
in example models, 55, 58–63
in piecewise growth models, 169–71

Raudenbush, S. W, 20, 25, 31, 32, 33, 35, 47–9
recursive structural equation models, 96, 119
reliability, 32–3, 34–5
residual variances, 5–6, 19
respecification, 87–9
restricted maximum likelihood (REML),
31, 36, 177
Rights, J. D., 42–3, 49, 50–4, 61
root design effect (DEFT), 11, 12
root mean square error of approximation
(RMSEA), 86

sampling variability, 10
second-order polynomial growth models, *175*,
177–80, **179**
Shaw, M., 43
Singer, J. D., 149, 152, 153, 154
single-level regression, 16, 18, 26, 41
spline growth models *see* piecewise linear
growth models
standard errors
multilevel modelling, 3–4, 9–12, *12*
structural equation modelling, 73
standardised path coefficients, 73, 83, 90
standardised root mean square residual
(SRMR), 86
standardised tracing rules, 107–15, *108*, **110**,
*110*, *112*, **113**, *114*, 232–40, *233*, *235*
Sterba, S. K., 42–3, 49, 50–4, 61
structural equation modelling (SEM), 68, 96
advantages of, 68–9
assumptions and requirements, 70–1
causality, 78, 90–1
covariance matrix as sufficient
statistic, 71, 96
degrees of freedom, 96–100, *98*,
104, 105–6, *106*
direct, indirect and total effects, 69,
75–8, *75*, 140–2, **140–1**
disturbances, 74, 83–4
equivalent models, 90–1
exogenous and endogenous variables,
74–5, 82, 83, 245–6
factor analysis, 68, 71, 80–1
inadmissible solutions, 90
just-identified models, 99, 118, 120, 122–4,
**132**, 135–7, *136*, **137**
limitations, 90–1
means and intercepts, 159–61
measurement errors, 68–9, 78, 83–4, 101–2,
104, 113–15
model modification, 87–9
multilevel SEM, 70
multiple group SEM, 69
overidentified models, 100, 104, 105
respecification, 87–9

sampling, 70–1
underidentified models, 90, 99
*see also* hybrid structural equation models;
    measurement/CFA models; path
    analysis/models; Wright's tracing rules
structural equation modelling (SEM)
    estimation, 84
    absolute fit indices, 86
    chi-square difference test, 85–6, 87, 88, 121
    hypothesis testing, 84–6
    incremental fit indices, 86
    maximum likelihood estimation, 70, 84
    model comparison, 87–8
    model fit, 84–8, 89
    model prediction, 89–90
    modification indices, 88–9, 121
structural equation modelling (SEM) example
    *see* hybrid structural equation models
    example
structural equation modelling (SEM) growth
    models, 69, 153–4
    degrees of freedom, 161–2
    linear growth models, 159, *159*, 161–4, **164**
    parameter interpretation, 163–4, **164**
    piecewise growth models, 169–74,
        *172*, **173**, **174**
    time-varying covariates (TVCs), 168–9
    *see also* growth curve modelling example
structural models, 81–3, *82*, 118
    *see also* hybrid structural equation models;
    path analysis/models
systems of equations, 96
    linking to path diagrams, 228–9, *228*
    in matrix notation, 225
    for measurement/CFA models, 106, *107*
    *see also* Wright's tracing rules

third-order polynomial growth models *see*
    cubic growth models
time-structured data, 152, 161

time-unstructured data, 153, 161
time-varying covariates (TVCs), 164–9
total effects, 69, *75*, 77, 140–2, **140–1**
tracing rules *see* Wright's tracing rules
trimming models, 88
Tucker–Lewis Index (TLI), 86
Type I error rates, 3–4, 8–9, 10, 78

unconditional growth models, 154–5, 158,
    163, 165, 183–4, 185
    *see also* growth curve modelling example
unconditional ICC, 47, 50
unconditional random-effects ANOVA model,
    32, 34–5, 50, 56
underidentified models, 90, 99
unidimensionality, 101
unstandardised path coefficients, 73
unstandardised tracing rules, 107,
    242–6, *246–7*

variance components, 19, 23–5, 41
    estimating, 31, 36, 177
    example of evaluating, 61–3
    growth curve models, 157–8,
        166–8, 170–1
    proportional reduction in variance, 48–9
    statistical tests of, 53
vertical scaling, 152

Willett, J. B., 148, 149, 152, 153, 154
within-cluster variance, 4, 7–8, 19
Wright, Sewall, 72, 91, 107
Wright's tracing rules
    standardised, 107–15, *108*, **110**, *110*, *112*,
        **113**, *114*, 232–40, *233*, *235*
    unstandardised, 107, 242–6, *246–7*

Yzerbyt, V., 78

zero-order growth models, 175–7, *175*